T0264986

LOCAL APPLICATIONS OF THE ECOLOGICAL

APPROACH TO HUMAN-MACHINE SYSTEMS

RESOURCES FOR ECOLOGICAL PSYCHOLOGY

A Series of Volumes Edited by:
Robert E. Shaw, William M. Mace, and Michael T. Turvey

LOCAL APPLICATIONS OF THE ECOLOGICAL

APPROACH TO HUMAN-MACHINE SYSTEMS

edited by

Peter Hancock
University of Minnesota

John Flach
Wright State University

Jeff Caird
University of Calgary

Kim Vicente
University of Toronto

CRC Press
Taylor & Francis Group
Boca Raton   London   New York

CRC Press is an imprint of the
Taylor & Francis Group, an **Informa** business

First Published by
Lawrence Erlbaum Associates, Inc., Publishers
365 Broadway
Hillsdale, New Jersey 07642

Transferred to Digital Printing 2009 by CRC Press
6000 Broken Sound Parkway, NW Suite 300, Boca Raton, FL 33487
270 Madison Avenue, New York, NY 10016
2 Park Square, Milton Park, Abingdon, Oxon, OX14 4RN, UK

Library of Congress Cataloging-in-Publication Data

Local applications of the ecological approach to human-machine systems /
  edited by Peter Hancock ... [et al.].
      p.   cm.
  Includes bibliographical references and index.
  ISBN 0-8058-1379-9 (acid-free paper). -- ISBN 0-8058-1380-2 (pbk.)
  1. Man-machine systems. 2. Human engineering.   I. Hancock, Peter
A., 1953-   .
  TA167.L63 1995
  620.8'2--dc20   94-47100
                                                       CIP

Publisher's Note
The publisher has gone to great lengths to ensure the quality of this reprint
but points out that some imperfections in the original may be apparent.

To Ron

Contents

Contributors

Wendy Arnone
Department of Applied Psychology
New York University

Reinoud J. Bootsma
UFR STAPS
University of Aix-Marseille-II

Beth Crandall
Klein Associates Inc.

V. Grayson CuQlock-Knopp
Human Engineering Laboratory

John Flach
Department of Psychology
Wright State University

Steven B. Flynn
Northeast Louisiana University

Jacqueline Grosz
Faculty of Human Movement Sciences
Free University

John P. Hansen
System Analysis Department
Risø National Laboratory

Marian Heinrichs
Center for Research in Learning, Perception and Cognition
University of Minnesota

Robert R. Hoffman
Department of Psychology
Adelphi University

Gary A. Klein
Klein Associates, Inc.

Peter N. Kugler
Department of Computer Science
Radford University

Gavan Lintern
Aviation Research Laboratory
University of Illinois

Michael D. McNeese
Human Engineering Division
US Air Force Armstrong Laboratory

Dan R. Montello
Department of Geography
University of California, Santa Barbara

J. (Bob) Mulder
Faculty of Aerospace Engineering
Delft University of Technology

Herbert L. Pick, Jr.
Institute of Child Development
University of Minnesota

Richard J. Pike
U.S. Geological Survey

Gary E. Riccio
Department of Kinesiology
University of Illinois

Rolf Th. Rysdyk
Faculty of Aerospace Engineering
Delft University of Technology

Leon D. Segal
NASA Ames

William Schiff
Department of Applied Psychology
New York University

Gerda Smets
Department of Industrial Design
Delft University

Kip Smith
Human Factors Research Laboratory
University of Minnesota

Thomas A. Stoffregen
Department of Psychology
University of Cincinnati

C.N. Sullivan
Institute of Child Development
University of Minnesota

W.B. Thompson
Department of Computer Science
University of Utah

C. van der Vaart
Faculty of Aerospace Enginnering
Delft University of Technology

Piet C.W. van Wieringen
Faculty of Movement Sciences
Free University

Rik Warren
Engineering Research Psychology
Armstrong Laboratory

Leslie A. Whitaker
Psychology Department
University of Dayton

Michael F. Young
University of Connecticut

Preface

This is the second of a two-volume set on ecological approaches to human factors. Volume I takes a global perspective. It attempts to lay a theoretical foundation for an ecological approach. We think that this is an important exercise. The importance of a firm theoretical base cannot be overestimated. However, the demands of design and application often require a degree of pragmatism. And, in fact, perhaps the best way to present the theoretical base is through a "case-based' presentation. That is, the theoretical base can be illustrated through examples of how it can be applied to specific problems. That is the fundamental goal of this volume. It is a collection of applications that illustrate the importance of an ecological perspective.

An important area of application for the ecological approach has been the control of locomotion. In fact, it might be argued that human performance issues associated with driving cars and landing aircraft were instrumental to the development of an ecological approach. Schiff and Arnone begin Volume II with a discussion of driving. This chapter provides a nice comparison and contrast of traditional information-processing-based approaches to driving with the ecological approach. They show how the two approaches can compliment each other to provide a more comprehensive framework for understanding human performance in driving. Flynn and Stoffregen consider the problem of tractor roll-over. They develop a framework for modeling stability properties of the human-machine system. Flach and Warren consider the problem of low-altitude flight. They use a state-space representation to illustrate the fundamental role of higher order variables in determining performance boundaries. They also discuss the information in optic flow that is available to help the pilot close the control loop. Grotz, Rysdyk, Bootsma, Mulder, van der Vaart, and van Wieringen also consider an important aspect of flight—the flare landing. The flare maneuver is very critical to safe landings. However, the informational basis by which the skilled pilot is able to execute this maneuver remains a question mark. Grotz et al. consider *tau* as a potential source of information for flare. Finally, Riccio reminds us that the feedback loops that allow skillful control of locomotion are not exclusively closed through the visual modality, and that control of locomotion includes internal control loops to stabilize posture of the human with respect to the vehicle. It is likely that higher order properties defined across modalities are critical to control of posture and to skillful control of locomotion. He considers the implication of this fact for the design of simulators to train skilled flight performance.

A defining characteristic of the ecological approach is the importance it gives to understanding the coupling between perception and action. One technological environment in which understanding this coupling is critical is in the area of telepresence and virtual reality displays. Smets reviews the work that the industrial design group at Delft University has done on this problem. She nicely illustrates how ecological theory can be the inspiration for unique and innovative solutions to important problems in the design of displays and controls for the support of telepresence. Hansen examines design of displays for more complex, process control environments. He outlines a number of principles for building representations that allow direct perception of the constraints within this type of workspace.

Three chapters examine the problem of spatial orientation and recognition. Whitaker and Grayson-CuQlock-Knopp study the strategies of expert orienteers in order to design better decision aides for navigation in natural terrain. Pick, Heinrichs, Montello, Smith, Sullivan, and Thompson look at the problem of topographical map reading. This chapter demonstrates how field and laboratory research can be combined to develop a more comprehensive understanding of a problem. Hoffman examines expertise in aerial photo interpretation. He argues that experts see these photos in terms of dynamic geobiological events rather than in terms of static dimensions.

Much of the work that has been characterized as ecological has focused on perceptual and motor skills. However, a few adventurous pioneers are beginning to breach the more "cognitive" frontier of decision making and problem solving. Klein and Crandal have found that perception plays a critical role in expert decision making in naturalistic settings. Experts "see" good alternatives early, and rather than generating several alternatives, experts tend to do a deep evaluation of the single alternative to determine if it will satisfy the demands of the situation. This evaluation often involves a "mental simulation" in which the experts are able to visualize the potential success or failure of the option under consideration. Young and McNeese consider the evidence that human performance in problem solving is very sensitive to contextual variables. They discuss some methodological tools for a situated approach to cognition.

The final two chapters turn from "human" factors to "organization" or "crew" factors. There is an increased awareness within the human factors community of the critical role the organization plays in shaping the performance of individuals within it. The system is no longer a single operator and its tools. Rather, the success of a system depends on cooperation and coordination of many operators (and many machines). Segal considers the information topology for coordination within teams of operators and how various design choices can modulate this topology. Kugler and Turvey consider the implications of dynamic systems theory for understanding the coupling between micro and macro factors in large-scale industrial systems.

It should be apparent that this volume contains a diverse range of topics. What is the unifying theme? What is the rationale for collecting such a menagerie under the heading of an "ecological approach"? It is not a shared allegiance to Gibson. Although Gibson's ideas have been influential for many, there are a number of authors who would bristle at the characterization of "Gibsonian." But there are some justifications for collecting these authors together. There is a shared skepticism of reductionistic approaches that are based on the assumption that data collected on elemental tasks under sterile laboratory conditions can be integrated into a comprehensive theory of human performance. There is a shared commitment to study the human environment as a unified, dynamic system. There is a belief that the contribution of the human agent can be best understood in the context of the emergent properties that arise from interaction with an environment. Thus, there is a movement away from bottom-up reductionistic approaches toward more top-down holistic approaches to the problem of human performance. Finally, there is a commitment to a basic science of behavior that is more than puzzle solving: commitment to a basic science that is informed by problems arising in complex human-machine environments, that will support the design of new technologies, and that will foster and enrich the growth and development of the human factor.

As with any production of this scope, contributions are made by many who do not get credit as authors or editors. In particular, our sincere thanks is extended to Rob Stephens, Jonathan Sweet, Jeff Preston, Shannon Skistad, Kevin Connolly, and Becky Hooey. Thanks to the Ecological Psychology Seminar class at Wright State University for editorial comments on early versions of many of the chapters. Particular thanks to Bart Brickman, Rob Hutton, and Charlie Garness for their editorial comments.

John Flach   Peter Hancock   Jeff Caird   Kim Vicente

RESOURCES
for ECOLOGICAL PSYCHOLOGY

Edited by
Robert E. Shaw, William M. Mace, and Michael Turvey

This series of volumes is dedicated to furthering the development of psychology as a branch of ecological science. In its broadest sense, ecology is a multidisciplinary approach to the study of living systems, their environments, and the reciprocity that has evolved between the two. Traditionally, ecological science emphasizes the study of the biological bases of *energy* transactions between animals and their physical environments across cellular, organismic, and population scales. Ecological psychology complements this traditional focus by emphasizing the study of *information* transactions between living systems and their environments, especially as they pertain to perceiving situations of significance to planning and execution of purposes activated in an environment.

The late James J. Gibson used the term *ecological psychology* to emphasize this animal-environment mutuality for the study of problems of perception. He believed that analyzing the environment to be perceived was just as much a part of the psychologist's task as analyzing animals themselves, and hence that the "physical" concepts applied to the environment and the "biological" and "psychological" concepts applied to organisms would have to be tailored to one another in a larger system of mutual constraint. His early interest in the applied problems of landing airplanes and driving automobiles led him to pioneer the study of the perceptual guidance of action.

The work of Nicolai Bernstein in biomechanics and physiology presents a complementary approach to problems of the coordination and control of movement. His work suggests that action, too, cannot be studied without reference to the environment, and that physical and biological concepts must be developed together. The coupling of Gibson's ideas with those of Bernstein forms a natural basis for looking at the traditional psychological topics of perceiving, acting, and knowing as activities of ecosystems rather than isolated animals.

The purpose of this series is to form a useful collection, a resource, for people who wish to learn about ecological psychology and for those who wish to contribute to its development. The series will include original research, collected papers, reports of conferences and symposia, theoretical monographs, technical handbooks, and works from the many disciplines relevant to ecological psychology.

Series Dedication

To James J. Gibson, whose pioneering work in ecological psychology has opened new vistas in psychology and related sciences, we respectfully dedicate this series.

Chapter 1

Perceiving and Driving: Where Parallel Roads Meet

William Schiff and Wendy Arnone

New York University

1.0   Introduction

Perceptual psychology has followed several roads to the present state of the art. The road emerging from classical sensory psychophysics has had a major influence on the practice of human factors and human engineering. This avenue assumes that our senses basically serve as energy transducers, responding to energy variables such as wavelength and intensity of light, frequency of sound, and so on. Following this sensory transduction, perception is then considered as an elaborative process by which information stored in memory becomes integrated with current sensory input. With this sort of theoretical framework, automobile driving then emerges as a complex psychomotor process in which prior knowledge is combined with current sensory input. In this light it would make sense to inform human factors practitioners about the sensitivities and limitations of our sensory (and motor) apparatus.

It would seem likely that it is largely from this model that current driving measures have elected to screen drivers for sensory sensitivities (e.g., color and acuity), for basic perceptual discrimination (e.g., coded sign shapes), and to ask (cognitive) questions pertaining to rules of the road or state laws.

Another road to the current state of the art has emerged from Gibson's functional psychophysics. As applied to automobile driving, such an approach reaches back at least to 1938, to a seminal paper with Crooks (Gibson & Crooks, 1938). This approach suggested that rather than being mechanisms to provide energy-level data to the central nervous system, the senses might be conceptualized as information-gathering systems that pickup useful environmental information permitting humans to perform ongoing real-world tasks. Perception would be a process involving the pickup of information. The level of description of that information was not the gram–centimeter–second level of classical physics, but that of higher level goals, vectors, openings, and obstacles. Human factors practitioners would be appropriately informed as to the nature of the higher order information used in driving, and driving screening measures would examine how well drivers pick-up and utilize the higher order dynamic information relevant to driving. This approach to perceiving and to driving automobiles became merged with Gibson's later analysis of flying aircraft (Gibson, 1950) to yield a parallel path to topics of perception and driving that examined dynamic optical information as a basis for vehicle control.

This chapter examines these two perceptual approaches to the driving task, discussing relevant literature emerging from both. We then examine the human factors literature for traces of both approaches in an attempt to "integrate" the parallel roads to current knowledge.

1.1   A Historical Perspective

1.1.1   The Classical Psychophysical Approach to Perception, and Its Impact on Automobile Driving Research and Application

Assumption and tradition may be as underwater rocks are to sailors; they may be passed over without notice, or they may force notice by interference with smooth sailing. Like any discipline, perceptual psychology has its many assumptions and traditions. Among the broadest and most captivating has been the persisting view that perception of events is based on constructed concatenations of static retinal images. From this view, perception of motion events may be treated as a sort of secondary process, our understanding of which will rather automatically occur once we have understood static scene perception. This may be termed the static snapshot assumption of classical psychophysics (Schiff, 1986, pp. 52–54; Turvey, 1977). We can find it in both older and newer accounts of perception (e.g., see Coren & Ward, 1989). It has also been assumed in various traditional theoretical approaches to perceptual psychology, that an essential basis for static scene perception must be the registration of simple stimulus energies by the senses. The macroassumption here was traditionally that the proper measurement scales for these energies were those of classical physics, such as wavelength, frequency, intensity, time, and so on. This aspect of the tradition may be termed the sensory *sensitivities* assumption (e.g., see Rumelhart, 1979).

These twin sets of assumptions have been quite influential in the study of perception and automobile driving. By the time the automobile appeared on the scene, these assumptions were so ingrained in perceptual psychologists' thinking that they were seldom, if ever, questioned, and came to permeate the study of and theorizing about how we gather information about the world, including information used in guiding an automobile. True, the brilliant Hermann von Helmholtz used such assumptions to develop rather comprehensive and still well-respected theories of how our visual and auditory systems may permit us to perceive the colors, sounds, and even sizes and distances of objects in the world. The respected theorists using these assumptions provided part of the "establishment view" attacked by Gibson's radically innovative approach to perceiving (e.g., see Reed, 1988). Gibson suggested that the assumptions were wrong-headed and were leading sensory-perceptual psychologists astray (e.g., see Gibson, 1950, 1966, 1979). He thought such assumptions and traditions led us to examine the wrong information (the energy variables of psychophysics) when thinking about perception, which always led us to the view that the world and its events underdetermine perception, and theoretically forced us to believe that the perceiver-knower must supplement or enrich such meager "sense data" from our knowledge or memory. Thus, the study of perception would be largely an analysis of the perceivers' contributions to perception, rather than the contributions of the light patterns reaching the visual system, or the sound patterns reaching the auditory system. Gibson believed the perceiver had a role in producing percepts, but it was the role of sampling the information in the world, not refashioning from memory transformations of simple sense data reaching sensory surfaces.

One would think that such a radical departure from tradition and assumptions would not go unnoticed. But it seems as if the message does not always reach intended audiences. One can pick up textbooks in sensation and perception (e.g., Coren & Ward, 1989), perception and cognition (e.g., Matlin, 1988), and find the same assumptions, often unstated, with no qualifiers. If the assumptions are questionable, one would not gather so from a reading of most perception texts. There are, of course, exceptions (e.g., Bruce & Green, 1990). A similar state of affairs exists in the human factors field. Respected texts having fairly comprehensive treatment of perception as it may relate to their fields often bypass Gibson's alternative approach (e.g., McCormick, 1957; McCormick & Sanders, 1976; Sanders & McCormick, 1987; Wickens, 1984).

When we think of driving, we should think first of the issues of motion perception, because the driver is often in motion, and parts of his traffic world are too. Perception and judgment of such motions are surely central to perceptual issues in driving. But if one looks in the perceptual texts, one usually finds motion perception treated as something of an afterthought, perhaps a consequence of the static snapshot assumption, with motion often viewed as another illusion. Phenomena of illusory motion have a fully researched history, from Wertheimer's Gestalt treatment of phi and other related simple motions, to more recent ones, such as Wallach's kinetic depth effect and the familiar Ames rotating trapezoidal window illusory motion phenomena (e.g., see Schiffman, 1990). It is of interest that most perceptual texts (and technical works as well) seem to convey the impression that most of the motion registered by the human visual system is continuous rotation at a constant velocity around a vertical axis at a fixed distance from the eye. This sort of motion is perhaps most frequently examined in traditional theories of motion perception and in perceptual texts. The driver sees very few instances of this sort of motion. This assumption has strong historical roots in Gestalt psychology, which shares the same view of illusions. It may also get a boost from the fact that much of the motion we perceive in man-made devices (motion pictures, video pictures, commercial signs, computer displays) is indeed phi-based motion, rather than the natural optical motion occurring when we walk, drive, or fly.

The remnants of classical psychophysical assumptions are to be seen in human factors texts and journals as well as in perceptual texts. For example, McCormick's justifiably popular introductory text in Human Factors Engineering (McCormick & Sanders, 1976) includes a major section concerned with "man in motion" in which illusions during self-motion are given major attention. The relatively small section on vehicle driving (5 pages in 1976 edition) refers to sensory components, treats information input to the vehicle driver as objects (other vehicles, road signs, etc.), and concludes that "the driver 'processes' the information somehow, leading up to his control responses" (McCormick & Sanders, 1976, p. 399). Further material acknowledges the role of perception of one's own speed. The section also treats reaction time, which includes time required for recognition and motor responses, and the mechanics of vehicle control devices (pedals, etc.). This rather brief treatment of such an important human factors topic as driving may be contrasted with an earlier discussion of classical psychophysics extending over 8 pages, including static visual acuity, color discrimination, dark adaptation, luminance contrast, and so on. This example is meant only as an indication of the continuing tutorial impact of assumptions and traditions, and not as a criticism of this particular book; it is but one example of the state of the fields of perception and human factors as they may impact on automobile driving research and applications.

Along with the classical psychophysical view that sensory sensitivity factors are foundational (although an inadequate basis for perception), we find a related assumption that because sensory phenomena cannot account fully for perception, our senses frequently lead us to error—the illusory assumption of the classical psychophysical view (e.g., see Coren & Girgus, 1978; Leibowitz, 1965, pp. 42–49). Indeed, it has been thought for many decades that if sensation is an accurate reflection of the external world, given the facts of illusions, that is, mistaken percepts, the basis for these errors must be postsensory, that is, perceptual. This set of related theoretical underpinnings may not only leads to possibly exaggerated "bells-and-whistles" treatments of illusions in perceptual textbooks, journals, and museum exhibits dealing with perception (e.g., see Schiff, 1988), it also may lead the applied researcher to illusory hypotheses. A recent example from the automobile driving literature may serve to demonstrate this point.

A leading researcher in the field of perception and driving analyzed possible causal factors potentially responsible for the very serious problem of motorists' crashes at railroad crossings (Leibowitz, 1985). Over 700 persons are killed each year in the United States alone while trying to drive across railroad crossings. A prime hypothesis advanced to partly explain such accidents was the size-velocity illusion. In this illusion (usually demonstrated with abstract two-dimensional laboratory displays) larger objects tend to be perceived as moving more slowly than smaller objects moving at the same velocity, in the same trajectory. From these findings it might indeed follow that motorists who are used to judging car-sized objects approaching transversely (approximately perpendicular to the line of sight) might misperceive the locomotive's approach speed and fatally misjudge whether there was time to safely cross the tracks before the train arrived at the crossing.

There are additional assumptions encased in this hypothesis, namely, that velocity is the variable used to make such decisions (e.g., Schiff, Oldak, & Shah, 1992), and that small magnitude errors such as those found in the illusion would be sufficient to prevent an error on the cautious side. What about the vast majority of living motorists? Is not their perception of velocity of locomotive-sized objects also in error? Yes, the classical psychophysical theorist might reply, but their more conservative decision strategies lead them to stop safely, even though they may misperceive the train's approach velocity. Thus, higher level mechanisms are invoked to rescue error-prone percepts—a view achieving great popularity with psychologists as different as Piaget (e.g., see Schiff, 1983) and Coren (Coren & Girgus, 1978). Again, we make no strong claims concerning the empirical viability of the hypothesis, but only note how readily the illusion explanation pops up from the stew of theoretical assumptions carried by researchers.

Another major source of research hypotheses related to perceiving and driving may be traced to the concern with sensory sensitivities as the basis for perceiving. When driving errors occur (as in the case of many accidents), a prime candidate for presumed causes is an inadequately functioning sensory system, failing sensory acuities, sensitivities, and the like. The history of driving research is replete with searches for driving errors that may be predicted by static or dynamic acuity defects (e.g., Burg, 1966), nighttime myopia (e.g., Leibowitz & Owens, 1977), visual field restriction (Ball, 1990; Keltner & Johnson, 1987; Owsley & Sloane, 1987), contrast sensitivity losses, and so on. Again, we do not claim these are unimportant as possible causal factors in traffic accidents; indeed, some are likely the roots of certain sorts of traffic accidents. We merely note the assumption-driven focus on sensory sensitivity factors as a major basis (besides illusions) for driving errors.

In addition to the impact of assumptions on driving research, one may consider their roles in the makeup of licensing examinations, administered by hundreds of thousands each year in the United States alone. Such tests are apparently meant as minimal screening devices for the required knowledge and skill to operate a potentially lethal machine under increasingly difficult traffic conditions. Although the licensing tests vary from state to state, most dwell on the same aspects of driving knowledge and performance and will be briefly and generically considered. The core of such tests dwells primarily on knowledge of state laws pertaining to driving and more general rules of the road, as tapped by multiple choice questions. Many include pictures or static diagrams in which the examinee must indicate which car has the right of way, the meanings of particular road sign shapes or colors (e.g., the octagonal stop sign), and verbal questions regarding safe vehicle operation (e.g., the number of car lengths required to stop a car moving at particular velocities). These tests are sometimes administered and scored by microcomputers, undoubtedly increasing their scientific credibility. One must note that driving "knowledge" is not necessarily functional knowledge. For example, one cannot begin to know how many times you may have exceeded posted speed limits, rolled slowly through stop signs at empty intersections, or accelerated to "beat" a yellow traffic light. Again, note that the nature of the knowledge tested in the written portion of the test is not functional; one may know the formal answer to the car-length question, and still it may have no bearing on a driver's actual performance behind the wheel of a car. The ability to correctly select the alternative meaning of a sign shape given a substantial period of time, and given a look at the alternatives, is not the same as responding to it quickly in a complex traffic situation.

Moreover, the licensing test typically screens for static visual acuity, color discrimination, and possibly stereopsis. This is another example of the role of psychophysical assumptions when applying perceptual knowledge to a practical issue. It is interesting because we again see minimal functional appropriateness of these measures (see Flach, Lintern, & Larish, 1990, and Owen, 1990, for a related treatment of functionality in perceptual research issues). No doubt that if a testee cannot see well enough to discriminate the microfeatures of a static acuity grating, checkerboard, Landoldt circle, or Snellen letter, that person may have vision too poor to drive safely. But the converse does not hold, because persons able to read a chart may still have problems with moving targets, their relative speeds and directions, and likely trajectories.

Unfortunately, driving is not merely a matter of discriminating fine detail. The issue is not what make of truck (indicated by the small letters emblazoned on its hood) is bearing down on you at high speed on a collision course, but that there *is* one, and that you are required to initiate specific evasive maneuvers. Large-scale research confirms that static acuity measures are not significantly related to driving accidents, and even dynamic acuity measures, which tap the fine visual-motor integrity of the visual system, are only minimally related to driving accidents (e.g., see Burg, 1971). However, one should remember that those with the worst vision have likely been screened out of the driving population by the tests used.

Color vision as tested in licensing tests is similarly irrelevant to driving. No relationship has appeared in the literature between color sensitivity and accidents (e.g., Allen, 1969; Cooper, 1990; Richards, 1966). This is fortunate because 8% of the male population is color defective, yet likely drive as safely as those with full color discrimination capability. Traffic engineers have added blue to the green lights, yellow to the red lights, and red-green defectives have little difficulty in using color codes found in driving.[1]

Again, the real issue is whether the beginning driver responds appropriately to such traffic signals, based on whatever kind of contextual information is available, and not the psychophysical sensitivity of the sensory system. Stereopsis may be related to maneuvers while parking. Whereas mishaps in such locations are bothersome for insurance companies as well as other drivers, they seldom result in serious injuries!

Precise measurement of these sensory sensitivities appears to bear no relationship or minimal relationship to safe motor vehicle operation; this fact is borne out by so much research that it is difficult to be skeptical (see Burg, 1971). But the few sensory sensitivities that may prove of predictive value in averting driving accidents, for example, bilateral field restriction, are seldom if ever tested in licensing examinations (see Keltner & Johnson, 1987).

Finally, there is the road test. This is a possible ultimate criterion for whether the testee has sufficient knowledge and skill to drive in public. Such tests are typically conducted on closed courses, or in low-density, low-speed traffic in an area neighboring the licensing site. They are typically conducted in full daylight, on dry pavement, and last between 10 and 30 min (shorter tests seem to be increasingly the rule due to cost factors). They are little more than minimal skill screens for minimally skilled beginning drivers, most of whom have spent less than 50 hours behind the wheel. A major component of such road tests, one that is often carefully scored and weighted, is the parallel parking test. (How many people have been injured, maimed, or killed while parallel parking? Note that today, most parking is not even parallel, but diagonal, and located in grocery store and shopping mall parking lots.) The testee who cannot maintain minimal control of the vehicle may be screened out by such subtests, but they hardly seem functional given the brief nature of the road test. Then, the novice driver is usually asked to

---

[1]The first author is a red-green defective, has held licenses in 3 states, has driven 43 years without an accident, and has never had even a near-miss due to color defect.

turn right and to turn left (whether he or she signals is very important); the novice driver must stop and start at a traffic light, stop at a stop sign (i.e., obey all traffic signs), turn into the lot, and stop the car smoothly before a license is granted. Thus, the road test, too, appears to be a minimally functional coarse screening test that bears little functional relationship to driver skill, alertness to potentially dangerous situations, ability to deal effectively with traffic environment stressors, and driver alertness.

Numerous attempts have been made to construct more suitable tests for driver licensing or relicensing. Almost all measures presumed requisite "basic abilities" that should apply to a wide range of driving situations. From DeSilva's early (1935) attempts to much more recent ones (e.g., Olson, Butler, Burgess, & Sivak, 1982), similar notions from the classical psychophysical view have dominated. All such tests rest on assumptions that it is possible (and desirable) to predict perceptual aspects of driving performance from some combination of mostly static vision tests (e.g., contrast sensitivity, acuity, color vision) plus general abstract measures of a more perceptual nature (e.g., depth perception—again, using static displays). One recent foray into integrated visual function tests was sponsored by the National Highway Traffic Safety Administration (NHTSA) (Shinar, 1977) and included tests for static acuity under normal, low, and glare illumination conditions; dynamic acuity; central movement in depth; central and peripheral angular movement; and field of view. The most useful measure for predicting overall accidents was acuity under low illumination levels, but dynamic visual acuity predicted daytime accident rates best. None of the measures accounted for large amounts of accident variance.

Although specific road tests have been devised that appear more closely related to driving skill (e.g., Olson et al., 1982), it is not clear that the skills chosen are the most appropriate ones; nor is it clear that road performance tests are practical, when one takes into account their administrative cost and the need for many specialized facilities. More recently, there have been rumblings about computer-video driving tests, with California being a likely site for testing (see Viets, 1990). Concerned agencies and individuals continue to call for the development and validation of functional driving tests (e.g., NHTSA, 1989). Several are being used for research (e.g., Schiff & Oldak, 1993), but none has yet emerged in a licensing bureau.

1.1.2   The Ecological Approach to Perception, and Its Impact on Automobile Driving Research and Applications

In 1938, Gibson and Crooks published a unique paper (Gibson & Crooks, 1938; also reprinted in Reed & Jones, 1982). It adopted the then-popular field-theoretical approach favored by Gestalt psychologists, in general, and Lewin (1936), in particular. The paper presented an analysis of the perceptual and control characteristics of driving that diverged markedly from what had gone on before, and much of which has occurred since. There was no mention of sensory acuity, illusions, or constructive processes in which the mind concatenated driving events from sensation-like bits and pieces. It may have marked the beginning of the application of "flow models of perceiving" (e.g., see Johansson, von Hofsten, & Jansson, 1980). There was no mention of wavelengths of light, sensory sensitivities, classical thresholds, or simple motor acts involved in vehicle operation. Indeed, the authors proclaimed "the task of the automobile driver is so predominantly a perceptual task, and that the overt reactions are so relatively simple and easily learned, that the analysis has to be carried out on a perceptual level and with concepts more appropriate to this requirement" (Gibson & Crooks, 1938, p. 453). Every student of perception and driving, and the human factors integration of these two bodies of knowledge, could profit from a careful reading of this paper. Whereas Gibson gave up the particular theoretical approach (valanced field theory) in his later works dealing with locomotion and perception (e.g., Gibson, 1950, 1958, 1966, 1979), the early basic approach was quite similar and was a harbinger of what was to come later. Its pioneering theoretical analysis of automobile driving contained concepts that were translated more or less intact (although becoming more specific) into Gibson's general approach to perceiving.

A nonexhaustive summary of the major concepts of Gibson and Crooks's (1938) paper follows:

1. The major focus of driving is locomotion-like walking or running, but it is accomplished from within a tool-like vehicle. The function of driving is usually to arrive at a destination, and the locomotor course is continually guided by vision, primarily, and modified to avoid collisions with other cars or obstacles and to follow particular paths.
2. The visual field of the driver is special, being selectively attuned to locomotor-relevant aspects. The most important part of the

environment is the road. The road has borders and consists of fields of possible paths through which one may pass unimpeded. These paths may be termed *fields of safe travel*, and they move with the car and driver through terrestrial space. The point of reference is not stationary objects in the environment, but the driver himself.

3. Steering involves keeping the vehicle headed down the middle of the field of safe travel. The field of safe travel contains a minimum stopping zone, which varies in length with the speed of the vehicle, its braking characteristics, and road surface conditions. The forward margin of the minimum stopping zone is behind the forward boundary of the field of safe travel.

4. The driver's awareness of his own velocity does not consist of estimates in miles per hour, but of the distance within which he could stop. (Note that this concept may be translated to time or a spatiotemporal estimate.)

5. A frontal shearing off of the field of safe travel which reduces it to or below the minimum stopping zone produces the experience of imminent collision and an immediate braking response.

6. The field zone ratio is the ratio of depth of the total visible field of travel to the minimum stopping zone. It is an habitual index of cautiousness, but may decrease when the driver is in a hurry. When speed is increased so that the forward margin of the minimum stopping zone nears the size of the field, driving begins to feel "dangerous." When the field of safe travel contracts because of a perceived obstacle encroaching on the zone, deceleration is proportional to the rate of recession of the forward margin of the field of safe travel toward the minimum stopping zone.

7. In high speed driving, the minimum stopping zone extends further down the road ahead of the vehicle than in slower driving. Good roads and relatively light traffic offer an extended field of safe travel, and the person who tends to drive faster usually does so. The driver must look farther ahead to search for potential obstacles or impediments to continue travel at that speed. The field of safe travel is modified by moving obstacles (traffic flow) and static obstacles. It is also modified by symbols, such as road signs, markings, traffic lights, and brake lights of other cars. If legal signs are unrealistic (e.g., speed limits on curves that are far lower than speeds with which the

curves can be safely negotiated), they will tend to be ignored,[2] or the driver will strike a compromise between the natural constraints of the road configuration and car maneuvering capabilities and those recommended by traffic control signs. It is these factors rather than legal taboos per se (maximum speed laws) that tend to govern the driving speeds of drivers, although the visible presence of police increases the likelihood of adherence to posted laws.

8. When an obstacle suddenly cuts off the field of safe travel inside the minimum stopping zone, an entirely new field of travel may open up, for example, the shoulder of the road in emergency situations. The field of safe travel is limited by a number of factors, including obstacles that encroach on it, the distances at which daytime vision and nighttime vision become inadequate due to purely optical factors, the margins of road illuminated by headlights, constriction of the field by glare from other's headlights, or horizon cropping, as by brows of hills one is traversing. All these should yield deceleration when they diminish the field-zone ratio.

9. Overdriving brakes or headlights leads to contraction of the objective field of safe travel until it approaches the minimum stopping zone, or does so without a corresponding contraction of behavioral capabilities. Learning safe driving habits involves semi-automatic perceptual-motor habits in which a safe margin is maintained between the minimum stopping zone and the field of safe travel. (Gibson, 1966) later called this the "margin of safety," which is found in much of the literature on driving [e.g., AAA, 1985a; Leibowitz, 1965, p. 51; Schiff & Oldak, 1990; Schiff, et al., 1992].)

10. Clearance lines radiate from obstacles, that is, a negatively

---

[2]Both authors have driven extensively in states restricting speeds on high-speed divided highways to 55 mph. In spite of the legal taboos, average traffic flow rates on these roads range between 65 to 70 mph—unless police are evident. Here is a contemporary instance in which Gibson's point about goals (desire to complete long-distance trips in a reasonable time span) and natural laws— constraints of roads, traffic (flow rates) and vehicle capabilities versus man- made laws and posted speed limits—hits home. Legislators who support unrealistically low speed limits (see Pevsner, 1991) might do well to travel their own roads at "legal" speeds for 4-5 hours (instead of flying about!) before voting on such legislation—which may encourage ignoring reasonable as well as unreasonable traffic laws.

valanced halo of avoidance. The more injurious potential an obstacle may have, the more it is avoided and the greater the extent of its clearance lines. (Consider the degree to which an auto driver tries to steer clear of an adjacent semi-trailer truck, versus a small sport roadster!) For a moving obstacle, the clearance lines radiate from the point where said obstacle will be located when one's own car comes closest to it, that is, the point of potential collision. Thus, one may actually steer toward the present location of a walking pedestrian with the foreknowledge that he or she will be well beyond the present location when one's vehicle arrives at the "trail." The greater the speed of a moving obstacle, the further ahead its clearance lines are projected. The correct perception of such clearance lines involves the projection of other vehicles along their paths of travel, as they may be jointly involved. Thus, passing another car traveling in the same direction, with another car approaching in the adjacent lane, requires an "estimate" (quote is Gibson & Crooks's) of the speed of obstacle(s) plus an "estimate" of the speed of one's own car.

> *Although the term estimate is misleading since there seems to be no conscious process calculation involved. Here is a case of a highly complex situation, involving relationships between two speeds of movement, which would not be an easy problem to solve with pencil, paper, and formulae. But for the skillful driver, the perceptual field-situation may be immediate, clear, and (let us hope) accurate. Complex "estimates" of speed and location are represented in the experience of the driver only by the simple seeing of an open or closed field of safe travel. (Gibson & Crooks, 1938, p. 465; emphasis in original; see also Reed & Jones, 1982, p. 129)*

Here Gibson foresaw the "computational issue" found in the recent time-to-arrival literature (e.g., McLeod & Ross, 1983; Schiff & Detwiler, 1979; Schiff & Oldak, 1990; Tresilian, 1990; Schiff et al., 1992).

11. Driving skill consists of the organization within the field of view of a correctly bounded stopping zone for a full range of speeds, road and traffic conditions, and fields of safe travel, precisely fitted to actual and potential obstacles in the field of view at any instance. This field must conform to the objective possibilities for safe locomotion. The motion

behaviors required to maintain a safe and constant ratio between the field of safe travel and the minimum stopping zone and to keep the car in the middle of the field are the basic operations of driving an automobile.

12. Traction is perceived. (Note that there are visible proprioceptive components of skidding, so that skidding is not a kinesthetic experience, but involves specific kinds of directional motions that have optical as well as kinesthetic consequences.) The centrifugal force producing a skid functions as a potential obstacle encroaching on the concave side of a curved field of travel; losing traction may thus be perceptually projected onto the road ahead and may cause adjustments of speed or course to prevent loss of traction.

In this conceptual analysis, world and driver were not considered as separate entities, but as interactive and reciprocal aspects of perception and action (see Lombardo, 1987). The driver's world is a phenomenological construct, having properties relevant to the operation of the vehicle through a moving maze of interspaces and objects. As the driver proceeds down the road, all this is represented in the optic array and produced by the vehicle's and driver's motion. As the vehicle proceeds forward, more of the world, including changes in vistas (Gibson, 1966, p. 207), is revealed via "disocclusion"; whereas world details flow by and disappear at the periphery or under the visible edge of the car's hood via progressive "occlusion." A glance in the rear-view mirror reveals the reverse backward flow, not only specifying where the car has been, but also a view of other cars flowing by in the opposite direction or coming up from behind.

From aspects of this complex flow of detail, we perceive our own motion through the world, termed *egomotion* (Warren, 1976; Warren & Wertheim, 1990). Egomotion has been studied extensively as a source of information used to guide aircraft, automobiles, or other vehicles (see Figure 1.1).

The driver co-perceives the stable aspects of the world (road, signs, trees, bridges, etc.) as well as his or her own course of motion relative to those stable aspects. The driver also perceives moving objects within the stable framework, including other vehicles, pedestrians, and so on. It is the continually changing relations among these aspects of the dynamic flow pattern that provide the critical information for vehicular guidance among features of the stable world and flowing traffic. As signs, houses,

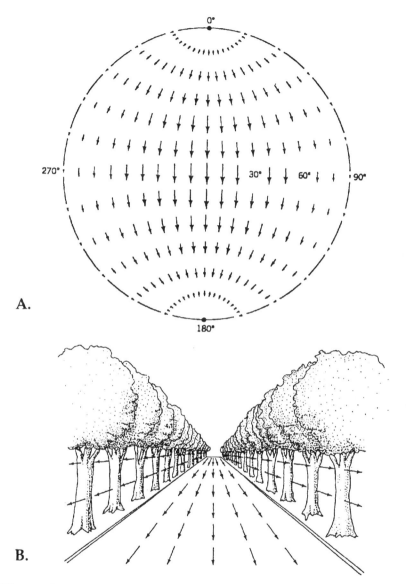

Figure 1.1. **A.** *The flow of velocities in the optic array reflected from the surface of the earth. This is the flow pattern obtained with locomotion parallel to the earth in the direction of the pole at the top of the graph. The vectors are plotted in angular coordinates. This is a view from above.* **B.** *Motion perspective for a perceiver moving in a straight line down a country road and fixating at the horizon. (A) From: Gibson, J.J., Olum, P., & Rosenblatt, F.*

bridges, and like are approached, they swell in projected angular size and change in projected angular shape as a visual system approaches them. The mathematics of these optical changes have been spelled out by several researchers and theoretical writers (e.g., Carel, 1961; Cutting, 1986; Lee, 1976; Warren & Wertheim, 1990). As moving vehicles (e.g., cars and trucks) approach, their projected forms dilate more slowly if moving in the same direction, but at a slower speed. If they are moving on a course different from a driver's, their dilating projections in the optic array will have a nonsymmetrical shape change (e.g., see Cutting, 1986; Gibson, 1958; Schiff, 1965). These differences and changes in relative rates of "image" dilation (or contraction) provide potential information for drivers concerning when and if objects reach particular locations in the driver's world.

Gibson's conception of this complex optical flow field of the driver was developed more elaborately and elegantly when he applied it to the perceptual information likely used by aircraft pilots during his work in aviation psychology during World War II (Gibson, 1950). Of course, the analysis of the driver's optical flow is simpler than that for pilots, because eye height above the ground changes dramatically when flying, but is virtually stable when driving. Also, the vehicular motion parameters of pitch, roll, and yaw—which substantially complicate the pilot's optical flow field—are essentially absent in non-crash auto-driving situations. And finally, whereas the vehicular maneuvers of the driver lead to a far more constrained optical situation for the driver than the pilot, environmental clutter is generally far greater for the driver than for the pilot.

1.2   The Ecological Approach to Perception and Action and Its Impact on Automobile Driving Research and Applications

A number of hypothesized parameters of the information flow field just described were analyzed by Gibson, his students, colleagues, and others as to their sufficiency and necessity in reproducing perceptual experience and guidance of action. Some of these are now introduced.

1.2.1   Egomotion

Perhaps the first aspect of driving, from the driver's point of view, is movement of the self. Unlike "naked" (nonvehicular) egomotion, driving also involves the perception of the driver's "vehicular envelope" (the car) moving through a stable environment. This basic form of perceived vehicular motion has been carefully researched by a number of researchers (e.g., Warren, 1976; Warren & Wertheim, 1990), following its initial description by Gibson (1950, 1958). More recent research and theory has been aimed toward the practical goal of producing apparent egomotion artificially via various simulators used for training, research, and assessment.

Egomotion is usually defined as displacement of the perceiver with respect to an environment (Warren & Wertheim, 1990, p. 7). Egomotion is specified by, and is typically perceived primarily by, virtue of optical flow, as shown graphically in Figure 1.1. This optical flow emerges from the changing light pattern projected to a point in space so that it may be picked up by a visual system, and the pattern is produced by structure in the environment that reflects (or emits) light varying in brightness and color. We generally call this structure *ground*, *surfaces*, and *objects*. Because the visual system is in motion relative to these stable environmental features during egomotion, the light pattern flows or streams as one moves or drives along (see Koenderink & van Doorn, 1975). Thus, it is the observer's own motion that produces the optical transformations of the entire optic array and the resulting flow patterns. In Gibson's words: "Flow of the ambient array specifies locomotion, and non-flow specifies stasis. By flow is meant the change analyzed as motion perspective (Gibson, Olum, & Rosenblatt, 1955) for the abstract case of an uncluttered environment and a moving point of observation. A better term would be flow perspective, or streaming perspective" (1979, p. 227).

The resulting flow pattern may be described in terms of vectors— not the general vectors of human movement described in 1938, but vectors in the optic array produced by such movements. These vectors specify both the magnitude and direction of egomotion, such as how fast and where the observer is going. Resulting vector fields are the hypothetical basis for the driver's perceiving, acting, and sometimes judging (Warren & Wertheim, 1990, p. 15). Gibson (1950) and his colleagues (Gibson et al., 1955) rendered the now neoclassical concepts for such travel over ground surfaces such as those over which drivers maneuver their vehicles.

Some recent mathematical characterizations of such a "canonical" flow field for drivers (and other egomotion perceivers) were provided by Warren and Wertheim (1990, p. 15):

$$\vec{p_i} = -s_i \ (\vec{t} - (\ t \ . \vec{d_i}) \ \vec{d_i}) - \vec{R} \ . \ \vec{d_i} \tag{1.1}$$

$$\frac{\vec{du_i}}{dt} = [u_i \cdot (\vec{u_i} \cdot \vec{w}) - p_i - 1 (\vec{u_i} \cdot v)] \cdot \vec{u_i} \tag{1.2}$$

The direction of the driver's egomotion can be derived from the pattern of vectors in the flow field, as shown in Figure 1.1b (e.g., see Bruce & Green, 1990, p. 281; Lee & Lishman, 1977, p. 281; Schiff, 1986, p. 247).

The magnitudes of flow vectors may specify how fast the driver is going ("egospeed" or "ego velocity"), but he or she is especially sensitive to changes in velocity, rather than absolute velocities of flow. Two different flow variables (global flow rate and local edge rate) seem usable for detecting egomotion velocity, depending on how the driver (or pilot) samples the optic array. Note that as shown in Figure 1.1, flow velocities of particular elements vary depending on their position in the field of view. The flow velocities resulting from elements are zero at the horizon and at the "aim point," which Gibson termed the *focus of expansion* and others termed the *pole* of the optic flow field (Bruce & Green, 1990, p. 330). Flow velocities are greater the closer elements on the surface on a road surface are to the self. Further, flow velocities expand or dilate from the road straight ahead, but are lamellar at the sides of the windshield or side windows if the driver looks straight ahead (Owen, 1990, p. 44). The flow velocities of the particular elements in the array also vary as a function of their vertical distance from the eye, and as a function of their sizes, if bounded contours. A driver might sample the overall "global" flow rate of an entire flow field, which would comprise a weighted average of the various flow rates. A pilot, who flies at different altitudes above the ground, is more likely to utilize global flow rates to discriminate his or her own velocity variations, because of the great variety of factors influencing element flow velocity (e.g, see Owen, 1990; Warren & Wertheim, 1990;). However, the driver's eye height remains approximately constant, and projected optic element size varies relatively little.

Denton (1980) identified another potential source of optical information for egomotion velocity, termed *edge rate*. The edge rate of flow is the number of optical texture units traversed by some portion of the self or self envelope per unit of time. The driver might pick up edge rate at the leading edge of the car's hood, for example; whereas the pilot could do likewise at the leading edge of the plane's cabin window or the wing.[3]

The driver's constant eye height (about 4 ft for most cars) permits him or her to use either flow velocity or edge rate, although many drivers seem to use edge rate to assess and control changes in velocity of egomotion. In trucks, eye height may be twice that of the car driver, but note that to determine absolute (metric) velocity from flow velocity of either kind, one must calibrate the flow rate in mph or kmph, most likely via the speedometer. Moreover, drivers, alternately driving both cars and trucks, may have temporary effects of miscalibrated egomotion velocities, until they recalibrate from the lower or higher eye height.

Flow rate, as Gibson and Crooks noted in their 1938 paper, is most likely used to detect changes in ongoing velocity, is derived from changes in the rate of change (of global or edge rate velocities), and is one of many higher order derivatives thought to govern locomotion and egomotion (see Koenderink & van Doorn, 1975). Thus, changes in the rate of flow velocities provide potential information for our a driver's acceleration and deceleration, as well as maintenance of a constant speed on the road—without the need to constantly look at the speedometer. Recently, the widespread use of cruise-control devices in passenger cars obviates the use of constant flow rate to maintain highway speeds while cruising low traffic density roads.

Denton (1980) utilized this notion in a clever application to traffic control. To slow autos approaching potentially dangerous exits or roundabouts from high-speed highways, stripes were painted on the road perpendicular to the path of travel, such that the closer one came to the curved danger point, the closer the lines were spaced. Even drivers who may have adapted to a high rate of travel (adaptation to flow rate;

---

[3]Note the edge rate is responsible for the vivid "illusion" of deceleration experience as one's plane accelerates and gains altitude during takeoff. Note that for the Gibsonian position, the illusory percept is information driven and proves nothing about the insufficiency of information for veridical perception. One may attend to informative global flow-rate information or (in this case) misinformative edge-rate information.

see Schmidt & Tiffen, 1969) perceive themselves as accelerating as they traverse the more closely packed striped pavement and apply their brakes to counteract the ("illusory") perceived acceleration. Accidents at such locations were substantially reduced as speeds were reduced by drivers under the influence of illusory acceleration.

Egospeed must be calibrated via a speedometer before it can be used to control metric velocity, although no such calibration is required to use it to maintain, increase, or decrease one's own speed. It is a simple matter to maintain one's own egomotion velocity without using metric instrumentation by maintaining a constant rate of change in optic flow, even though metric estimates may be in error. Schiff and his colleagues (1992) found that driving ego velocities tend to be underestimated at typical highway speeds, with egospeed underestimated by about 10 mph at 52 mph, 14 mph at 62 mph, and 18 mph at 72 mph, that is, on the order of 20%–25% (see Figure 1.2). These judgments were made from egomotion films, however, and it is not certain that the results pertain in real egomotion ("vection").

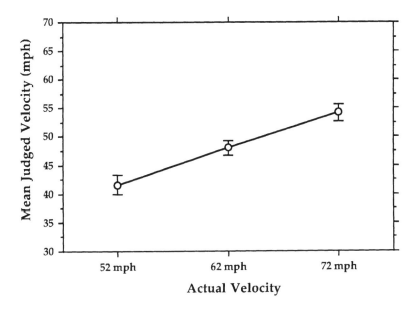

Figure 1.2. *Judged velocity as a function of actual ego velocity.*

When the driver stops his or her car, egomotion flow field velocity reduces to zero magnitude. Upon acceleration, the flow velocity accelerates and changes accordingly. Thus, the regulation of forward motion and the "threading" of the car along a desired path is largely accomplished via the driver's attention to changes in the flow field produced by the car's own motions (Spurr, 1969). Note that there is no known measure of utilization of such flow information in extant tests of drivers' performances. Although it is known that as drivers gain experience, they tend to monitor farther and different parts of the flow field (e.g., see AAA, 1985b; Nagayama, Morita, Miura, Watanabem, & Murakami, 1979). "

Within the optic array is what Gibson (1950) termed the *focus of expansion*, that is, the point in the optic pattern from which all flow vectors radiate. Gibson initially conceived of the f.o.e. as the "aim point" used when pilots approach runways, when drivers follow lead cars, or when a perceiver approaches any target. A lively debate has occurred in the f.o.e. literature concerning the utility of f.o.e. information in controlling locomotion or vehicular motion, some from empirical grounds (e.g., Johnston, White, & Cummings, 1973; Kaufman, 1968), and some from theoretical grounds (e.g., see Cutting, 1986; Regan, 1985; Regan & Beverly, 1982; Torrey, 1985). The main issue concerned cases in which the observer might look at some point in the world toward which he or she was not moving while the observer, him- or herself, was also in motion. An example would be looking at an oncoming car in an adjacent lane while continuing to drive straight ahead in one's own lane.

In addition to the f.o.e., the automobile driver typically utilizes other aspects of the flow pattern. For example, Figure 1.3 is an abstract rendition of a "dip" in the road along which one is driving, specified optically by "curving" local deformations in the continuous flow pattern during egomotion. The driver may perceive that this dip does not permit traveling at his or her current speed and may slow down accordingly. In driving, the "dip" shown in Figure 1.3, might lead to acceleration and "leaping" for a stunt driver, but lead to braking for a conservative motorist. A red traffic signal is a learned symbol initiating braking; a smooth, dry, clear, straight road permits driving faster than a rough, wet, crowded, twisting road. Drivers must learn most of these relations, including the one in which a "clear" railroad crossing at which lights are flashing and a locomotive "slowly" approaches does not permit driving across. Such relationships are conceived by Gibson and his adherents as linking perception and action—they specify the action potential of the environment-perceiver interface, not just the physical world (e.g., see Warren, Young, & Lee, 1986).

In addition to these aspects, the driver's flow field may contain incursions, such as perturbances produced by other vehicles' or pedestrians' entrances into the field of view. In such cases, the flow produced by the terrain would be progressively occluded by the boundaries of the incurring object. When an object crosses the boundary of peripheral vision, the driver may perceive and act (braking, steering, blowing horn, uttering expletives), depending on the perceived confluence of the driver's and other's respective paths.

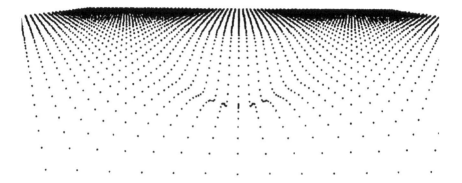

Figure 1.3. *Depiction of either a rigid depression which will soon pass underneath the observer, or a non-rigidity in the ground that travels with the observer such as might be produced by the downwash of a helicopter. Dimple based on the Witch of Agnesi.*

Note that in the ecological theoretical treatment, the shape of the object, its nature, and its likely location or path are also specified by another set of vectors, those produced by the occlusion of background texture or by its visibly discriminable contours or patches of color/brightness differences that comprise its visible surfaces. If the intruding object is a vehicle, its translatory motion would be approximately linear, and it would be perceived as in front of that part of the environment it progressively occludes and disoccludes (e.g., see Kaplan, 1969). If the intruding object is a pedestrian, its event/object nature (a walking person) is specified by a complex set of motion vectors, studied via computer-generated programs that capture the motion configurations specifying such objects and events (e.g., see Cutting & Kozlowski, 1977; Cutting, Proffitt, & Kozlowski, 1978; Johansson, 1973). Further, if the current rates of motion would lead the driver's path (of the self-envelope) to intersect with the path of the potential obstacle, the driver would brake, steer, or accelerate to change that state of affairs (see Arnone, 1991a, 1991b). In other words, the driver would perceive the confluent "times-of-arrival" of him- or herself and the other. The corrective actions of braking, steering, and so on, would be further monitored by the driver to achieve a satisfactory resolution of the intersection of self-envelopes, that is, being at the same place at the same time. This then would constitute a complex perception of, and avoidance of, impending collision—a major topic researched primarily by researchers adopting the ecological approach to perceiving and acting.

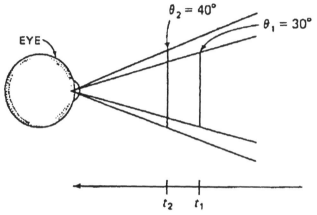

Figure 1.4. *Time-to-collision (T_c) is optically specified as a relative rate of angular size over time. From: Perception: An applied approach (p. 256) by Schiff, W., Copyright © Houghton Mifflin. Reprinted by permission.*

1.2.2   Time-to-Collision Time-to-Arrival *Tau*

One of the early problems studied within the ecological framework which bears strongly on the task of automobile driving is that of perceiving time-to-contact or time-to-collision (T_c). T_c is the time remaining before contact between the self and another object or surface, with either the self, other object, or both in motion. This information may be available from a number of sources, including the optical dilation of an object's contours during approach or rate of distance change. For example, two cars approaching each other at a closure rate of 60 mph at a distance of 500 ft would have a T_c of 5.7 sec. If a driver may perceive him- or herself to be on a collision course with another object, he or she may utilize optical (as well as acoustical) information to avoid such collisions (e.g., see Gibson, 1958; Lee, 1976; Schiff, 1965; Schiff

& Detwiler, 1979; Schiff & Oldak, 1990). Paradoxically, this involves perceiving the endpoint of an event (approach, then collision) before it occurs. This class of percepts was of particular interest to Gibson because it was a prototypical example of the pickup and utilization of optical information over time, rather than an instantaneous "snapshot" as conceived in traditional theories.

In his 1958 paper on animal locomotion, Gibson stated:

> *Approach to a solid surface is specified by a centrifugal flow of the texture of the optic array. Approach to an object is specified by a magnification of the closed contour in the array corresponding to the edges of the object. A uniform rate of approach is accompanied by an accelerated rate of magnification. At the theoretical point where the eye touches the object the latter will intercept a visual angle of 180 degrees; magnification reaches an explosive rate in the last moments before contact. This accelerated expansion in the field of view specifies imminent collision, and it is unquestionably an effective stimulus for behavior in animals with well-developed visual systems. In man it produces eye blinking and aversive movements of the head, even when the stimulus is a harmless magnification of a shadow on a translucent screen. At lesser intensities this "looming" motion; as it might be called, presumably yields lesser degrees of aversion, or a slowing down of approach. The fact is that animals need to make contact without collision with many solid objects in their environments: food objects, sex objects, and the landing surfaces on which insects and birds alight (not to mention helicopter pilots). Locomotor action must be a balance between approach and aversion. The governing situation must be a balance between flow and non-flow of the optic array at the moment when the contour of the object or the texture of the surface reaches that angular magnification at which contact is made. (p. 188)*

Shortly after this pioneering paper was published, one of Gibson's students, William Purdy, generated the mathematics of the stimulus information for time-to-contact or time-to-go, and it appeared in his thesis (Purdy, 1958) and in a theoretical treatise, including a study of humans estimating time-to-contact (T_c) with an apparently approaching surface (Carel, 1961). The simple formula for T_c is as follows, given the spatial change in optical separation (i.e., angular size change) of any two points or contours over time (t):

$$T_c = \frac{\theta_1}{\theta_2 - \theta_1 \, / \, t_2 - t_1}$$

(note: $\Delta\theta / \Delta t$)

(1.3)

Formula 1.3 calculates T_c from angular size changes over time. As shown in formula. 1.3, the relative rate of increase in angular separation of points or contours on the approaching object's surface is the reciprocal of the time-to-go.

The effectiveness of such information for producing avoidant behaviors (indicating impending collision was being perceived) was explored in various animal species and humans (Schiff, 1965; Schiff, Caviness, & Gibson, 1962) and by numerous other researchers (e.g., see Schiff, Benasich, & Bornstein, 1989, Schiff & Detwiler, 1979, for brief critical reviews). But it was Lee (1976) who developed the hypothesis that perceiver-actors as diverse as automobile drivers, diving birds (Lee & Reddish, 1981), long jumpers (Lee, Lishman, & Thompson, 1982), and fast-ball sports participants (Lee, Young, Reddish, Lough, & Clayton, 1983) use *tau* information to control their visual-motor behaviors. Lee's analysis of *tau* (τ) utilization in automobile braking is based on the notion that the driver may utilize the inverse of the proportional rate of dilation to control brake pressure, because, as derived from Purdy, *tau* = 1/rate of dilation, for example, between the headlights of an approaching car. Then the issue becomes one of applying sufficient pressure, to cancel dilation prior to contact. A driver's deceleration is adequate to avoid collision if $\dot{\tau}$ is greater then -1/2 (where $\dot{\tau}$ is the derivative of τ with respect to time), and a driver may maintain a safe following distance by maintaining $\dot{\tau}$ at a marginal value greater than one-half (i.e., 1/2) (see Lee, 1976, for details of derivation). Recent critiques by Tresilian (1991) and Wann, Edgar, and Blair (personal communication 1992) noted some problems regarding the adequacy of this formulation.

Numerous studies have been performed in attempts to discover whether drivers or other perceivers can indeed use *tau* information to perceive and judge time-to-contact, with a consensus that they can do so with high accuracy if the temporal value is about 2 sec or less. But when the time remaining before contact reaches 3 sec or more, estimates made by pushbutton or verbal means indicate increasing underestimation as time-to-contact increases (e.g., see Cavallo & Laurent, 1988; McLeod & Ross, 1983; Schiff & Detwiler, 1979; Schiff & Oldak, 1990; Schiff et al., 1992). The amount of underestimation is considerable, with most perceivers estimating time-to-contact at about 60% to 70% of its actual value, regardless of whether perceivers are motionless and approached by an oncoming object or car, or whether they are in motion approaching a static target.

These findings derive from a variety of situations, including artificial animated events (e.g., Schiff & Detwiler, 1979), filmed real events (McLeod & Ross, 1983; Schiff & Detwiler, 1979; Schiff & Oldak, 1990; Schiff et al., 1992), and real approach events in three-dimensional space (Cavallo & Laurent, 1988; Kebeck & Landwehr, 1991). Also of interest, females tend to underestimate arrival time more than males (especially older females, see Schiff et al., 1992). Cavallo and Laurent (1988) suggested that such underestimation decreases with driving experience, although their results were confounded by gender selection and possibly by risk taking (see Schiff & Oldak, 1990). They used French drivers, often noted for their aggressive driving styles, relative to American drivers (see Altman, 1989, p. 33). Further, the trajectory of approach is quite important in determining whether arrival time is underestimated, and by how much (Schiff & Oldak, 1990). Transverse trajectory motions are underestimated far less, if at all, whereas radial (head-on) approaches are underestimated. Thus, cars approaching an intersection from a side road may be perceived or judged as likely to arrive *after* they actually would, leading to possible decision errors by a motorist approaching the intersection at a 90° path relation to the side-road trajectory.

A theoretical issue has pervaded this research, that is, whether observers are using dilation information (e.g., *tau*) to estimate when approach will terminate in contact, or whether they are obtaining distance information and velocity information and "computing" time-to-contact by neural division ($Tc = d/v$). Although there is no definitive answer to this question (e.g., see Tresilian, 1991), there are strong indications that drivers usually use the rate of dilation information (or gap-closure rate information; see Bootsma & Oudejans, 1990) rather than metric forms of distance and velocity information as other researchers have suggested. For example, Schiff et al. (1992) showed films of approaching cars and had observers push a button to indicate when they thought an approaching car would arrive had they not lost sight of it, and verbally estimate the velocities and distances of the approaching cars seen in an identical set of films. The correlations between the two sets of judgments were essentially zero; when an observer gave a "seat of the pants" estimate of when the car would arrive, indicated by pushing a button, it was unrelated to metric judgments of distance/velocity of the same approach event. Yet, much of the research in the human factors field relies on verbal distance and velocity estimates of autos to reach decisions about human perceptual performance in the driving field (e.g., Olson & Sivak, 1986; Scialfa, Guzy, Leibowitz, Garvey, & Tyrrell, 1991; Scialfa, Kline, Lyman, & Kosnik, 1987; Sivak, 1987). Again, it is of some interest that driving assessment measures do not attempt to use arrival-time measures in any form, although they are acknowledged as important (e.g., Olson & Sivak, 1986). However, a recent study has shown that left turn margins of safety are preferable to arrival time estimates used for the same purpose (Arnone, 1994).

1.3   Human Factors and Automobile Driving: Can Parallel Roads Meet?

It should be clear from the foregoing that there are at least two distinct approaches to perceptual theory that lead to different sorts of research practices, selection of different variables to study, and resulting in two universes of discourse that appear to be parallel and nonintersecting. But this is likely a dangerous obstacle to innovations in the field of automobile driving research and to the design of better vehicles, equipment, roads, and ways to measure which is better. Rather than adopt a "good guy–bad guy" approach to the field, and lose the fruit of perceptual research because of honest differences in conceptual approaches, it may be time to bring the parallel universes together.

Practitioners in the engineering and human factors fields should more carefully choose research methods and measures in driving research, and should more carefully consider the ecological validity of their procedures and stimuli. Ecological validity does not mean doing everything in the laboratory or test situation just as it is done in the real world. Such an approach would be so expensive and time-consuming as to be almost impossible, although real road and driving track studies are widespread (e.g., see Gawron & Ranney, 1988; Olson & Sivak, 1986; Sivak, 1987). What this refers to is selecting those processes and situations that typify driving and utilizing stimuli similar to those picked up and used while driving (e.g., see Forbes, 1972, p. 37; Neisser, 1976).

Thus, static acuity measures used in driving research and testing have close to zero ecological validity. Dynamic acuity measures have a little more, but perhaps not very much more. Again, the driving task does not consist of discriminating fine details in patterns—even moving patterns. Some human factors researchers recognized this fact (Forbes,

1972; Rockwell, 1972). The driving task seems to be largely a complex set of vector detection and control processes. Studies and measures that reflect that reality will likely be far more successful than those that continue to ignore the ecological validity issue and test drivers for verbally accessed knowledge rather than performance and for irrelevant perceptual processes (Fox, 1988). Recent state-of-the-art reports may be found in Gale, Brown, Haslegrave, Taylor, and Moorehead (1991), Schiff and Oldak (1993).

On the other hand, sensory capabilities may indeed limit performance in actual driving situations, and wherever they can be economically measured, measuring them seems quite a good idea, however "sensory" rather than "global" they might appear. But in devising measures of such sensitivities, it would be more ecologically valid to embed them in a driving-like task. Familiar context may aid detection and discrimination (or so research has shown). And on the other side of the coin, cluttered environments (which almost all driving environments are) may hinder even a good visual system trying to locate a sign, signal, or other detail. Thus, unless one measures sensory capabilities in environments having characteristics similar to those of driving environments, one will never know whether one is under- or overestimating the functional utility of a system that may do well (or poorly) in the static and uncluttered environment of the medical practitioner's office. It is likely that if the impairment is extreme, the ecologically invalid measure will still detect it. But it is the marginal driver who is of most concern. It may be easier to detect marginal driving skills using dynamic realistic displays, rather than the usual battery of multiple choice question, static, and overly simplified perceptual measures and isolated sensory-sensitivity measures (e.g., see Schiff & Oldak, 1990, 1993).

Good engineering and human factors design is also important, as spelled out by Norman (1988). Even the good skilled driver with good vision may become a road hazard if he or she has to look for vehicle controls that all look and feel alike, that are in the wrong place, or that display information in a way that takes longer than necessary to "decode" (see Gale, Freeman, Haslegrave, Smith, & Taylor, 1986, 1988).

Thus, in closing, it appears that researchers engaged in the study of automobile driving should incorporate information from both the classical sensory psychophysics and ecological approaches, as well as findings concerned with human factors and design. Driving involves the integrated activities of sensing, perceiving, deciding, and acting. Before any actions or decisions regarding such actions can be made, the human factors scientist must become aware of phenomena in the perceiving and driving realm. Perceiving involves the capability of the driver to detect relevant phenomena. Sensing also involves the ability to pick up different sorts of information from the dynamic environment. For example, an individual may be fully capable of detecting an oncoming car (i.e., possess "normal" vision), but through some circumstance misjudge or misinterpret information such as closure rate, causing a collision. Or, the individual may pick up and interpret the information correctly, but as a result of poor engineering/design be unable to manipulate appropriate controls in a timely fashion.

Each of the aforementioned approaches to automobile driving research has provided a uniquely traveled road to the current state of the art. It is time for them to consider the findings from other approaches and unite the parallel universes of discourse and knowledge.

Acknowledgments

The authors of this chapter gratefully acknowledge the generosity of Michael Palij and Doris Aaronson in its preparation.

1.4   References

AAA (1985a). *To drive at night*. Washington, D.C.: The AAA Foundation for Traffic Safety.

AAA (1985b). *Visual perception in driving*. Washington, D.C.: The AAA Foundation for Traffic Safety.

Allen, M. (1969, September). Vision and driving. *Traffic Safety*, pp. 8–10.

Altman, J. (1989). *Berlitz: Blueprint France*. Lousanne,Switzerland: MacMillan S.A.

Arnone, W. (1991a). *An examination of response times with different input devices to unknown targets in driving scenes*. Unpublished manuscript.

Arnone, W. (1991b). *The effect of illumination level on response latency to target events in driving scenes*. Unpublished manuscript.

Arnone, W. (1994). Assessment of time to arrival and margins of safety gaps for left turns in younger and older drivers. Unpublished doctoral dissertation, New York University.

Ball, K. (1990). Cognition. In Conference on Research and Development Needed to Improve Safety and Mobility of Older Drivers. U.S. Dept. of Transportation, *NHTSA Conference Proceedings* (pp.

30–40), Washington, D.C.

Bootsma, R., & Oudejans, R. (1990). *Visual information about time-to-collision between two objects.* Unpublished manuscript.

Bruce, V., & Green, P. (1990). *Visual perception: Physiology, psychology, and ecology.* Hillsdale, NJ: Lawrence Erblaum Associates.

Burg, A. (1966). Visual acuity as measured by dynamic and static tests: A comparative evaluation. *Journal of Applied Psychology, 50,* 460–466.

Burg, A. (1971). Vision and driving: A report on research. *Human Factors, 13,* 79–87.

Carel, W. L. (1961). *Visual factors in the contact analog* (No. R61ELC60). Ithaca, NY: General Electric Company Advanced Electronics Center.

Cavallo, V., & Laurent, M. (1988). Visual information and skill level in time-to-collision estimation. *Perception, 17,* 623–632.

Cooper, P. (1990). Differences ages, different risks: The realm of accident. In J. Rothe (Ed.), *The safety of elderly drivers: Yesterday's young in today's traffic* (pp. 85–133). New Brunswick, NJ: Transaction Books.

Coren, S., & Girgus, J. (1978). *Seeing is deceiving: The psychology of visual illusions.* Hillsdale, NJ: Lawrence Erlbaum Associates.

Coren, S., & Ward, L. (1989). *Sensation and perception.* New York: Harcourt Brace Jovanovich.

Cutting, J. (1986). *Perception with an eye for motion.* Cambridge, MA: MIT Press.

Cutting, J., & Kozlowski, L. (1977). Recognizing friends by their walk: Gait perception without familiarity cues. *Bulletin of the Psychonomic Society, 9,* 353–356.

Cutting, J., Proffitt, D., & Kozlowski, L. (1978). A biomechanical invariant for gait perception. *Journal of Experimental Psychology: Human Perception and Performance, 4,* 357–372.

Denton, G. (1980). The influence of visual pattern on perceived speed. *Perception, 9,* 393-402.

DeSilva, H. (1935). *Research on the scientific investigation of driving skill* (No. XS-F2-U-25). Amherst: University of Massachusetts, F.E.R.A.

Flach, J., Lintern, G., & Larish, J. (1990). Perceptual motor skill: A theoretical framework. In R. Warren & A. Wertheim (Eds.), *Perception and control of self motion* (pp. 327–355). Hillsdale, NJ: Lawrence Erlbaum Associates.

Forbes, T. (Ed.). (1972). *Human factors in highway traffic safety research.*

New York: John Wiley Interscience.

Fox, M. (1988). Elderly drivers' perceptions of their abilities compared to their functional visual perception skills and their actual driving performance. *Physical Occupational Therapy in Geriatrics, 7,* 13–49.

Gale, A., Freeman, M., Haslegrave, C., Smith, P., & Taylor, S. (Eds.). (1986).*Vision in Vehicles I. Proceedings of the First International Conference on Vision in Vehicles.* North Holland: Elsevier.

Gale, A., Freeman, M., Haslegrave, C., Smith, P., & Taylor, S.(Eds.). (1988) *Vision in Vehicles II. Proceedings of the Second International Conference on Vision in Vehicles.* North Holland: Elsevier.

Gale, A, Brown, I., Haslegrave, C., Taylor, S., & Moorehead, I. (Eds.). (1991) *Vision in Vehicles III. Proceedings of the Third International Conference on Vision in Vehicles.* North Holland: Elsevier.

Gawron, V., & Ranney, T. (1988). The effects of alcohol dosing on driving performance on a closed course and in a driving simulator. *Ergonomics, 31,* 1219–1244.

Gibson, J. J. (1950). *The perception of the visual world.* Boston: Houghton Mifflin.

Gibson, J. J. (1958). Visually controlled locomotion and visual orientation in animals and men. *British Journal of Psychology, 49,* 182-194.

Gibson, J. J. (1966). *The senses considered as perceptual systems.* Boston: Houghton Mifflin.

Gibson, J. J. (1979). *The ecological approach to visual perception.* Boston: Houghton Mifflin.

Gibson, J. J., & Crooks, L. (1938). A theoretical field-analysis of automobile-driving. *American Journal of Psychology, 51,* 435–471.

Gibson, J. J., Olum, P., & Rosenblatt, F. (1955). Parallax and perspective during aircraft landings. *American Journal of Psychology, 68,* 372–385.

Johansson, L. (1973). Visual perception of biological motion, and a model for its analysis. *Perception and Psychophysics, 14,* 201–211.

Johansson, L., von Hofsten, C., & Jansson, G. (1980). Event perception. *Annual Review of Psychology, 31,* 27-64.

Johnston, I., White, G., & Cummings, R. (1973). The role of optical expansion patterns in locomotor control. *American Journal of Psychology, 86,* 311–324.

Kaplan, G. (1969). Kinetic disruption of optical texture: The perception of depth at an edge. *Perception and Psychophysics, 6,* 193–198.

Kaufman, L. (1968). Research in visual perception for carrier landing: *Studies of the perception of impact point based on shadow graph techniques* (sgd-5265-0031, Suppl. 2). Great Neck, NY: Sperry-

Rand Research Center.

Kebeck, G., & Landwehr, K. (1991). *Optical magnification as event information.* Unpublished manuscript.

Keltner, J., & Johnson, C. (1987). Visual function, driving safety,and the elderly. *Ophthalmology, 94,* 1180–1188.

Koenderink, J. (1990). Some theoretical aspects of optic flow. In R. Warren & A. Wertheim (Eds.), *Perception and control of self motion* (pp. 53–68). Hillsdale, NJ: Lawrence Erlbaum Associates.

Koenderink, J. J., & van Doorn, A. J. (1975). Invariant properties of the motion-parallax field due to the movement of rigid objects with respect to an observer. *Optica Acta, 22,* 773–791.

Lee, D. (1976). A theory of visual control of braking on information about time-to-collision. *Perception, 5,* 437–459.

Lee, D., & Lishman, J. (1977). Visual control of locomotion. *Scandinavian Journal of Psychology, 18,* 224–230.

Lee, D., Lishman, J., & Thompson, J. (1982). Regulation of gait in long jumping. *Journal of Experimental Psychology: Human Perception and Performance, 8,* 448–459.

Lee, D., & Reddish, P. (1981). Plummeting gannets: A paradigm of ecological optics. *Nature, 293,* 293–294.

Lee, D., Young, D., Reddish, P., Lough, S., & Clayton, T. (1983).Visual timing in hitting an accelerating ball. *Quarterly Journal of Experimental Psychology, 35A,* 333–346.

Leibowitz, H. (1965). *Visual perception.* New York: MacMillan.

Leibowitz, H. (1985). Grade-crossing accidents and human factors engineering. *American Scientist, 73,* 558–562.

Leibowitz, H., Owens, D. (1977). Nighttime driving accidents and selective visual degradation. *Science, 197,* 422–423.

Lewin, K. (1936). *Principles of topological psychology.* New York: McGraw-Hill.

Lombardo, T.J. (1987). *The reciprocity of perceiver and environment.* Hillsdale, NJ: Lawrence Erlbaum Associates.

McCormick, E. (1957). *Human factors engineering.* New York: McGraw-Hill.

McCormick, E., & Sanders, M. (1976). *Human factors in engineering and design.* New York: McGraw-Hill.

McLeod, R., & Ross, H. (1983). Optic-flow and cognitive factors in time-to-collision estimates. *Perception, 12,* 417–423.

Matlin, M. (1988). *Sensation and perception* (2nd ed.). Boston: Allyn & Bacon.

Nagayama, Y., Morita, T., Miura, T., Watanabem, J., & Murakami, N.

(1979). *Motorcyclists visual scanning pattern in comparison with automobile drivers'* (Tech. Paper Series No. 790262). Society of Automotive Engineers.

National Highway Traffic Safety Administration (NHTSA). (1989, August 23-24). In Conference on Research and Development Needed to Improve Safety and Mobility of Older Drivers. Washington, D.C.: *U.S. Dept. of Transportation.*

Neisser, U. (1976). *Cognition and reality.* San Francisco: Freeman.

Norman, D. (1988). *The psychology of everyday things.* New York: Doubleday.

Olson, P., Butler, B., Burgess, W., & Sivak, M. (1982). *Toward the development of a comprehensive driving test: Low-speed maneuvering. Highway* (Rep. No. UM-HSRI-82-4). Ann Arbor: Safety Research Institute, University of Michigan.

Olson, P., & Sivak, M. (1986). Perception-response time to unexpected roadway hazards. *Human Factors, 28,* 91–96.

Owen, D. (1990). Perception and control of change in self-motion: A functional approach to the study of information and skill. In R. Warren & A. Wertheim (Eds.), *Perception and control of self motion* (pp. 289–326). Hillsdale, NJ: Lawrence Erlbaum Associates.

Owsley, C., & Sloane, M. (1987). Contrast sensitivity, acuity, and the perception of "real world" targets. *British Journal of Ophthalmology, 71,* 791–796.

Pevsner, D. (1991, September 29). It's time to stop poking along at 55. *New York Times,* p. 13.

Purdy, W. (1958). *The hypothesis of psychophysical correspondence in space perception.* Unpublished doctoral thesis, Cornell University, Ithaca, NY.

Reed, E. (1988). *James J. Gibson and the psychology of perception.* New Haven, CT: Yale University Press.

Reed, E., & Jones, R. (Eds.). (1982). *Reasons for realism.* Hillsdale, NJ: Lawrence Erlbaum Associates.

Regan, D. (1985). Visual flow and the direction of locomotion. *Science, 223,* 1064–1065.

Regan, D., & Beverly, K. (1982). How do we avoid confounding the direction we are looking and the direction we are moving? *Science, 215,* 194–196.

Richards, O. W. (1966). Motorist vision and the driver's license. *Traffic Quarterly, 3,* 3–20.

Rockwell, T. (1972). Skills, judgment, and information acquisition in driving. In T. Forbes (Ed.), *Human factors in highway traffic safety research* (pp. 133-164). New York: Wiley Interscience.

Rumelhart, D. (1979). *Introduction to human information processing.* New York: Wiley.

Sanders, M., & McCormick, E. (1987). *Human factors in engineering and design.* New York: McGraw-Hill.

Schmidt, F., & Tiffen, J. (1969). Distortion of drivers' estimates of automobile speed as a function of speed adaptation. *Journal of Applied Psychology, 53,* 536–539.

Schiff, W. (1965). Perception of impending collision: A study of visually directed avoidant behavior. *Psychological Monographs General and Applied, 79* (11, Whole No. 604).

Schiff, W. (1983). Conservation of length redux: A perceptual-linguistic phenomenon. *Child Development, 54,* 1497–1503.

Schiff, W. (1986). *Perception: An applied approach.* Acton, MA: Copley.

Schiff, W. (1988b). Unpacking the Ames room: What the Transactionalist's brain doesn't tell the ecological psychologist's ear. *Newsletter of the International Society for Ecological Psychology, 3,* 6–9.

Schiff, W., Benasich, A., & Bornstein, M. (1989). Infant sensitivity to audiovisually coherent events. *Psychological Research, 51,* 102–106.

Schiff, W., Caviness, J., & Gibson, J. (1962). Persistent fear responses inrhesus monkeys to the optical stimulus of "looming." *Science, 136,* 982–983.

Schiff, W., & Detwiler, M. (1979). Information used in judging impending collision. *Perception, 8,* 647–658.

Schiff, W., & Oldak, R. (1990). Accuracy of judging time to arrival: Effects of modality, trajectory, and gender. *Journal of Experimental Psychology: Human Perception and Performance, 16,* 303–316.

Schiff, W., & Oldak, R. (1992). *Functional screening of older drivers using interactive computer-video scenarios* (Final Rep.). Washington, DC.: AAA Foundation for Traffic Safety.

Schiff, W., Oldak, R., & Shah, V. (1992). Aging persons' estimates of vehicular motion. *Psychology & Aging, 7,* 518–525.

Schiffman, H. (1990). *Sensation & perception: An integrated approach* (3rd ed). New York: Wiley.

Scialfa, C., Guzy, L., Leibowitz, H., Garvey, P., & Tyrrell, R. (1991). Age differences in estimating vehicle velocity. *Psychology & Aging, 6,* 60–66.

Scialfa, C., Kline, D., Lyman, B., & Kosnik, W. (1987). Age differences in judgments of vehicle velocity and distance. *Proceedings of the*

Human Factors Society, 1, 558–561.

Shinar, D. (1977). *The effects of age on simple and complex visual skills.* Paper presented at the annual convention of the WesternPsychological Association, Los Angeles, CA.

Sivak, M. (1987). Human factors and road safety. *Applied Ergonomics, 18,* 289–296.

Spurr, R. (1969). Subjective aspects of braking. *Automobile Engineer, 59,* 58–61.

Torrey, C. (1985). Visual flow and the direction of locomotion. *Science, 227,* 1064.

Tresilian, J. (1990). Perceptual information for the timing of interceptive action. *Perception, 19,* 223–239.

Tresilian, J. (1991). Empirical and theoretical issues in the perception of time to contact. *Journal of Experimental Psychology: Human Perception and Performance, 17,* 865–876.

Turvey, M. (1977). Contrasting orientations to the theory of visual information processing. *Psychological Review, 84,* 67–88.

Viets, J. (1990, January). High-tech plans to test drivers. *San Francisco Chronicle, p. A3.*

Wann, J., Edgar, P., & Blair, D. (1992). *Personal communication.*

Warren, R. (1976). The perception of egomotion. *Journal of Experimental Psychology: Human Perception and Performance, 2,* 448–456.

Warren, R., & Wertheim, A. (Eds.). (1990). *Perception and control of self motion.* Hillsdale, NJ: Lawrence Erlbaum Associates.

Warren, W., Jr., Young, D., & Lee, D. (1986). Visual control of step length during running over irregular terrain. *Journal of Experimental Psychology: Human Perception and Performance, 12,* 259–266.

Wickens, C. (1984). *Engineering psychology and human performance.* Columbus: Merrill, OH.

Chapter 2

Perceiving and Avoiding Rollover in Agricultural Tractors

Steven B. Flynn

Northeast Louisiana University

Thomas A. Stoffregen

University of Cincinnati

2.0   Introduction

Tractors tip over, injuring or killing their operators. Tractor overturns account for nearly half of reported tractor fatalities, or about 175 deaths and injuries per year in the United States (Goldberg & Parthasarathy, 1989; Murphy, Beppler, & Sommer, 1985). Tractor overturns account for a large proportion of total farm fatalities as well (Goldberg & Parthasarathy, 1989). Clearly, tractor turnover is a problem warranting attention from the human-factors community.

2.1   The Problem

Tractor tipping is a problem of vehicular orientation: Is the tractor upright or recumbent relative to its environment? People perceive and control the orientation of a wide variety of vehicles such as automobiles, motorcycles, boats, airplanes, and spacecraft, as well as tractors. But the problem of rollover per se (i.e., as an isolated event rather than as a consequence of some other event, such as a collision) is peculiar to a relatively small group of vehicles including tractors and certain recreational or off-road vehicles (three-wheeled all-terrain vehicles, or ATVs). Rollover is problematic for these vehicles for two reasons: They have relatively high centers of mass (cf. Robertson, 1989), and they often operate on rough, sloping ground.[1] Rollovers are dangerous and may also be common with tricycles that are popular among the elderly.

2.2   The Facts

Humans (and other animals) control the dynamic orientation of their bodies. Typically, humans are adept at this; people rarely fall over despite a wide range of postural disturbances, and despite voluntarily entering relatively unstable states (e.g., leaning). Orientation control is robust, allowing people to perform tasks as varied in their postural demands as gymnastics, in which large, potentially destabilizing forces are encountered, and brain surgery, which requires the maintenance of an extremely stable sensory platform and thus has low tolerance for even slight disturbances. This orientation control capability is not limited to bodies. After a period of adjustment, humans are capable of maintaining the orientation of extended-body dynamical systems such as bicycles, skateboards, and the like.

For stance, the body can be described as a lever with the fulcrum at the point of contact with the ground (the feet) and with forces acting at the center of mass. Mechanically, a standing human is an inverted pendulum (cf. Stoffregen & Riccio, 1988). An inverted pendulum is passively unstable; in the absence of compensatory effort, perturbations in the system will tend to increase in magnitude (the same thing happens when one tries to balance a pencil on its end). When an inverted pendulum is not upright, gravitoinertial force (GIF) produces torque on the system that tends to move it further from upright.[2] This increases the torque, which in turn increases the tendency to fall, and so on (Figure 2.1). This means that humans depend on continuous compensatory muscular action to maintain an upright stance; a system

---

[1]These two criteria also obtain for off-road motorcycles. Orientation is certainly a problem for motorcycles, and for any two-wheeled vehicles (and for unicycles, for that matter), but it is not clear that the appropriate term for loss of orientation with these vehicles would be *rollover*. There are a number of important differences in the passive dynamics of one- or two-wheeled vehicles and those having more than two wheels. There are corresponding differences in the perception and control of orientation for these types of vehicles.

[2]We define *upright* as alignment with the direction of unstable equilibrium (for passive bodies such as tractors). This is the direction in which the center of gravity is above the base of support such that there is no torque acting at the base of support (Stoffregen & Riccio, 1988).

that is unstable when passive is rendered dynamically stable via muscular effort. An upright tractor, by contrast, is not an inverted pendulum. Its wider base of support makes it passively stable over a wide range of angular disturbances.[3] If the tractor system is perturbed (for example, if one wheel leaves the ground), GIF tends to reestablish the 3-point base of support.

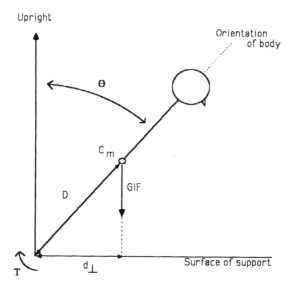

Figure 2.1. *Rotational dynamics of human standing posture. GIF = the gravitoinertial force vector, D = distance from point of contact with the ground to the center of mass, d = moment arm, T = torque, θ = angular deviation from upright, Cm = center of mass.*

The efficacy of the human postural control system implies a high level of perceptual sensitivity to the dynamics of orientation. That is, highly refined postural tuning indicates an exquisite ability to perceive

---

[3]We use *disturbance* to refer to a force that is applied to the system, such as an angular torque that would be produced in hitting a bump, and *perturbation* to refer to the effects of a given disturbance on the system's motion and or stability (cf. Riccio & Stoffregen, 1988). This is because a given disturbance (input) can produce different perturbations (output) depending on the momentary state of the system. This is discussed at greater length later.

the information needed for tuning. Empirical evidence also indicates high sensitivity to orientation information (e.g., Martin, 1990; Riccio, Martin, & Stoffregen, 1992).

2.3   The Conundrum

From a human factors perspective, this information leads to a conundrum. The conundrum concerns resolving The Problem with The Facts, as summarized in the following statements:

1. People demonstrate a refined ability to manage the orientation of both their bodies and various extended-body systems every day.
2. Tractors, on occasion, roll over.

The conundrum is deepened by the following:

1. The consequences of tractor rollover generally are much more serious than the consequences of losing one's balance; despite this, people appear to be more sensitive to the dynamics of stance than the dynamics of tractor operation.
2. Tractors, being tripods, are inherently stable. The human body, being mechanically an inverted pendulum, is inherently unstable. Despite this, controlling the orientation of the human body is apparently much easier than controlling the orientation of a tractor, at least when the tractor is operating at risk for rollover.

Exploring possible resolutions of the tractor conundrum constitutes the goal of this chapter. The following candidate explanations are considered:

1. Rollover occurs because there is no information available (in the ecological sense; Gibson, 1979/1986) specifying the stability state of the tractor system.
2. Rollover occurs because the tractor operator is not attuned to information specifying impending rollover.
3. Rollover occurs because the human perception/action system is ill equipped for the control of passively stable systems (Riccio, 1991).

Most existing analyses of tractor operation are based on information-processing approaches to perception and control (e.g., Bottoms, 1983, Goldberg & Parthasarathy, 1989; Murphy et al., 1985). Our analysis of tractor rollover is developed from an ecological approach to the perception and control of orientation (Owen & Lee, 1986; Riccio et al., 1992; Riccio & Stoffregen, 1988, 1990, 1991; Stoffregen & Riccio, 1988, 1990, 1991).

2.4   Candidate Explanation #1: Rollover occurs because there is no information available specifying the stability state of the tractor system.

The term *information*, as employed by ecologists, refers to patterns of energy that are specific to environmental circumstance (Michaels & Carello, 1981). Candidate Explanation #1 asserts that such patterns specifying tractor rollover are not available to the tractor operator. To evaluate this claim, tractor dynamics must be analyzed to determine if tractor stability states impart structure to ambient energy arrays.

2.4.1   The Dynamics of Tractor Stability

As previously stated, tractors are passively stable; they tend to re-establish a 3-point base if perturbed. This passive stability does not extend to all possible perturbations, however. The tractor is passively stable only so long as its center of mass remains within the triangle defined by its wheels (Figure 2.2a).[4] When the tractor is rotated so that the center of mass is no longer within this triangle, the tractor ceases to be passively stable and rolls over (Figure 2.2b).

There are thus two important differences in stability for humans and

---

[4]In the context of angular perturbations, the tractor's base of support is defined as the triangle connecting the left and right rear wheels with the point of articulation of the front axle with the tractor body. Modern tractors often have front wheels that are some distance apart, suggesting that they might have 4-point bases of support. However, the front axle is articulated so that it can rotate relative to the body of the tractor (this allows all four wheels to remain in contact with the ground when the ground slope is different for the front and rear wheels). Up to the limits of front axle articulation, then, the effective point of support for the front of the tractor is the articulation point.

tractors. The upright human is passively unstable, and thus requires constant, internally generated effort to compensate for perturbations. The tractor, by contrast, is passively stable for most perturbations; however, it has no internal source of torque (no feet and legs) to compensate for those perturbations that destabilize it. Under safe operating conditions the operator does not have to do anything to maintain tractor orientation, because the tractor maintains its orientation passively and spontaneously. Under unsafe conditions the operator has available only a few compensatory actions (e.g., sudden turns, fuel cut-off), and these are limited and indirect in their effects on roll-axis orientation.

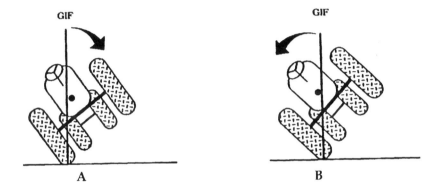

Figure 2.2. *Passive behavior of the tractor depends on the location of the center of mass relative to the direction of GIF. A. The tractor will passively return to an upright position (all wheels on the ground). B. The tractor will passively roll over.*

What factors cause tractors to roll over? The angle at which a stationary tractor will roll over is determined by the angle of the tractor with respect to ground slope (Goldberg & Parthasarathy, 1989; Murphy et al., 1985). On a sufficiently steep slope the center of mass of the stationary tractor will be outside the base of support, and the tractor will roll over. The torque required to cause rollover is determined by the momentary angular position of the tractor with respect to the GIF vector (cf. Cabe & Pittenger, 1992; Stoffregen & Riccio, 1988). This is expressed in Equation 1 (Figure 2.3).

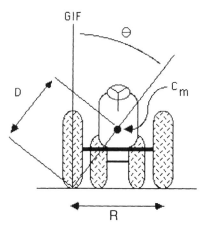

Figure 2.3. *The dynamics of tractor tipping in the roll axis (graphical illustration of Equation 1).*

$$T = [W (R/D)] \sin \theta \qquad (2.1)$$

where T is the torque required to turn over a tractor at a given moment, W is the weight of the tractor, R is the distance between the rear tires, D is the distance, in the roll plane, from the center of mass to the point of contact between a rear tire and the ground, and θ is the angle between the center of mass and GIF with the vertex at the point of contact. For a given tractor R/D and W are constant, and so Equation 2.1 can be simplified as follows:

$$T = k (\sin \theta) \qquad (2.2)$$

where k is a constant. As the instantaneous tilt of the tractor increases, the torque required to push it into instability decreases; a tractor that is highly tilted relative to GIF can be pushed over by a very small disturbance. On the other hand, a large torque is needed to overturn a tractor whose rear axle is perpendicular to GIF; a small disturbance will have little influence on its stability. This illustrates how a given disturbance may be destabilizing at one moment but not destabilizing at another; the perturbation resulting from a disturbance is a function of instantaneous tractor tilt relative to GIF. This is illustrated graphically in Figure 2.4.

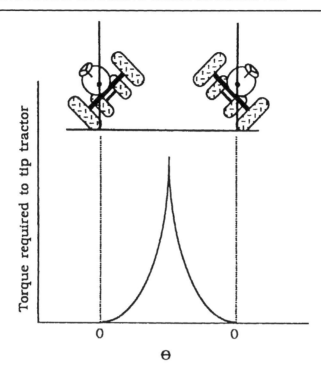

Figure 2.4. *The torque required to push a stationary tractor's center of mass "through" the direction of the GIF vector in the roll plane as a function of θ, the angle between GIF and a line connecting the center of mass with the point of rotation (see Figure 2.3). The two zero points on the horizontal axis indicate rotation about the left and right rear wheels of the tractor. This is a graphical representation of Equation 2.2.*

Of course, tractors do not roll over when they are stationary; they roll over while they are being used, while they are moving. For the moving tractor it is useful to define rollover with respect to a given perturbation rather than an absolute angle. This allows us to emphasize the fact that different factors interact in determining the likelihood of rollover. The factors that influence the stability of a moving tractor are (a) linear velocity, (b) the slope of the ground relative to the GIF vector operating on the tractor, (c) the turning radius (steering angle) of the tractor relative to the ground slope, and (d) the roughness of the ground, either in general or in the case of individual bumps.[5] For turnovers in

---

[5]These four parameters were identified by Murphy et al. (1985).

the roll axis the bumps that matter are those that are encountered solely by the uphill rear wheel. Symmetrical bumps will not influence roll stability, whereas bumps to the downhill wheel will move the tractor toward rather than away from stability. The reverse is true for holes or depressions in the ground (for clarity these will not be considered in our analysis).

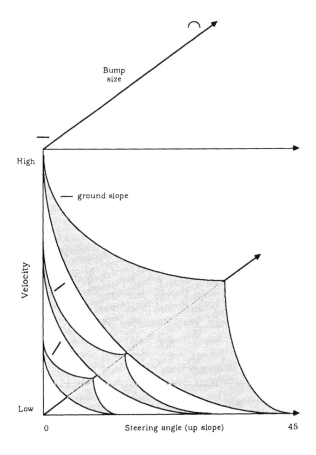

Figure 2.5. *A four-dimensional state space for dynamic tractor stability. The shaded areas are the stability separatrices for different values of ground slope. Points within a separatrix (closer to the origin) correspond to states in which the tractor is passively stable. Points beyond a separatrix correspond to states in which the tractor is passively unstable and will roll over (see Figure 2.2).*

2.4.2   A Control Space for Tractor Stability

Our analysis focuses on the factors influencing tractor stability. We represent these in an n-dimensional control space that is similar to a state space (cf. Marmo, Saletan, Simoni, & Vitale, 1985; McGinnis & Newell, 1982). The control space represents relations among the different variables that influence the tractor's stability. One such space is defined by three axes corresponding to the tractor's linear velocity, the roughness of the ground, and the steering angle (Figure 5; ground slope will be addressed shortly). At any given moment the values of these variables can be plotted in the control space, with a single point representing the instantaneous values of all three variables. Some of these points will correspond to stable tractor states (i.e., ones from which the tractor will not roll over), whereas others will correspond to unstable states. If we plot enough of these points, we find that there are distinct regions of the control space that correspond to tractor stability and instability. There will be a discrete boundary, or separatrix, between the regions of stability and instability (Figure 2.5).

The form and location of the separatrix will be influenced also by the magnitude of ground slope; this is shown in Figure 2.5 by presenting separatrices for three different slope magnitudes (the arrangement of dimensions in the state space is arbitrary, so long as they are orthogonal).

A turn introduces centripetal force to the system. This acceleration sums with other forces to yield the single GIF vector (Stoffregen & Riccio, 1988). One effect of centripetal force is to change the direction of GIF; the GIF vector is no longer parallel with gravity (Figure 2.6). Changing the direction of GIF with respect to the vehicle and the ground is equivalent to changing the orientation of the ground and the vehicle with respect to GIF.

Accordingly, the effects of turning can be incorporated in the ground slope dimension, reducing the dimensionality of the control space to 3 (Figure 2.7). This is preferable for several reasons, the most important being the connection to the real world. The stability of the tractor (and of the operator on it) is not determined exclusively by gravity. Rather, stability is determined by one's relationship to the direction of balance (for the operator) or the direction of unstable equilibrium or zero torque (for the tractor). This means that the three-dimensional control space of Figure 2.7 more closely describes the actual dynamical relations between the tractor and the physical/inertial environment. It has the additional advantage of having fewer dimensions, that is, of being more parsimonious.

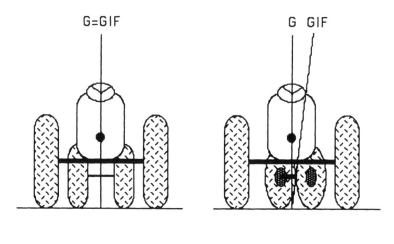

A   B

Figure 2.6. *Tilt defined with respect to the direction of GIF. A tractor that is upright relative to gravity can nevertheless be tilted with respect to GIF. This can occur during turns (B).*

Tilt, velocity, and bump size can change rapidly (and independently), and hence the position of the tractor relative to the boundaries of the region of stability can vary from moment to moment. A tractor operating under conditions of constant velocity, tilt, and bump size would occupy a single point in the control space over time. But individual points do not occur in isolation. The tractor "moves within the space" whenever there is a change in the value of one or more of the three dimensions. Constancy in velocity, tilt, and (especially) bump size is rare, so in the typical case the representation of the tractor over time in the control space will occupy multiple points. If we plot the tractor's position in the control space as velocity, tilt, and/or bump size are varied, we generate a line or path that describes the tractor's behavior over changes in the system's dynamics. These lines are referred to as trajectories (cf. Riccio & Stoffregen, 1988). Rollover occurs when a trajectory crosses the separatrix between the stable and unstable regions.

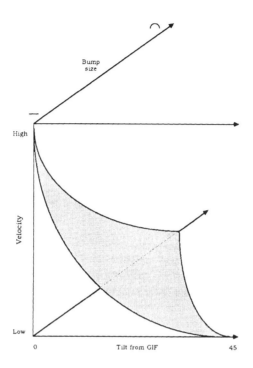

Figure 2.7. *A three-dimensional state space for dynamic tractor stability. The "steering angle" and "ground slope" dimensions from Figure 2.5 have been combined into a single dimension, Tilt from GIF. The resulting graph has a single stability separatrix.*

We assume that the majority of tractor overturns have their immediate cause in transient changes in ground roughness or steering angle, because these variables tend to change more rapidly than velocity or ground slope. Ground roughness is the most accessible example. For the uphill wheel, individual bumps will cause the system to move toward its stability separatrix. An individual perturbation, then, can cause the tractor to move in such a way that it will cross the stability boundary and roll over. That is, the disturbance will give rise to a trajectory that begins in the region of stability and crosses the separatrix into the region of instability. The key point is that this ill-fated trajectory begins somewhere within the region of stability. This means that there will be some period of time between the beginning of the perturbation and the moment at which the center of mass passes through the separatrix. At any given moment the amount of time remaining until this happens is a function of the tractor's instantaneous position relative to the stability boundary and the velocity at which it is approaching the boundary. This temporal interval can be referred to as the "time to contact" of the tractor's center of mass with its limits of stability.

2.4.3   Information for Impending Rollover

To reiterate: Candidate Explanation #1 suggests that rollover-specifying information may not be available to tractor operators; evaluating this explanation requires ascertaining whether the tractor stability state maps into ambient stimulus arrays. An important question, then, is to what should the operator be sensitive in order to maintain stability? The answer to this question constrains our search for potential stability information sources. Previous analyses focused on instantaneous tilt from upright as the most important variable (Goldberg & Parthasarathy, 1989). In contrast, our description suggests that it would be useful for operators to be sensitive to the spatiotemporal relation of the tractor to the separatrix between stable and unstable regions of the control space. The question is whether tractor dynamics engender structure in ambient energy arrays that specifies this relationship.

It is important to emphasize that the time-to-contact between the tractor and its stability limits is a real parameter of tractor operation. As a general concept, time-to-contact is relevant to many events that animals naturally control; studies have demonstrated that a wide range of behaviors are modulated on the basis of stimulus information about time to contact. Among these are wing furling in sea birds as they dive into the ocean (Lee & Reddish, 1981), stride length in long jumping (Lee, Lishman, & Thomson, 1982), the timing of paddle strokes in table tennis (Bootsma, 1988), and the avoidance of contact with an oncoming object (Stoffregen & Riccio, 1990). In these examples the contact is between two physical things—an object or surface and an animal. But there can also be time-to-contact of an animal with particular dynamical states; information about this may be used to control behavior in a manner similar to that observed in other contexts. For example, Martin (1990) demonstrated that human postural responses in stance are organized around the time-to-contact between the person's momentary position and the stability limits for stance (defined in terms of the maximum hip/ankle states from which falling can be prevented). Martin reported that the organization of humans' spontaneous compensatory postural actions was more highly correlated with temporal safety margins (time to instability) than with spatial safety margins (distance to instability).

Martin (1990) argued that a disturbance's effects on posture would depend on the subject's position with respect to the direction of balance at the moment of the disturbance. Disturbances that move the center of mass forward relative to the feet should be dangerous if the subject were already leaning forward, but relatively innocuous if the subject had been leaning backward. Martin's subjects stood on a force platform that could be moved beneath them, perturbing their posture. Subjects adopted a leaning posture, such that they were closer to the forward or backward limits of standing stability. Martin found that responses to disturbances were governed by the time-to-contact between the instantaneous position of the center of mass and the stability boundary toward which it was moving. Similarly, Cabe and Pittenger (1992) analyzed the effort required to bring an object into an upright, or balanced position, for example, standing a fallen log on its end. Using the same acceleration function described by Stoffregen and Riccio (1988), they identified a parameter of dynamic stimulation—"time-to-topple"— that is related to the time remaining in a rotational event before the object's center of mass passes through the direction of unstable equilibrium. In essence, Cabe and Pittenger's (1992) time-to-topple and Martin's time to contact with stability boundaries are the same. Both Martin (1990) and Cabe and Pittenger (1992) addressed instances in which animals wish to perceive impending crossing of stability boundaries so as to prevent this from happening. In this sense their work is relevant to our discussion of the perception of vehicular stability.[6]

Previous analyses of tractor rollover (Goldberg & Parthasarathy, 1989) described rollover kinematics in terms of angular velocity only. Here we discuss acceleration. Consider a tractor that has hit a bump and is tilting toward its stability separatrix. As the tractor rotates, the lever arm shortens and less torque is exerted on the center of mass (Figure 2.8; see also Figure 2.1). The result is that there is a pattern of decreasing acceleration that is uniquely related to instantaneous proximity to GIF. Changes in torque produce changes in acceleration, and so changing acceleration (sometimes called jerk) is characteristic of a

---

[6]Stoffregen and Riccio (1990) argued that these events are perceived not in terms of contact per se but in terms of time to reaction, which is especially salient for animals wishing to prevent "contact." This is similar to the τ-margin hypothesis proposed by Lee, Young, Reddish, Lough, and Clayton (1983).

tipping body. In general, the acceleration rate at any moment is uniquely related to a particular point of the movement arc.[7] Hence, there is a pattern of acceleration that is specific to the tractor's dynamic proximity to GIF, and thus the accelerative pattern may specify time-to-contact with the tractor's stability separatrix. It is not known whether humans are sensitive to the informative properties of these accelerative patterns.

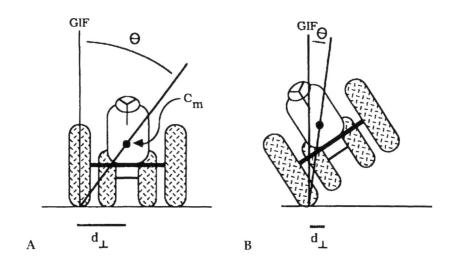

Figure 2.8. *The amount of torque acting on a tractor is proportional to the length of the lever arm (d).*

2.4.4   Dynamic Proximity to Stability Limits

Information about time-to-contact with stability boundaries is available only for an individual perturbation that will actually take the passive vehicle across those boundaries (for other perturbations the time to contact is infinitely large). There is not much time from the initial

---

[7]This is not true of instantaneous velocity. A force (disturbance) temporarily applied to a tractor will affect its subsequent velocity but not its subsequent acceleration (Stoffregen & Riccio, 1988).

disturbance until the vehicle crosses into the region of passive instability. In short, operators may be sensitive to impending contact with the stability separatrix but gain nothing from this. On its own, perceptual sensitivity is not sufficient to prevent tractor rollover; the perception is useless if the operator does not have enough time to take corrective action (Goldberg & Parthasarathy, 1989). It may be impossible or impractical to prevent rollover on the basis of time-to-contact for individual perturbations.

Goldberg and Parthasarathy (1989) reported a roll rate for side overturns of $75°$/sec under some conditions. For a vehicle that was operating at $15°$ from the stability boundary prior to the destabilizing disturbance, this would leave the operator with only 200 msec in which to detect and respond to the roll in order to prevent overturn. It may be the case that no functional correction is possible within this interval.[8] This in turn would mean that perception of impending instability would be without benefit. Also, the only actions an operator typically has available are cutting the throttle or turning the steering wheel; these have only a limited effect on roll-axis orientation. Thus, we are motivated to seek parameters of stimulation that could provide information about proximity to stability boundaries when the vehicle is not yet in a motion that will cross them.

Stability is sometimes discussed as if it were an all-or-nothing proposition, and it is true that at any given moment the vehicle is either on one side or the other of the stability boundary. But it may be more appropriate to consider stability as a relative, rather than absolute, quantity. A tractor that is operating near the stability boundary is arguably less stable than one that is far from that boundary. That is, its tendency toward passive stability is less robust when it is near the separatrix than when it is far from it. This is because the torque that is required to tip a tractor over is proportional to the momentary position of the center of mass relative to GIF (see Equation 2.2 and Figure 2.4). This "relative stability" phenomenon is shown in Figure 2.9, which represents stability as a basin in which the system resides (analogous to a marble in a bowl).

The deeper and more steeply walled the basin, the greater the force required to push the marble (system) out of the basin (upright orientation). Basin depth and wall steepness are not fixed, however;

---

[8]Goldberg and Parthasarathy (1989) discussed data suggesting that the total response times to rollover is more than 300 msec.

changes in certain variables (in this case, ground tilt alone is being considered) are reflected in concomitant changes in basin characteristics. As ground tilt deviates from zero, the basin representing upright orientation becomes more shallow, ultimately too shallow to restrain the "marble." At this point the marble rolls into another basin, representing the reorienting result of a rollover. Thus, although the system is nominally stable up until the moment of a rollover-inducing perturbation, it is arguably more stable when its "basin walls" are steep and deep than when the walls are shallow.

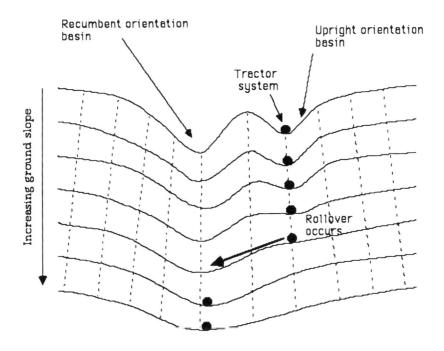

Figure 2.9. *Representation of relative tractor stability.*

Is there some way to detect the fact that the vehicle is dynamically close to instability, even though it is not yet moving to cross into it? In principle, the operator could perform a calculation involving bump size, velocity, and tilt from GIF. This would provide the required information, but it is doubtful whether humans perform such "perceptual calculations"; moreover, this approach would provide accurate information only when the operator was actively monitoring each of the relevant variables.[9] Implicit in this approach is the assumption that operators have independent perceptions of bump size, tilt, and velocity, and that the perception of stability derives from a calculation based on these independent percepts. But stability is an interaction among these parameters; the nature of the interaction may be qualitatively different from any of the parameters taken independently.

Stability is defined in terms of the position of the vehicle's center of mass relative to the separatrix between stable and unstable regions of the state space. Our hypothesis is that there may be patterns of angular motion of the center of mass that specify its proximity to the separatrix. These kinematic patterns are determined by the dynamics of the system and therefore enjoy a unique specificational relationship to system stability (Stoffregen & Riccio, 1988). This relationship holds independent of variations in linear velocity (translational motion), terrain properties, and so on. It should be possible, therefore, to detect dynamic stability on the basis of these kinematics alone, without prior or independent sensitivity to other parameters of motion.

Disturbances that cause a rear wheel to leave the ground may be rare. Alternately, they may be so powerful as to always lead to rollover. Either case would render unavailable or useless information about such events. It is therefore desirable to identify, if possible, a parameter of vehicle motion that provides information about dynamic proximity to the stability separatrix in general, that is, across all operating situations. When the vehicle is tilted (in the roll axis) relative to GIF, there will be an asymmetry in the effects of equivalent disturbances on the two rear wheels, even when both are on the ground. A given disturbance (e.g., bump, or change in velocity or steering angle) will cause a larger perturbation when it affects the uphill wheel (Pu), and a smaller perturbation when it affects the downhill wheel (Pd; Figure 2.10). For a tractor that is upright relative to GIF, the ratio of perturbations for the two rear wheels (Pu/Pd) will have a value of 1.0. As the tractor approaches the stability separatrix, the magnitude of perturbations to the downhill wheel will approach zero, such that the value of Pu/Pd will approach infinity. High values of Pu/Pd correspond to dangerous conditions. Another way of stating this is that there will be an

---

[9]Murphy et al. (1989) suggested that such a calculation be performed electronically, with the results displayed to the operator.

asymmetry in the vehicle's response to disturbances, with the magnitude of the asymmetry providing information about dynamic proximity to the stability separatrix. Human sensitivity to information in asymmetric patterns of angular acceleration has not been assessed (Stoffregen & Riccio, 1988).

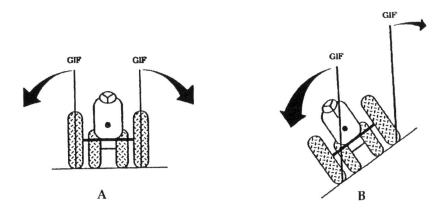

Figure 2.10. *Asymmetries in tractor perturbations. A. When the tractor is upright, equal disturbances to the two rear wheels will produce equal perturbations. B. When the tractor is tilted relative to GIF, equal disturbances to the two rear wheels will no longer produce equal perturbations. The magnitude of the asymmetry increases with increasing tilt up to the point at which the center of mass crosses the stability separatrix.*

The rate of change of the disturbance-response asymmetry as a function of tilt will not be linear. Rather, it will be governed by the changing acceleration function for inverted pendula that we described earlier. This means that the rate of change of the asymmetry could provide information about whether the system (the vehicle/ground interaction) was moving toward or away from the separatrix over time (i.e., across perturbations). This source of information might apply more to turns or changes in speed than to changes in ground slope or roughness. This is because it is unlikely that ground slope changes regularly in the required manner. So, in general, it seems that the operator would need to be sensitive to relatively constant values of the ratio, rather than to its rates of change. This may not be a problem, because the constant values, nevertheless, obtain in dynamic situations, as described earlier.

2.5 Candidate Explanation #2: Rollover occurs because the tractor operator is not attuned to information specifying impending rollover.

It is possible that humans would not have a spontaneous capacity to perceive the stability of an extended-body dynamical system such as a tractor. However, such a perceptual ability might be acquired. Perhaps tractors roll because operators have not become sensitive to rollover information. It certainly is possible for control of extended dynamical systems to develop through learning. For example, a spontaneous sensitivity to bicycle stability generally does not exist, yet, bicycle rollover is not problematic with skilled operators. How is it that people develop sensitivity to bicycle dynamics more so, apparently, than to tractor dynamics?

We offer two possible explanations for this. The first concerns qualitative differences between bicycle and tractor stability; this is discussed in Candidate Explanation #3. The second concerns perceptual learning (cf. Gibson, 1991). The hypothesis is that operators are not sensitive to rollover dynamics because they have never experienced them.

For most people, learning to ride a bicycle involves a period of adjustment. That is, most riders experience a period during which stability problems are common, resulting in both spills and frequent foot faults (i.e., placing one foot on the ground to counter a destabilizing torque). A similar period is experienced by infants as they learn to stand and walk; during this time, falls are frequent. Such periods may serve an important function: By repeatedly experiencing a loss of stability, new riders and standers can become familiar with patterns of stimulation associated with a loss of stability. In other words, falls provide an opportunity to become attuned to information specifying time to contact with stability limits.

It may be that sensitivity to stability limits requires experience with stability limits. Thus, most tractor operators, having never experienced a rollover, are not attuned to tractor stability information. More often than not, an operator's first opportunity to learn is also his or her last.

If stability limits must be experienced before they can be detected, this raises the possibility of simulator training as a prophylactic.

Perhaps potential operators could be required to attend "tractor school," part of which could involve simulated overturns. In this way, operators could be presensitized to rollover information. The issue of training is addressed further in the Research Issues section of the chapter.

2.6 Candidate Explanation #3: Rollover occurs because human perception/action systems are ill equipped for the control of passively stable systems.

As mentioned previously, the major difference between the control of stance and the control of tractor stability is that the human body is passively unstable for most postures, whereas the tractor is passively stable for most "postures" (orientation with respect to GIF). It may be that people are not able to perceive time to contact with stability limits for tractors because the *perception* of dynamic orientation is dependent on the *control* of orientation (Riccio, 1993; Riccio et al., 1992; Stoffregen & Riccio, 1988). That is, the dynamics of stability may be perceptually available only to observers who are engaged in active control of the system parameter in question. This claim has been advanced in the context of control of the human body, but it may be equally relevant to inanimate objects with which a person may interact, such as a tractor.[10] Tractor operators do not control roll orientation within the region of stability, and hence they may be incapable of perceiving the tractor's proximity to stability boundaries if control of orientation is needed to perceive orientation. Thus, the operator may be aware of nearness of his or her own body to its stability limits (because this is under his or her active control), yet at the same time be unaware of the relation of the tractor to its stability limits. The operator must perceive his or her own orientation simultaneously with and separately from that of the tractor; confusing one for the other can lead to disastrous results, as Stoffregen and Riccio (1988) argued in the context of aircraft orientation.

2.6.1 Other Vehicles

Elderly people sometimes ride specially designed tricycles as an

---

[10]Cabe and Pittenger's (1992) time-to-topple is defined explicitly in terms of the perceiver's own control actions.

alternative to bicycles or walking. These can be powered electrically or by pedaling. Tricycles are chosen because of their passive stability; less effort is required to control their orientation, and there is no need for continuous control (as is required on a bicycle). As with tractors, the center of mass of a tricycle tends to be high relative to the wheel base, and hence rollover may be a problem under some operational conditions. Tricycles normally operate on less tilted and smoother ground than do tractors; this should further increase stability. But with tricycles the mass of the operator is a much higher proportion of system mass than is the case with tractors. It may be that a major cause of rollovers with tricycles is movements of the operator that shift the center of mass of the tricycle system. Another possibility is that tricycle rollovers are caused by overly sharp turns, that is, when the operator employs a very short turning radius. The elevated center of mass of the tricycle-plus-rider makes tricycles relatively susceptible to this kind of turnover.

The operation of tricycles on flat, smooth, pavement means that there will be fewer and smaller perturbations in roll orientation than occur with tractors. Although this makes for a smoother ride, it could have the unfortunate effect of making the stability state of the tricycle less detectable by its operator. With fewer bumps and untilted pavement there will be fewer disturbances to orientation and less angular (roll) motion. A reduction in roll motion will attenuate the kinds of kinematic information about dynamic stability that we discussed. With fewer disturbances the system can be said to be more stable, but at the same time its dynamic proximity to stability boundaries may be less detectable. We return to this later.

All-terrain vehicles (ATV's) have dynamics that are similar to those of tractors, and this extends to the propensity to overturn (U.S. Consumer Products Safety Commission, 1986). All-terrain vehicles having four wheels are known to be unstable in the roll axis, as are certain models of off-road automobiles (Robertson, 1989). However, the three-wheeled ATV models are the most susceptible to rollover, resulting in a ban on their sale in the United States. These vehicles operate on tilted, bumpy terrain similar to that encountered by tractors. The risk of rollover for these vehicles may be greater than that for tractors. This is because recreational vehicles tend to be smaller and lighter in proportion to the power of their engines, and because they tend to operate at higher speeds and with "tighter" turns. Also, as with tricycles, the operator's weight is a relatively high proportion of total system weight for ATVs, such that shifts in the operator's position on the vehicle may materially affect its stability. Control of orientation, in these conditions, depends on the operator's ability to perceive the intimate interaction between the dynamics of his or her body and the dynamics of the vehicle. It is the interaction between these that will determine stability or instability.

2.7   Conclusion

Why do tractors occasionally roll over? The answer, we suspect, lies in the linkage between stability information and the operator. We discussed sources of information about the orientation stability for tractors and other vehicles. There appear to be patterns of motion (kinematics) that are specific to dynamic orientation and so provide information about stability. We discussed the fact that these patterns of motion are brought into existence by the vehicle's response to disturbances in orientation, such as results from bumps in the ground and from changes in velocity or turning radius. One implication of our analysis is that there may be a direct relation between the availability of information about stability and the number of disturbances to which the system is subjected. That is, if information about stability is created in the system's responses to disturbances, then disturbances may be a necessary precondition for the perception of system stability. This is consistent with Riccio's (1993; cf. Riccio & Stoffregen, 1991) analysis of the role of movement variability in the perception of orientation. One implication is that operator sensitivity to tractor stability may increase when, for example, the ground is bumpy. For tractors (and tricycles) there can be a tradeoff between perceivability and stability. Such a tradeoff need not obtain when orientation is controlled by the operator (e.g., stance, high-performance aircraft), that is, for systems that are passively unstable (Riccio & Stoffregen, 1988, 1991). This underlines the difference between perceiving the orientation of one's body and the orientation of a vehicle.

In analyzing tractor dynamics we did not explicitly include the physical dynamics of the human operator. For instance, the operator's mass is not taken into account in Equations 1 and 2. This omission simplifies the analysis and is considered acceptable because the operator's mass is trivial relative to that of the tractor. Ignoring the operator's physical dynamics may be a reasonable strategy when analyzing the dynamics of tractor rollover, but it may be unreasonable when considering the perception of tractor dynamics by the operator. We identified several parameters of tractor motion that provide information about dynamic stability. These motions will be transmitted to the operator's perceptual systems through a complex mechanical linkage. This is because the operator is not rigidly linked to the tractor. The tractor-operator interface (the seatpan and buttocks) constitutes a point of articulation in the tractor-operator system. Moreover, the operator's body is not rigid; there are many points of articulation (joints) within it. This means that motion-based stimulation will differ at different points on the body (and at different times). For example, stimulation of the vestibular and visual systems will not be identical to stimulation of somatosensory receptors in the trunk and buttocks (e.g., accelerometers placed on the head and hips will give different readings). Furthermore, the operator may not maintain the same orientation as the tractor. The operator generally will maintain his or her body in alignment with GIF (Stoffregen & Riccio, 1988), even when the vehicle is tilted with respect to this vector (Riccio & Stoffregen, 1991; cf. Riccio, this volume). That is, the operator acts to "null" tilting motions of the tractor so as to maintain his or her own state of balance.

These considerations may make it seem that informative parameters of tractor motion must be hopelessly washed out in the complicated stimulation of the operator's various perceptual systems. But this may not be so. The success of the operator's control of his or her own orientation suggests that the operator perceives (at least) the effects of the tractor's motion on his or her own bodily dynamics. Stimulation of the operator's perceptual systems is surely complex, but it is not random. The motion functions of the tractor, and their transmission through the various articulation points between the tractor and the operator and within the operator's body, are determined by the physical dynamics of the various masses that are involved. It is generally assumed that the perception of orientation is based on sensitivity to static quantities such as gravity (such that perception of orientation becomes more complicated in the presence of motion dynamics), and that different perceptual systems are sensitive to incompatible and independently variable frames of reference (Stoffregen & Riccio, 1988). However, both of these assumptions can be questioned (Riccio & Stoffregen, 1990, 1991; Stoffregen & Riccio, 1988, 1991). Our analysis suggests that perceivers are directly sensitive to the dynamics of their interactions with the physical world, in the context of bodily orientation, locomotion, vehicular control, and so on (Riccio et al., 1992; Stoffregen & Flynn, 1992; cf. Runeson & Frykholm, 1983).

2.8   Research Issues

We discussed several parameters of dynamic stimulation that make available information about the dynamic stability of the vehicle-in-motion. Many of these also obtain in, and provide information about, bodily orientation. Tests for sensitivity to these parameters, when they have occurred at all, have generally taken place in the context of the perception and control of bodily, rather than vehicular, orientation. It should not be assumed that sensitivity to information about bodily orientation necessarily implies sensitivity to similar sources of information about vehicular orientation. One reason for this is that changes in bodily and vehicular orientation generally take place around different axes of rotation. For a tractor operator, changes in bodily orientation typically occur around the axis of the hips or buttocks, whereas changes in orientation of the tractor take place around a point of contact between the tractor and the ground. Sensitivity to the latter might, therefore, require sensitivity to the location of axes of rotation that are outside the perceiver's body. Very little is known about humans' sensitivity to angular whole-body motion when the axis of rotation is outside the body (cf. Riccio & Stoffregen, 1991, pp. 222, 235). Existing devices that produce or simulate whole-body angular motion generally involve rotation around an axis that passes through the animal's body (Riccio et al., 1992); such devices may not be appropriate for studying operators' sensitivity to tractor roll motions.

Research on the perception of vehicular orientation often employs methodologies in which the human participant is restrained within a vehicle or simulator. This is often done in the interest of safety or under the assumption that bodily motion acts only as "noise" to complicate the perception of orientation. However, restraint can attenuate or eliminate many of the motion functions that characterize a vehicle in operational conditions (cf. Riccio & Stoffregen, 1990). Our analysis suggests that these motion functions may not be noise to be suppressed, but may constitute useful information about dynamic stability. Research paradigms that suppress or eliminate these motion functions may, then, have the unintended effect of reducing the fidelity or verisimilitude of the simulation and potentially of leading to excessively pessimistic conclusions about operator sensitivity to vehicle orientation dynamics.

Two types of experiments are desirable for evaluating the operational importance of the motion functions described in this chapter. One type should assess subjects' perceptual sensitivity to the motion functions, that is, their ability to perceive dynamic tractor orientation on the basis of the motions that characterize operational situations. The second type of study should assess subjects' ability to use these sources of information for the control of tractor orientation. In both types of experiments powerful whole-body motion devices will be required to generate faithfully the requisite motion functions. An appropriate device would place a seated subject at one end of a rod or board that could be rotated around its other end. Under computer control and with sufficiently powerful motors, the motion of the board could simulate the dynamics of tractor roll motion. That is, the angular acceleration of the device could be specific to its instantaneous position with respect to upright; the device could be made to move as an inverted pendulum. Such a device could also be made to simulate directions of unstable equilibrium (GIF) that deviate from the direction of gravity, thus providing a more accurate simulation of tractor operation during turns.[11] With such a system experimenters could simulate accurately trajectories of the "vehicle" within the state space that governs tractor orientation under operational conditions.

Such a device would be uniquely valuable for the study of the perception and control of vehicle orientation. One dependent variable would be participants' sensitivity to time to contact with the stability separatrix, echoing Martin's (1990) work with stance. Other dependent variables would include subjects' sensitivity to asymmetries in angular acceleration and to the rate of change of such asymmetries. Subjects' ability to utilize these sources of information could be assessed by giving them different forms of control over the device's motion. The dynamics of different control actions (e.g., changes in steering angle or in fuel flow) could be programmed into the simulation.

Our analysis has revealed new sources of information about dynamic vehicular orientation. These are available to operators (and passengers) independent of their sensitivity to variables that have previously been thought to be central to the perception of vehicle stability, such as translational velocity, ground roughness, and tilt relative to gravity (Goldberg & Parthasarathy, 1989; Murphy et al., 1985). Assessment of the importance of these new variables will require new experimental methodologies. These promise to provide important information about the perception and control of vehicles, and about relations between orientation of vehicles and orientation of the body.

---

[11]This has been achieved with a device that rotates around an axis passing through the body of the subject (Riccio et al., 1992).

Acknowledgments

The idea for this chapter was developed by Steven B. Flynn. The chapter grew out of discussions in a graduate course in human factors that was co-taught by Thomas A. Stoffregen and Joel S. Warm. We thank Dr. Warm for his support of this work.

2.9   References

Bootsma, R. J. (1988). *The timing of rapid interceptive actions*. Amsterdam: Free University Press.

Bottoms, D. J. (1983). The interaction of driving speed, steering difficulty and lateral tolerance with particular reference to agriculture. *Ergonomics, 26*, 123–139.

Cabe, P., & Pittenger, J. B. (1992). Time to topple: Haptic angular tau. *Ecological Psychology, 4*, 421–426.

Gibson, E. J. (1991). *An odyssey in learning and perception*. Cambridge, MA: MIT Press.

Gibson, J. J. (1986). *The ecological approach to visual perception*. Hillsdale, NJ: Lawrence Erlbaum Associates. (Original work published in 1979)

Goldberg, J. H., & Parthasarathy, V. (1989). Operator limitations in farm tractor overturn recognition and response. *Applied Ergonomics, 20*, 89–96.

Lee, D. N., & Reddish, P. E. (1981). Plummeting gannets: A paradigm of ecological optics. *Nature, 293*, 293–294.

Lee, D. N., Lishman, J. R., & Thomson, J. A. (1982). Regulation of gait in long jumping. *Journal of Experimental Psychology: Human Perception and Performance, 8*, 448–459.

Lee, D. N., Young, D.S., Reddish, P.E., Lough, S., & Clayton, T. M. H. (1983). Visual timing in hitting an accelerating ball. *Quarterly Journal of Experimental Psychology, 35A*, 333–346.

Marmo, G., Saletan, E. J., Simoni, A., & Vitale, B. (1985). *Dynamical systems: A differential geometric approach to symmetry and reduction*. New York: Wiley.

Martin, E. J. (1990). *An information-based study of postural control: The role of time-to-contact with stability boundaries*. Unpublished master's thesis, University of Illinois at Urbana-Champaign.

McGinnis, P. M., & Newell, K. M. (1982). Topological dynamics: A framework for describing movement and its constraints. *Human Movement Science, 1,* 289–305.

Michaels, C. F., & Carello, C. (1981). *Direct perception.* Englewood, NJ: Prentice-Hall.

Murphy, D. J., Beppler, D. C., & Sommer, H. J. (1985). Tractor stability indicator. *Applied Ergonomics, 16,* 187–191.

Owen, D. M., & Lee, D. N. (1986). Establishing a frame of reference for action. In M. G. Wade & H. T. A. Whiting (Eds.), *Motor development: Aspects of coordination and control* (pp. 287–308). Boston: Martinus Nijhoff.

Riccio, G. E. (1993). Information in movement variability about the qualitative dynamics of posture and orientation. K. M. Newell and D. M. Corcos (Eds.), *Variability and motor control* (pp. 317–358). Champaign, IL: Human Kinetic Publishers.

Riccio, G. E. (1995). Coordination of postural and vehicular control: Implications for multimodal perception of self motion. P. A. Hancock, J. M. Flach, J. K. Caird, & K. J. Vicente (Eds.), *Local applications of the ecological approach to human-machine systems.* Hillsdale, NJ: Lawrence Erlbaum Associates.

Riccio, G. E., Martin, E. J., & Stoffregen, T. A. (1992). The role of balance dynamics in the active perception of orientation. *Journal of Experimental Psychology: Human Perception & Performance, 18,* 624–644.

Riccio, G. E., & Stoffregen, T. A. (1988). Affordances as constraints on the control of stance. *Human Movement Science, 7,* 265–300.

Riccio, G. E., & Stoffregen, T. A. (1990). Gravitoinertial force versus the direction of balance in the perception and control of orientation. *Psychological Review, 97,* 135–137.

Riccio, G. E., & Stoffregen, T. A. (1991). An ecological theory of motion sickness and postural instability. *Ecological Psychology, 3,* 195–240.

Robertson, L. S. (1989). Risk of fatal rollover in utility vehicles relative to static stability. *American Journal of Public Health, 79,* 300–303.

Runeson, S., & Frykholm, G. (1983). Kinematic specification of dynamics as an informational basis for person-and-action perception: Expectation, gender recognition, and deceptive intention. *Journal of Experimental Psychology: General, 112,* 585–615.

Stoffregen, T. A., & Flynn, S. B. (1992). Visual perception of support surface deformability from human body kinematics. *Ecological Psychology, 6,* 33–64.

Stoffregen, T. A., & Riccio, G. E. (1988). An ecological theory of orientation and the vestibular system. *Psychological Review, 95,* 3–14.

Stoffregen, T. A., & Riccio, G. E. (1990). Responses to optical looming in the retinal center and periphery. *Ecological Psychology, 2,* 251–274.

Stoffregen, T. A., & Riccio, G. E. (1991). An ecological critique of the sensory conflict theory of motion sickness. *Ecological Psychology, 3,* 159–194.

U.S. Consumer Products Safety Commission. (1986). *U.S. consumer product alert: CPSC urges caution for three- and four-wheeled all-terrain vehicles.* Washington, DC: Author.

Chapter 3

Low-altitude Flight

John M. Flach

Wright State University
Armstrong Laboratory
Wright-Patterson Air Force Base

Rik Warren

Armstrong Laboratory
Wright-Patterson Air Force Base

3.0   Introduction

> *The vast majority of jet fighter aircraft crashes occur in the absence of any mechanical or operational failure in the aircraft, any obvious medical emergency on the part of the aircrew, or any deterioration of visibility due to weather or nighttime conditions (McNaughton, 1981). The pilot simply flies into the ground. In the early 1980s these "collisions with the ground" occurred in the United States at a rate of about two crashes per month, in each case inevitably costing the lives of the entire aircrew and destroying the aircraft (Miller, 1983). In spite of a substantial and superb Air Force investigation program in which each crash is probed to determine its cause and rapid dissemination of the findings provided to all flight personnel (see* Life Sciences, *most issues from 1980 to present), the frequency of such crashes does not appear to be diminishing either in absolute number or as a function of number of sorties flown.* (Haber, 1987, p. 519)

High speed low-altitude flight is an advanced form of flight requiring skilled pilots with special training. In its most taxing varieties and moments it demands precision and responsiveness of control almost to the level of prescience. Precision is demanded because a deviation of a few tens of feet from an intended altitude might mean clipping a tree or exposing oneself to an enemy. Responsiveness is demanded because, with high speeds and short distances, the "closure rate" to the ground or other terrain features is fast, and hence a pilot has little time to act to achieve objectives or to react to avoid collision or exposure. This fast changing spatiotemporal relationship of a pilot and the low-altitude environment forms a highly dynamic and unforgiving ecology. It is the unforgiving character of the dynamic ecology that imparts an element of ever present, if not always clear, danger to each moment of action or inaction and makes prescience desirable.

Prescience is, of course, impossible. However, it can be approximated by careful attention and educated perception. A trained and sensitive perceiver can exhibit remarkable "foresight" and thus be able to anticipate and forestall problems and also to capitalize on opportunities.

In order to better develop the foresight and insight of pilots by better training and sensitization, we need a body of perceptual facts and a good theory to guide their use. In this chapter, we selectively review the available facts on the perception and control of low-altitude flight with an emphasis on those perceptual aspects that seem most likely to support situation awareness and foresight. Further, we adopt an ecological approach to perception (Gibson, 1979) to provide a motivated structure to the facts and, more importantly, to provide a coherent theoretical basis for extracting meaning from the facts.

In the following sections we first examine the functional ecology of low-altitude flight. We then discuss the nature of visual information in general. Finally, we describe some specific sources of information that might be relevant for low-altitude flight together with empirical evidence examining human performance as a function of the availability of this information. This sequence is used for convenience only and is not meant to imply that perception is a one-way path from the environment, through the eyes, and into the head. Rather, ecological perception is characterized as a continuously traversed dynamic loop or cycle of perception and action. In the spirit of the Gestalt psychologists: the cycle is other than the sum of its parts, as discussed in Flach and Warren (1995).

3.1   Ecology of Low-Altitude Flight

The performatory demands of low-altitude flight are due to the fast-changing and hazardous spatiotemporal relationship of the pilot and the environment, that is, the dynamic ecology. This environment provides advantages as well as dangers. The term *affordances* is used for these consequences (both positive and negative) that the environment offers the pilot.

Analysis of the affordances of flight environments begins with a consideration of the *field of safe travel.* Following Gibson and Crooks

(1938), the field of safe travel is defined as the field of possible paths which the aircraft may take unimpeded. As noted by Gibson and Crooks, this field "exists objectively as the actual field within which the car (aircraft) can safely operate, whether or not the driver (pilot) is aware of it" (p. 454). For aviation, the sky becomes the field of safe travel and the ground surfaces (and extensions from it—trees, buildings, power lines, etc.) become important delimiters of the field of safe travel. Thus, the negative affordance of collision offered by the ground provides an important boundary for defining the field of safe travel.

One formalism that can be used to describe the dynamic "affordance space" or "problem space" of low-altitude flight is the state space (Smith & Hancock, 1993, discussed the state space for air traffic control). Figure 3.1 shows a characteristic state space for flight. The axes in this diagram represent two dynamic state variables. The vertical axis indicates the distance from the vehicle to environmental surfaces (e.g., the ground— i.e., AGL altitude, in which AGL refers to above ground level). The zero point of this axis represents contact with the surface. The horizontal axis represents the rate of change of distance to the surfaces (i.e., the closure or climb rate). In the left half of this space, the rate of change of distance is positive, so that the distance to surfaces is increasing (i.e., the plane is climbing). In the right half of this space, the rate of change of distance is negative, so that the distance to surfaces is decreasing (i.e., the plane is diving or losing altitude). It is typical to represent position on the horizontal axis and velocity on the vertical axis in state-space diagrams. However, to preserve the natural relation to altitude, the vertical axis in Figure 3.1 corresponds to position or above-ground-level altitude.

The parabolic curves within the diagram represent trajectories through the state space. Particularly, the state trajectories represent "minimum time" switching curves[1] for a dynamic system whose response is characterized by a second-order differential equation (this is only a loose approximation to the dynamics of actual aircraft). These parabolic curves represent critical dynamic boundaries within the problem space. Curve (a) in the diagram represents the maximum closure rates for particular distances at which there is a control that will result in "soft contact" with the surface, where soft contact means touching the surface with a low rate of change. All points outside this

---

[1]The switching curve shows the locus where an optimal controller would "switch" from full thrust to full braking in a bang-bang control that would take the vehicle from one altitude to another in minimum time.

curve are on trajectories where collisions with the surface are unavoidable. That is, for the particular combinations of altitude and descent rate represented by the points outside curve (a), even a command that generates maximum lift will not be sufficient to avoid collision with the ground. For example, curve (b) shows the response to a command for maximum lift from a point outside this boundary. Note that at zero altitude, the vehicle still will have significant velocity. This might represent a controlled flight into terrain as described earlier in the opening quotation.

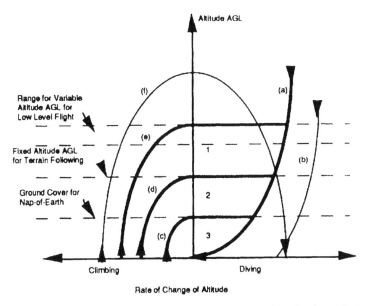

Figure 3.1. *A state-space representation showing critical action boundaries for different types of low-altitude flight: (1) Low-level flight; (2) Terrain following terrain avoidance, TFTA; (3) Nap of the earth flight, NOE.*

In addition to the zero point on the vertical axis that represents the ground, other points along the vertical axis represent important functional boundaries (i.e., ceilings) for special environments such as air combat and search and rescue. For example, these boundaries may represent the difference between exposure and occlusion from detection by an enemy. In these cases, positions below a boundary represent safe areas of the state space (out of sight from the enemy), and positions above a boundary represent dangerous areas of the state space (exposure to threats). Again, the parabolic curves (c, d, and e) represent trajectories that lead to contact with these altitude boundaries. Along these trajectories rate of change of altitude is positive, indicating that the aircraft is climbing. Each trajectory represents the boundaries for the maximum rates of climb at each altitude where "soft contact" with the ceilings is possible. Points outside these curves represent trajectories where "crashing" through these ceilings is unavoidable. Curve (f) represents a trajectory where, despite a maximum command to avoid contact, the aircraft breaks through the upper boundary on altitude before beginning its descent (to the left of the vertical axis the aircraft is gaining altitude; upon crossing the vertical axis the aircraft begins losing altitude). The areas within the curves, for example, the region bounded by curves (a & c), represent "safe" states where there are actions available that will keep the aircraft from crashing into the ground or breaking through the ceiling. These regions represent the fields of safe travel for various types of low-altitude flight.

It is convenient to distinguish three general flight environments within the state space that have different fields of safe travel:

1. In *low-level flight*, the aircraft flies at a relatively constant vector or level path. Specifically, undulations in terrain are ignored, so the altitude above ground level (AGL) and lateral distance to terrain features might be changing continuously. "Low" altitude here might mean 500 ft, and speeds might be 500 kts or less.

2. In *terrain-following-terrain avoidance* (TFTA) flight, the AGL altitude is ideally held constant. In a side profile, the altitude of the aircraft would closely match the profile of the terrain (excluding the smaller sized jagged features that would be impractical to try to follow).

3. In *nap of the earth* (NOE) flight, the aircraft follows the terrain profile quite closely in both vertical and horizontal dimensions. A helicopter will hug the ground and weave between hills and valley walls. Physically, altitude and other distances to terrain features can be quite low (e.g., altitude is 30 ft) and so also must speeds be low (e.g., 30 kts). Depending on the mission, the aircraft seeks either maximum cover using the terrain for protection from hostile forces, or seeks maximum pilot visibility for search and rescue features.

The state space representation in Figure 3.1 is only a caricature of the true problem space of low-altitude flight; however, it illustrates important aspects of affordances. The parabolic trajectories through this space represent higher order relational properties that reflect both dynamic constraints in terms of state variables as well as limitations on the action capabilities of the human-vehicle system (e.g., maximum lift).

This is critical to an ecological approach. Affordances, or meaningful properties, within ecologies are generally relational in that they are dually influenced by the environment and the acting system. Thus, the boundary of "too close" cannot be specified in terms of altitude or distance alone. Rather, too close, is a higher order variable that depends on the action capabilities of the human-vehicle system as well as on the state variables of distance and rate of change (e.g., Lee, 1976). This has important implications for the next section in which we consider the information required so that the pilot can "see" the boundaries that partition safe from dangerous areas of the state space.

The state space provides one representation of "what" the pilot needs to see. The state space provides a context in which the meaning of environmental states can be visualized. From a research perspective this problem must take precedence over the problem of information. The problem of information addresses the question of "how" pickup of meaning may be possible. However, from the perspective of skillful adaptation to ecologies, the question of what and how (of meaning and information) are intimately linked. This intimacy is reflected in the notion of direct perception in which Gibson (1979) proposed that there is a direct pickup of meaning (affordances) as a result of attunement to higher order invariants (information).

Further, the ecological approach would argue that in order to build an experimental program to study human performance, questions of meaning and information must take precedence. Understanding the nature of the natural problem space (i.e., meaning) and the information that is available are the bases for the intuitions that should guide the design of tasks and stimulation within empirical, experimental programs. Otherwise, it is unlikely that the experimental research will generalize to the natural environments of interest.

The boundaries on observer action constrain the space of behavior in the same way as the speed of light bounds the space of physical possibilities within the theory of relativity (see Hancock & Chignell, 1995, or Shaw & Kinsella-Shaw, 1988, for alternative uses of the light cone metaphor). Unfortunately, for human factors, there is no absolute boundary for behavior like the speed of light. The action boundaries on human-machine systems are continuously evolving with the development of new technologies and tools. Thus, the edge of our behavioral envelope is continuously changing. Where once our reach and grasp were constrained by the length of our arm and size of our hand, now teleoperated systems allow us to reach and grasp huge bodies floating in remote locations of space. Where once we were constrained to move along the surface of the ground at running speeds, now we move across the sky at Mach speeds. Nonetheless, it is essential that human factors professionals track the edge of this envelope so that their experimental and theoretical work can remain relevant to the natural ecologies within which we live and work. Meaning within the work or life space can only be understood with respect to these action boundaries. In the next two sections we first address the issue of visual information and second review some empirical work that evaluates human performance in vehicular control tasks.

3.2   Visual Information

The word *information* has so many meanings that before providing examples, we must give a definition and explain how it contrasts with other definitions. We also discuss key concepts that motivate and expand on the definition. By optical information or, synonymously, optical support, we mean that there is something about the pattern of light available to a pilot's eye(s) that stands in lawful correspondence with something to be perceived.

In the case of flight, the "something to be perceived" includes the state of the human-vehicle system in relation to the environment or, equivalently, the environment in relation to the action system. In the previous section, we described some important aspects of this relationship in terms of the state space in Figure 3.1. It is important to note that Figure 3.1 captures only a part of the things that might be perceived (a part that we feel is significant for controlling self-motion). Other things to be perceived might include features of the terrain (e.g., flat and smooth or rolling and rough), fixed objects such as landmarks, and moving objects such as other aircraft.

The phrase "pattern of light" has important implications. From the ecological perspective, information resides in the light. Or, to use a term that Gibson (1961) introduced, information is in the "ambient optic array." The ambient optic array can be visualized using a small mirrored ball (e.g., a Christmas ornament). Examine how the environment is reflected in this surface, and note how the reflections change as the ball is moved through the environment. The most obvious optical effect revealed by the mirrored ball when comparing no motion and displacement of the ball is a total sliding or flow of the projection of the environment on the surface of the sphere during self-motion. Assuming a rigid environment and rectilinear self-motion, this self-motion optic flow pattern is highly constrained.

The optical path followed by the projection of each environmental point is an arc of a great circle. The great circles for all the arc paths intersect at two points and form a pattern like the meridians or longitude lines on a globe with two poles. All flow is away from one point, called the *focus of expansion*, and toward the opposite point, called the *focus of contraction*. These points correspond respectively to the locations in the environment toward which the observer is heading and from which the observer is moving away.

Besides its path or direction, a flow element also has speed. Flow speed is zero at the poles and gradually increases along the meridian path and reaches a maximum at the optical "equator" that corresponds to the ground line currently directly below the observer and at right angles to the direction of self-motion. Although the optical equator is optically fixed midway between the two poles, it is important to keep in mind that the environmental line to which it corresponds is continuously changing.

Flow speed differs not only within a meridian but also across meridians. The meridian corresponding to the ground track of the self-motion contains the fastest flow. Hence, the point with the overall fastest flow is directly beneath the observer on the prime meridian. Along the other meridians, which for level travel correspond to ground lines parallel to the ground track, flow speed decreases the further a meridian is from the prime meridian and the ground line is from the ground track. This diminution in flow speed continues symmetrically on either side of the prime meridian until the horizon is reached and flow is zero. Thus, flow is always zero at the poles and also everywhere the horizon is visible. For level travel over flat ground planes, flow speed diminishes the further a ground point is from "directly below," hence, the zero flow at the horizon, because it corresponds geometrically to the projection of points lying infinitely far away on the ground plane.

The variation in flow speed within and across meridians is not symmetric with optical distance from the point directly below the observer. For level travel over a flat terrain, this nonisotropic variation can be expressed in the "roll-off" functions with respect to the maximum flow point directly below the observer (Gibson, Olum, & Rosenblatt, 1955; Warren, 1990). The speed rolls off as a cosine function across meridians but rolls off as a cosine-square function within a meridian.

These self-motion flow-pattern effects can be captured on other projection surfaces than a sphere. Gibson (1947, 1950; see also Langeweische, 1944) used simple line drawings of the flow vectors for views out the front, side, and bottom "windows" of an airplane traveling on a level path over a flat ground. For this case, the aim point is on the horizon, and the flow is zero everywhere on the horizon and is maximal directly below the aircraft. He also illustrated the case of approach to a runway. The flow is still zero everywhere on the horizon, but the aim point is now on the projection of the ground at a point exactly as many degrees below the horizon as the glide path of the aircraft forms with the ground. Flow is zero at the aim point, increases to a maximum value on a particular great circle at a point midway in arc length between the aim point and the horizon, and from there decreases to zero at the horizon. Langeweische (1944) also provided a vivid description of how this structured array provides information for guiding an approach to landing.

In addition to pictorial depictions of the flow field, Gibson also presented the first formal mathematical analyses (Gibson, et al., 1955). The general equations provided can be used to describe and analyze more complex scenarios than level travel over a flat terrain. The formal analyses continue (e.g., Koenderink, 1986, 1990; Lee, 1976, 1980; Zacharias, 1990), and still there is work to be done due to the richness of self-motion scenarios and terrains, especially pertaining to aviation. Smets (this volume), for example, pointed out the need to consider structures resulting from higher order derivatives within the flow patterns (e.g., acceleration).

The ambient optic array is considered to be a logically prior and more general "projection" surface than the retina. The ambient optic array is a full sphere, whereas the retina is functionally much less than a hemisphere. More importantly, the optic array continues to exist, even when there is no retina. It is thus defined independently of any particular optical system. It is also defined independently of any observer location. We may speak of the optic array at a fixed point in space or even as existing along a possible path over which an observer might move. Such a moving optic array is exactly what is needed for describing the optic projection structure available to a pilot in a moving airplane. The structure that appears within a moving optic array will be referred to as the "flow field."

The patterns within flow fields are in "lawful correspondence with" the relations between observer and environment. That is, the patterns are specific to the environmental surfaces and the motion of the observer relative to those surfaces. Thus, the flow field, the optical paths, and the speed variations described earlier are not just mathematical or analytical abstractions. They constitute patterns that stand in lawful correspondence to behavioral events. Working out these lawful correspondences is the problem of ecological optics.

The idea of "lawful correspondence" or "specificity" is central to the ecological approach to information. It is this correspondence or specificity that permits the observer to respond adaptively to his environment. One way to think about this "lawful correspondence" is as a "consistent mapping" (Flach, 1987; Flach, Lintern, & Larish, 1990). Within the ecological approach the term *invariant* is more typically used to indicate a lawful correspondence. However, whichever term is used, *consistent mapping* or *invariant*, the fundamental point is that the structure is a property of the stimulus event (in this case of the light). It is not in the head or the "eye" of the observer. If the structure is not present (e.g., a variable mapping exists), then skill will not develop. In other words, no amount of information processing can compensate for a lack of structure within the stimulus event.

On the other hand, the phrase "available to a pilot's eye(s)" emphasizes that the lawful correspondence is necessary but not sufficient for skilled adaptation to the environment. In addition, the observer must be attuned to the information or must be able to act in ways necessary to access the information. Thus, although there may be information that specifies a safe approach to a runway (e.g., see Langeweische, 1944), not all observers have learned to pick up this information.

The introduction of "eye(s)" into the discussion of information requires consideration of retinal flow. The retinal array differs from the optic array in that it corresponds to only a small subsample of the optical array in spherical extent. This can severely limit the availability of information as combat pilots know all to well. In order to get information about events behind their aircraft, they must turn their heads and "check six."

The necessity to shift gaze to cover all information sources is only one consequence of retinal pickup systems. Control of gaze means that an item in the environment can be fixated, even though it is a flowing item relative to the optically fixed aim point. The optic array, nonflowing aim point now becomes a retinally flowing point. An eye fixation causes a rotation of the eye, and this in turn adds a rotational field to the optic flow field. The resulting retinal flow pattern thus has different flow path and fixed point characteristics than the pure optic flow pattern. But, a new class of information emerges, namely, information about direction of gaze and attitude of the eye. The consequences of the distinction between retinal and optic flow for information definition and pickup are just beginning to be explored (e.g., Cutting, Springer, Braren, & Johnson, 1992). For example, does the visual system in some sense "parse" the retinal field into its pure flow and rotational constituents, or does it directly capitalize on the new patterns in the retinal flow? Exploring these questions promises to be a rich field of inquiry.

Gibson also emphasized that information was obtained rather than passively received. Pilots are not passive receivers of imposed stimulation. They actively seek out information. For example, a pilot may "pop up" after a low-level approach to better see the target. For Gibson, observers purposively move to explore and adapt to their environment. Indeed, his abandonment of traditional passive approaches to perception came from studying flight training during World War II. His radical reformulation of visual information and its adequacy came about precisely to explain how people fly (Gibson, 1947, 1950; Gibson et al., 1955) and how animals in general move about their environment (Gibson, 1958, 1966, 1979).

It is important to note that Gibson was not denying or invalidating the existence of passive perception or imposed stimulation, for surely they are facts of everyday life. He was calling attention to a neglected aspect of perception that he argued was more prevalent and important, namely, *control*. In our own investigations into the control of flight, we find it useful to combine obtained and imposed stimulation in the same display. In real life, a given locomotion command does not always produce the intended or customary effect. For example, turning one's car's steering wheel to the right usually produces a right turn. However, if one hits a patch of ice, the direction and speed of turning might be alarmingly different from what was expected. Similarly, in an airplane, a nose-up pitch attitude at sufficient speed produces a climb. However, the same speed and attitude might result in a drop in attitude in the presence of a strong vertical wind sheer. The point is that there may be a specific mapping from motion to the information in optical flow, but there is not a specific mapping from actions (i.e., command input) to either motion or optic flow. Thus, control of motion requires that the loop between perception and action be closed by information.

In sum, information refers to a mapping from structure within optical flow fields to relations between an observer and an environment. This information provides the foundation for adaptive behavior or control. Information in this sense "closes the loop." It provides the coupling that intimately links perception and action. It provides the feedback and feedforward necessary for stable control. Thus, a desired trajectory through the environment is achieved by acting so as to produce a particular optical transformation (Gibson, 1958; consider also Powers's, 1978, claim that "behavior is the control of perception").

3.3   Examples of Optical Concomitants of Self-motion

In this section we examine several examples of optical concomitants or correlates of self-motion. Optical concomitants are typically identified based on analytic descriptions of patterns within the flow. These patterns can be defined globally, over the entire flow field (e.g., global optical flow rate). They can also be defined locally, over some critical "landmarks" within the optical flow field (e.g., horizon, focus of expansion) or within the retinal flow field (e.g., differential motion parallax). This section is organized around some of the dimensions that might be important to pilots in controlling low altitude flight. First, we consider perception of speed and altitude. Next, we consider time-to-contact as a higher order variable that may be more directly available and more relevant to the control of locomotion than either speed or altitude. Finally, we consider information for heading or the direction of self-motion. The perception and control of heading has been the subject of important theoretical debates concerning the nature of information and its role as feedback in the control of locomotion.

Each section begins with analytic descriptions of the available structures within flow fields that correlate with the dimensions of interest (e.g., speed, altitude, etc.). This is followed by a brief review of some of the empirical work to evaluate humans' ability to use the available structures for controlling locomotion.

3.3.1   Speed

3.3.1.1 Ecological Optics.

Warren (1982) identified two aspects of the optic array that may affect the perception of the speed of self-motion: optical edge rate and global optical flow rate.

Optical edge rate is defined as the rate at which local discontinuities (e.g., edges) cross a fixed point of reference in the observer's field of view. Speed in edges per second can be calculated by timing how many edges pass a reference point (e.g., a wing tip) per second. Edge rate is dependent on ground speed and texture density. When traveling over an evenly textured surface (i.e., constant or stochastically regular spacing between discontinuities), increases in edge rate are directly proportional to increases in speed. However, if the surface texturing is not regular, that is, there are systematic changes in the distances between texture elements (i.e., discontinuities) on the surface, then edge rate will vary inversely with the distance between discontinuities.

Global optical flow rate is an alternative source of information for the speed of self-motion. Warren (1982) identified global optical flow rate as an extension of the analysis of the optic flow field performed by Gibson et al., (1955). Warren (1982) noted that the expression for the angular velocity of any point in the optic array could be separated into three independent components. Two of these components (azimuth and declination angle) are purely local factors of the position of a specific point on the ground surface in relation to the observer. For example, as in the description of the flow on our mirrored ball earlier, discontinuities directly below the observer change optical positions (i.e., flow) at the highest rate. Discontinuities that are near the horizon flow at much slower rates. These two components account for the roll off function that was described earlier—with angular speed maximal below the observer and decreasing as a cosine function across meridians and as a cosine-square function along a meridian.

The third component is a global multiplier that affects every point in the optic array. This multiplier, the ground speed divided by the altitude, was identified as global optical flow rate. The rate of change in visual angle for every discontinuity in the optical flow field increases in proportion to increases in the speed of self motion and decreases in proportion to increases in the distance from the surface. Thus, if altitude above the ground is constant, changes in global optical flow rate are directly proportional to the speed of self-motion. However, if altitude is changing, then global optical flow rate may vary independently from changes in velocity.

Thus, two structures within the flow field have been identified that are lawfully related to the speed of self-motion. Yet, neither structure is unambiguously related to speed. Optical edge rate is confounded by changes in the spacing of texture elements on a surface, and global optical flow rate is confounded by changes in altitude.

3.3.1.2 Human Performance.

Denton (1980) demonstrated the influence of edge rate perception on driving behavior. Denton painted stripes at the approaches to traffic circles so that the distance between edges decreased at an exponential rate. Thus, self-motion at a constant speed over these stripes resulted in an accelerating edge rate. A marked reduction in speed of approach resulted from this manipulation. Thus, Denton was able to take advantage of the ambiguity between edge spacing and speed of self-motion to induce drivers to slow down to safe speeds when approaching these intersections. It is noteworthy that accident rates at these intersections were also reduced.

Owen and Warren (1982) implicated global optical flow rate as a possible contributing factor to an increase in damage to wheel structures resulting from taxiing at excessive speeds in 747 aircraft. They observed that pilots in the 747s were trained in 707s and 727s. An important difference in the 747 is the height of the cockpit. The cockpit of the 747 is roughly twice as high above the surface when taxiing than either of the other two aircraft. If pilots were controlling taxi speed using global optical flow rate (which depends on height), this could explain the excessive taxi speeds. Because the height was doubled, pilots in 747s would be moving at roughly twice their normal taxi speed in order to produce the amount of global optical flow that they were familiar with from the smaller aircraft.

A series of experiments were performed to examine the optical information for the speed of self-motion by Owen and his associates at the Aviation Psychology Laboratory at Ohio State (Owen, Warren, Jensen, Mangold, & Hettinger, 1981; Owen, Wolpert, Hettinger, & Warren, 1984; Owen, Wolpert, & Warren, 1983; Tobias & Owen, 1984; Wolpert, 1987). Several important findings can be summarized from Owen's work. When subjects judged perceived change in velocity, both edge-rate and flow-rate information were useful to the observer, but edge rate information tended to dominate. The effects for edge rate and flow rate were additive with reduced effects found for global optical flow rate at lower levels of flow rate. Also, the usefulness of edge rate information was found to interact with display duration.

Larish and Flach (1990) examined magnitude judgments of the speed of self-motion. Principal independent variables were edge rate, global optical flow rate, and the type of texture (grid or dot). Results indicated that edge rate and global optical flow rate had additive effects on magnitude judgments with edge rate accounting for the larger proportion of variance. Effects were independent of texture type.

The available research appears to indicate that humans do use both edge rate and global optical flow rate when controlling or making judgments about their own speed. However, we must be cautious about generalizing from this research to actual flight control tasks. First of all, judgments of speed on the ground (Denton's [1980] work and Owen and Warren's [1982] observations with regard to taxiing) may not generalize to flight, because altitude does not vary dynamically in these tasks. A second concern is that all the research on flight-related judgments used flat-colored, computer graphic displays, in which texture was defined with respect to lines or points. Texture is impoverished relative to natural surfaces that are richly textured and on which texture is nested, so that the units of texture may vary with distance from the texture. That is, at high altitudes texture may be defined by fields and roads, but at lower altitudes texture may be defined by ears of corn or gravel on the runway surface. A third concern is that all of the work of Owen and the study by Larish and Flach (1990) used passive judgment tasks as opposed to active control tasks. Finally, based on the earlier analysis of the ecology of low altitude flight as illustrated in Figure 3.1, it is questionable whether speed, per se, is a relevant dimension for judgment or control. The boundaries that define fields of safe travel are not defined relative to a fixed speed. Certainly, speed is a consideration in defining critical boundaries within the state space. And speed is definitely a critical control variable as far as the aerodynamics of flight for fixed-wing aircraft are concerned (because lift is a function of air speed). However, in all of these instances speed interacts with other variables (e.g., distance from the surface, heading, attitude). This leaves open the possibility that information may be available with regard to these interactions. These higher order invariants may be both more relevant to the problem of flight control, and they may be more unambiguously specified within the optic array.

3.3.2   Altitude

3.3.2.1 Ecological Optics.

Three sources of information for altitude were identified within the optic array: global optical density, splay angle, and depression angle.

Global optical density has been defined by Warren (1982) as "the number of ground elements required to span one eyeheight distance." Formally, optical density (OD) is defined as follows:

$$OD = z/g$$

where z is the altitude of the observer, and g is the extent of a ground texture element (i.e., the distance between discontinuities on the surface). Thus, for a constant texture size, changes in altitude will result in proportional changes in optical density. As altitude increases, optical density will increase, and vice versa. Optical density is a global optical variable, in that it is an index of the general optical "sparseness" within the field of view. It is not associated with specific texture elements. At low altitudes the average distance between elements will be greater than at higher altitudes as illustrated in Figure 3.2.

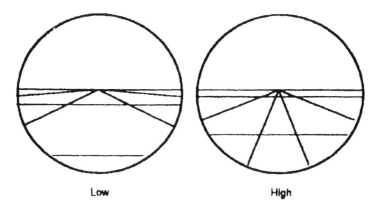

Figure 3.2. *The global optical density is an index of the sparseness of texture within the field-of-view. At low altitudes the density is low and at high altitudes the density is high.*

Optical splay was also identified as a source of information for altitude by Warren (1982). Warren cited Biggs (1966), who noted that when an observer maintains a constant distance to a line on the ground plane (e.g., the curb of the road), despite shifting optical positions of the individual points composing the line, the optical position of the line was invariant. For a straight line parallel to the direction of motion, the invariant optical position can be defined in terms of the angle at the vanishing point formed by the line and a second line perpendicular to the horizon along the ground trace of forward motion, as shown in Figure 3.3. This angle is defined by the equation:

$$S = \arctan\left(\frac{Y_g}{z}\right)$$

where S is the splay angle, Y_g is the lateral displacement of the line from the perpendicular, and z is the altitude (eye height) of the observer. The equation describes the projection onto the frontal plane for an observer moving parallel to the ground. For rectilinear motion over a flat ground plane, splay angle is constant when altitude is constant.

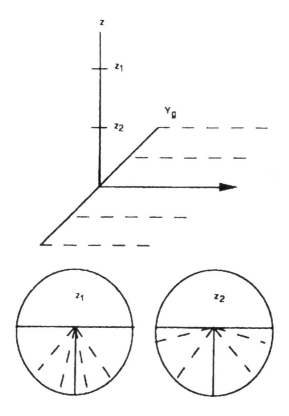

Figure 3.3. *Optical splay angle for a common texture viewed from two different eyeheights (z).*

The rate of change in splay angle is specified by the following equation:

$$\dot{S} = \left(\frac{\dot{z}}{z}\right) \cos S \, \sin S + \left(\frac{\dot{Y}_g}{z}\right) \cos^2 S$$

The first term [$\left(\frac{-\dot{z}}{z}\right)\cos S \sin S$] indexes change in splay as a function of changes in altitude [$\dot{z}$]. The negative sign indicates that as altitude

decreases, splay angle will increase, and vice versa. The term [$\frac{-\dot{z}}{z}$] specifies fractional change in altitude or change in altitude scaled in eye heights. This term indicates that the relation between change in altitude and change in splay angle will depend on the initial altitude. At high altitudes (large z), any given change in altitude would result in a smaller change in splay angle than when initial altitude was lower. As noted by Warren (1988), "sensitivity of the display [optical splay rate] varies inversely with altitude: the lower the altitude, the more change in visual effect for equivalent altitude change commands. At very low altitudes

this optical activity is dramatic even *optically violent*" (p. A121). The [$\frac{-\dot{z}}{z}$] term is independent of optical position. It scales the rate of change for every splay line in the field of view. For this reason, it has been termed *global perspectival splay rate* (Wolpert, 1987). The sine and cosine terms index the dependence of splay rate on optical position. For splay lines with 0° splay angle (perpendicular to horizon at the expansion point) and ±90° splay angle (the horizon), the rate of change will be zero. From these minima, the rate of change in splay angle for a given fractional change in altitude will increase to a maximum at a splay angle of ±45°.

    The second term in the equation for change in splay angle [$\left(\frac{\dot{Y}_g}{z}\right)\cos^2 S$] indexes change in lateral distance [$\dot{Y}_g$] from the observer to the line element such as might result from a lateral movement of the observer. For straight-ahead forward motion there is no change in lateral distance, and this term has no impact on the optical splay angle. For this reason, this term has not typically been included in analyses of splay. However, lateral displacements have sometimes been included in the events that have been simulated to study altitude control. Thus, it is important to understand the effects from this term. The first half of the

term [$\frac{\dot{Y}_g}{z}$] specifies lateral displacements scaled in eye heights. Changes in lateral distance will result in proportional changes in splay angle. The

second half of the term [$\cos^2 S$] indicates how change in splay angle varies as a function of the optical position of a particular line element. This term shows that change in splay will decrease from a maximum for the texture line directly below the observer (S=0°) to a minimum at the horizon (S=±90°). It is important to note that whereas changes in altitude will have symmetrical effects on lines spaced equal distances to each side of the observer, lateral motions will cause a reduction in splay angle for lines in the direction of the motion and an increase in splay angle for lines in the opposite direction from the motion. In other words, changes in altitude result in changes in splay angle that are symmetric around the motion path, whereas changes in lateral position will have asymmetric effects.

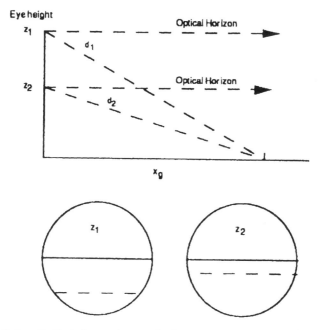

Figure 3.4. *Optical depression angle for a common texture element viewed from two different eye heights (z).*

Optical depression angle provides yet another potential source of information for changing altitude. Optical depression angle [δ] is the angular position below the horizon of a texture line perpendicular to the direction of motion as illustrated in Figure 3.4. This angle can be expressed as a function of altitude (z) and the principal distance on the ground from the observer to the texture element [x_g]:

$$\delta = \arctan\left(\frac{z}{x_g}\right)$$

For rectilinear motion over a flat ground plane, the rate of change of the optical depression angle will be:

$$\dot\delta = \left(\frac{\dot z}{z}\right)\cos\delta\,\sin\delta - \left(\frac{\dot x_g}{z}\right)\sin^2\delta$$

The first term [$\left(\frac{\dot z}{z}\right)\cos\delta\,\sin\delta$] shows the contribution of changes in the observer's altitude on the optical depression angle. The relation between depression angle and altitude is qualitatively identical to the relation between splay angle and altitude, with the exception of sign. As with splay angle, the rate of change in depression angle scales with fractional changes in altitude. Also, as with splay angle, the rate of change of depression angle will depend on the initial optical position of a texture element. Rate of change of depression angle will be zero at depression angles of 90° (directly below the observer) and 0° (horizon) and will be maximum at a depression angle of 45°.

The difference in sign indicates that splay angle increases with decreasing altitude, whereas depression angle decreases with decreasing altitude. This difference can be misleading. Splay angle is indexed to the line of sight, but depression angle is indexed to the horizon. Actually, both splay angle and depression angle are components of an expansion of texture that is associated with approach to a surface. This expansion is further discussed in terms of *tau* or time-to-contact.

The second term in the equation for rate of change of depression angle [$-\left(\frac{\dot x_g}{z}\right)\sin^2\delta$] indexes changes in depression angle as a result of forward motion of the observer. In the first part of this term, $\dot x_g$ is proportional to the speed of the observer. The term [$-\left(\frac{\dot x_g}{z}\right)$] is forward speed scaled in eye heights. As noted in the previous section on speed, this term has been identified as global optical flow rate. Thus, the rate at which depression angle changes is affected by both altitude and forward speed. The remaining part of this term [$\sin^2\delta$] accounts for changes in depression angle rate due to the initial optical position of a texture element. Rate of change of depression angle due to forward motion will be minimum at the horizon and will increase to a maximum at a point directly below the observer. Thus, the lower the texture element is in the forward field of view, the greater will be the rate of change in depression angle for a given speed of observer movement.

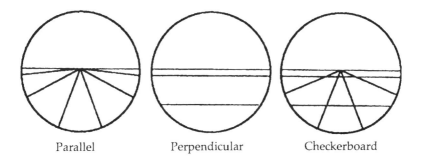

Parallel   Perpendicular   Checkerboard

Figure 3.5. *Three texture types that have been used to isolate the information for the control of altitude.*

3.3.2.2 **Human Performance.**

Wolpert, Owen, and Warren (1983) compared observers' ability to detect loss in altitude in a simulation of flight with constant forward speed using three types of texture, as shown in Figure 3.5: parallel (meridian) texture, perpendicular (lateral) texture, and square (checkerboard) texture. Parallel texture was chosen to isolate the information available from optical splay, and perpendicular texture was chosen to isolate the information available from global optical density. The results indicated that observers were best able to detect loss in altitude with parallel texture only. Performance was nominally worse with checkerboard texture and was significantly worse with parallel texture. A number of similar studies were summarized by Wolpert (1983, 1987; Wolpert & Owen, 1985). Wolpert (1987) noted that in these studies "loss of altitude scaled in eyeheights proved to be the functional variable, performance improving over increasing levels of that variable. In contrast, ground-unit-scaled loss in altitude showed a minimal effect over the different levels" (p. 24). Because the rate of change of optical splay is directly

related to change of altitude scaled in eye heights [$\frac{-\dot{z}}{z}$], whereas optical density is related to change in altitude scaled in ground units, splay was nominated as the effective source of information for judging change in altitude. At that time, no analysis of depression angle had been made. This conclusion must be reconsidered in light of the analysis of change in depression angle which shows that change of visual angle in the perpendicular texture displays also scales with fractional change in altitude.

Johnson, Tsang, Bennett, and Phatak (1989) employed a strategy similar to that of Wolpert et al. (1983) to isolate the optical information available for control of altitude. They used three texture types; parallel (meridian) texture only, which isolates optical splay; perpendicular (latitude) texture (with a single meridian line roadway to indicate flight path), which was intended to isolate optical depression angle; and square texture, which contains both optical splay and optical depression information. Unlike Wolpert et al., who measured performance in a passive psychophysical judgment task, Johnson et al. used an active control task. Johnson et al. introduced disturbances in both the vertical and lateral axes. Subjects were to minimize the effects of the vertical disturbance using a single-axis velocity control. Subjects' control actions had no effect on the lateral disturbance. The lateral (side-to-side) disturbance was introduced to prevent subjects from using local information such as the position of a meridian texture line on the bottom of the display (e.g., distance from the corner of a rectangular display) to control altitude. In apparent contradiction to the results of Wolpert et al., Johnson et al. found superior performance (lower tracking error) with the perpendicular (optical depression angle) and square textures. Higher tracking error was found for the meridian display, which contained the most salient information with regard to splay.

A second study by Johnson and his colleagues (Johnson, Bennett, O'Donnell & Phatak, 1988) examined active control of altitude in a hover task. In this task, Johnson et al. included disturbances on three axes: altitude [z], lateral [$\dot{Y}_g$] (visible only in parallel texture), and fore-aft [x_g] (visible only in perpendicular texture). Performance was examined for numerous texture types, four of which were of particular interest for the present discussion: parallel, perpendicular, square, and dot. The results showed equivalent performance (both in terms of tracking error and correlated control power) for perpendicular, square, and dot textures. Performance with the parallel (splay only) texture showed greater tracking error and lower correlated control power. Again, this result is in apparent contradiction to the findings of Wolpert et al. (1983).

Two differences between Johnson et al.'s studies and the earlier work of Wolpert et al. were the inclusion of disturbances on axes other than the altitude axis and the use of an active control task. Wolpert (1988) employed an active altitude regulation task with disturbances in altitude and roll (subjects only controlled altitude). Note that a roll disturbance affects the optical activity of both parallel and perpendicular texture, but not the angular relations of splay and depression angle. Wolpert found that "altitude was better maintained over parallel texture than over square or perpendicular texture" (p. 17). Wolpert found that whether or not the roll disturbance was included had no effect on performance. Flach, Hagen, and Larish (1992) also measured performance in an active control task with disturbances similar to those used by Johnson et al. (1988), except that whereas Johnson used a hover task, Flach et al. used a task with forward velocity so that the fore-aft disturbance, implemented as a variable headwind, affected forward velocity, not position. Flach et al.'s results were consistent with Wolpert's. Performance was best with parallel and checkerboard textures—both of which contain splay information. Performance was poor in the horizontal texture conditions contrary to Johnson et al.'s results.

Wolpert (1988) also included optical flow rate as a variable in his study. He found a performance decrement for increasing levels of global optical flow rate. This is consistent with previous research by Wolpert and Owen (1985). They used global optical flow rates corresponding to walking speed (1 eye height/sec) and very low flight (.25 and .5 eye heights/sec) and found that detection of descent over square texture deteriorated with increasing global optical flow rates. This is interesting in light of the optical analysis presented earlier. Optical splay angle is independent of global optical flow rate. However, optical depression angle is dependent on global optical flow rate.

Global optical flow rate [$-\left(\frac{\dot{x}_g}{z}\right)$] changes as a function of altitude (z). However, changes in global optical flow are not specific to altitude.

Global optical flow rate is directly proportional to forward velocity[$-\dot{x}_g$] and inversely proportional to altitude. This ambiguity was previously noted in the optical analysis of Gibson et al. (1955).

It is interesting to note that the global optical flow rates examined by Wolpert (1988) and Flach et al. (1992) were all greater than .5 eye heights/sec. However, the optical flow rates examined by Johnson et al.

(1988, 1989) ranged from 0 for the hover task to .25 eye heights/sec. Thus, in the Johnson et al. studies, the optical flow rates were far lower than in previous studies. Also, in each of the studies discussed earlier, the texture that isolated the most effective optical information (whether splay or depression-angle) always yielded performance that was superior to (though not typically significantly superior to) the display that combined the two sources of information (checkerboard or dot texture). Wolpert et al. (1983) and Wolpert (1988) found that performance was better with parallel (splay only) texture than with checkerboard texture. Johnson et al. (1988, 1989) found that performance was better with perpendicular (depression-angle only) texture than with checkerboard or dot textures. Also, Warren (1988) found that altitude control was superior with parallel-only texture than with parallel-plus-superimposed dot texture. Why does the combination of multiple sources of information result in performance degradation?

Perhaps, the optical activity resulting from forward motion (global optical flow rate) interacts with the optical activity for changes in altitude. In splay only textures, global optical flow rate is invisible so that there should be no interference. If the rate of forward motion is slow or altitude is high, then the contribution of global optical flow will be small so that interference will be small. But if global optical flow rate is high and can be seen in the display (perpendicular texture elements or dots are present in the display), then the noise created by this optical activity will make it difficult for the observer to disambiguate changes in altitude from changes of vertical position. To test this hypothesis, Kelly, Flach, Garness, and Warren (1993) examined active control of altitude as a function of both texture (parallel, perpendicular, checkerboard, and dot) and global optical flow rate (0, .25, 1, and 4 eye heights/sec). Results showed that performance in terms of both RMS error and correlated control power was always worse with perpendicular texture (no splay). However, the magnitude of the effect increased with global optical flow rate. That is, consistent with the earlier hypothesis, the presence of increasing levels of global optical flow interfered with the ability to control altitude. When splay information was present, however (i.e., parallel, checkerboard, and dot textures), performance was independent of global optical flow rate.

Johnson et al.'s results remain an anomaly. Even for zero global optical flow rate, Kelly et al. found performance with parallel (splay) texture to be superior. Johnson et al. found superior performance with perpendicular (depression angle) texture. One last consideration might be the use of local information within the flow field to control altitude. A local strategy might be to maintain an invariant relationship between a specific texture element and a landmark in the field of view (e.g., the edge of the display or in operational environments the edge of the windscreen or a smudge or local discontinuity on the windscreen). Thus, the pilot might try to keep the lowest perpendicular texture element in the field of view a fixed distance above the bottom edge of the screen. Using such a strategy, it should be impossible for the observer to disambiguate a change in altitude from any other motion that affects the relative optical position of the edge. Consistent with this observation, Johnson et al. (1988) found high levels of crosstalk in their subjects' control responses, such that there was a relatively large amount of control power correlated with the fore-aft disturbance. On the other hand, Flach et al. (1992) and Kelly et al. (1993) took care to minimize local strategies by using circular frames so that no local cues were available in terms of corners or edges. Also, in the Kelly et al. study the edge of the field of view was created by goggles worn by the subjects so that the frame was not fixed, but moved with the head. In these studies there was little or no crosstalk. That is, control power was not correlated with the nonaltitude disturbances.

To summarize, it seems that splay plays a very important role in the perception and control of altitude. However, other sources of information are available and can be used. These sources include other global variables such as optical density and depression angle as well as local variables such as the relative position of particular discontinuities within the field of view. Johnson told us that helicopter pilots are sometimes trained to maintain altitude in hover by picking out an object in their forward field of view and keeping that object at a fixed position on their windscreen. Note that this strategy will only work in a hover (when one is moving across the surface everything flows) and that depending on where the object is in the field of view, this will result in some crosstalk as a result of fore-aft motions. Johnson and Phatak (1989) modeled this local control strategy and found close agreement between the model and human performance in their altitude control studies.

3.3.3   Time-to-Contact

3.3.3.1 Ecological Optics.

Change in splay angle and depression angle discussed in the previous sections is actually a component of an expansion of texture associated with approach to a surface. The rate of the expansion is a function of fractional change in altitude $[\frac{\dot{z}}{z}]$. The optical changes in splay or depression angle (and by implication fractional change in altitude) are not unambiguously related to either speed or altitude. This was recognized by Gibson et al., (1955), who wrote that:

> Ground speed and altitude are not...independently determined by optical information. A more rapid flow pattern may indicate either an increase in speed or a decrease in altitude. Length of time before touching down, however, seems to be given by the optical information in a univocal manner. (p. 382)

This observation anticipates a more complete analysis by Lee (1976, 1980). This analysis shows that the time-to-contact (*tau*) with a surface is specified as the inverse of the rate of optical expansion, which in our notation would be $[\frac{\dot{z}}{z}]$. So, if one is losing altitude at 10 m/sec and he or she is 50 m from the surface contact will be in 5 sec. The optical analyses suggest that this time is specified unambiguously in the flow field. Grosz et al., (this volume) provide the derivation of *tau* and discuss time-to-contact information in the context of landing in the following chapter. Thus, we will not review the research on time-to-contact further. However, we would like to point out that *tau* provides an example of a possible source of information that may be both more directly and unambiguously available and more relevant to the control decisions required for safely guiding an aircraft in low altitude flight, than either of the "lower order" variables speed or altitude. As we have seen, these lower order variables may not be unambiguously specified by the information in optic flow. Judgments of distance or speed may be secondary to judgments of time. For example, Haber (1987) asked:

> When flying straight and level at 100 feet at 500 knots, how long can pilots spend looking down inside the cockpit at a computer display, without risking collision with the ground? How much shorter does this time become if a pilot is making a 6-g level turn at the same speed and altitude above the ground? How long does it take to hit the ground from 100 feet in a 1 deg descent? How much of that time is needed for the pilot to react, and how much for the plane to react after the pilot has done so? (p. 530)

Answering these questions requires both an understanding of the action constraints of pilot and aircraft as well as the information constraints that specify time-to-contact.

3.3.4   Heading

3.3.4.1 Ecological Optics.

In describing the flow on the mirrored ball, it was observed that all flow is away from one point and toward the opposite point. The focus of optical outflow or the focus of expansion is optically coincident with the point in the environment toward which the observer is heading. Likewise, there is a focus of optical contraction that is diametrically opposite the focus of expansion and that is optically coincident with the place in the environment from which the observer is "fleeing." Gibson (1958) discussed these sources of information with respect to the control of self-motion. For a predator to catch a prey, the predator must so move as to keep the focus of optical expansion within the optical contour of the prey's body. For a prey to avoid a predator, the prey should move, allowing for some jinking maneuvers, so as to keep the focus of optical contraction within the optical contour of the predator's body. For a plane to make a safe landing, the focus of expansion needs to be kept on the desired touchdown point (Gibson, 1950, Figure 58, p. 128).

Langeweische (1944) showed that the position of the focus of expansion not only specifies the point on the ground where the aircraft is heading, but also that the angular distance of the focus of expansion below the horizon (depression angle) specifies the glide slope of the aircraft as illustrated in Figure 3.6. This relation between aim point and horizon has been refered to as the *H-distance* or *H-angle* (Berry, 1970; Hasbrook, 1975; Lintern & Liu, 1991) Thus, a safe approach is optically specified when the focus of expansion and the threshold of the runway are aligned in the field of view at the appropriate distance below the horizon. This will work as long as the horizon corresponds to the point of optical infinity. If the horizon is at a different optical locus due to a sloping ground plane or to a lack of texture (e.g., at night or in heavy fog), then this strategy may lead to an unsafe approach. In a more gross sense, the position of the focus of expansion with respect to the horizon (above, below, or on the horizon) may be an important source of information for maintaining level flight.

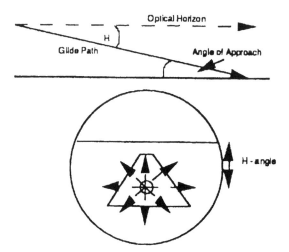

Figure 3.6. *The depression angle between the focus of expansion and the horizon specifies the angle of approach.*

The focus of expansion is not simply a local discontinuity within the flow pattern, but is globally specified. In principle, a small area of the optic array not explicitly containing the focus of optical expansion permits the geometrical extrapolation of the flow lines to their point of common intersection, the heading point. This is illustrated in Figure 3.7 in which the rectangle represents a window. The lines within the window represent the optical traces or optical flow paths of arbitrary texture points due to rectilinear self-motion of the observer. The dotted lines outside the rectangle represent the backward extrapolation of the optical flow paths. All these extrapolated lines intersect at the same optical location that specifies the aim point in the environment. Hence, even samples of the optic array not containing the expansion point contain information about heading.

The focus of expansion identifies the environmental point toward which the observer is heading at any instant. But instantaneous heading is only one possible meaning of the word *heading*. Lee (1976) pointed out that we can be heading around a curve. On curves, the instantaneous heading is not altogether meaningful. As one turns a car or aircraft, the instantaneous aim vector might momentarily point at a pedestrian or a mountain. But this does not mean that one is headed for the pedestrian of mountain; he or she is heading around a curve. For car driving, Lee showed that the optical contours corresponding to the edges of a curved road are optically coincident with the optical flow paths if one's control is such that he or she is staying on the curved road.

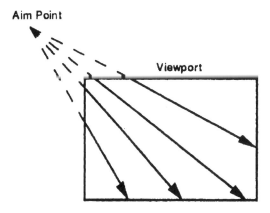

Figure 3.7. *The focus of expansion is specified even when it is not within the viewport.*

For aircraft, there are no road contours, but we speculate that the existence of curved flow lines, that is, the optical flow traces or "smears" formed over a short but not instantaneous period, is informative about the curved path of the aircraft. The greater the optical curvature, the tighter the turn. With aircraft, of course, there is additional curved path information contained in the relationship of visible elements of the aircraft and the horizon due to bank angle.

Thus, it is important to consider the focus of expansion as a global pattern within the flow field over time, as opposed to a single discontinuity instantaneously present within the field. It has also been observed that the discontinuity in the flow field may not correspond to discontinuities within the retinal flow field. Particularly, if the observer moves his or her eyes to track a particular texture element on the surface. That element will be fixed within the retinal field, and the point of expansion will flow across the retinal field. Cutting et al. (1992) suggested that observers may use local information within the retinal field as information for heading. They suggested that:

> Given an object to fixate and to pursue, in most environments near objects will generally move faster than, and in the opposite direction from far objects. A mobile observer need only shift his or her gaze opposite to the most rapid flow in the retinal array to find the eventual instantaneous aimpoint. (p. 46)

This local retinal cue for heading is called differential motion parallax. Much empirical and theoretical debate has centered on the availability and utility of global expansion within the optic array versus local differential motion parallax within the retinal array as the basis for judgments of heading (see Cutting et al., 1992, or Warren & Hannon, 1990, for recent reviews). This debate may be less important for high speed low-altitude flight, however, as indicated by the following quote from Calvert (1954; cited in Cutting et al., 1992):

> When a cinematographic film is taken of the face of a driver of a vehicle starting from rest, it is found that he scans the visual field only when the vehicle is moving slowly. As the speed increases, the driver scans less and less, until finally he begins to stare fixedly in the direction in which he wishes to go, usually at the aiming point if he can see it, or at some point close to it if he is unable to see it. The higher the speed the more difficult the task, the stronger the tendency to fixate, and the less likely the pilot is to glance at objects far removed from his path. (p. 240)

Thus, as Cutting et al. (1992) noted, when the eyes are riveted to the aim point, "optical and retinal flow fields are the same because the eye undergoes no rotation" (p. 70). For now, we must remain open to the possibility that both global information (in terms of an expansion field) and local information (in terms of differential motion parallax) are potential sources of information for heading.

3.3.4.2 Human Performance.

Kraft (1978) examined landing approach as a function of manipulation of the visual horizon. Kraft found that when the position of the visual horizon was altered by simulating a rising terrain behind the runway, pilots flew lower approaches. This is consistent with the hypothesis that pilots use the H-angle. In this case the "true" H-angle with respect to the optical horizon (optical infinity) would be smaller than the apparent H-angle as seen relative to the visual horizon created by the sloping terrain. Lintern and Liu (1991) also found that accurate approaches could be made when only the aim point and the horizon were visible in a simulated landing scene. Additionally, they found that when the H-angle was distorted by simulating up-sloping or down-sloping terrain beyond the runway, changes in flown glide slope were consistent with the hypothesis that pilots use the H-angle.

Research on the psychophysics of perception of heading have used different methods reflecting theoretical assumptions about the locus of information within the optical or retinal flow fields. R. Warren (1976) and W. Warren and colleagues (Warren & Hannon, 1988, 1990; Warren, Mestre, Blackwell, & Morris, 1991; Warren, Morris, & Kalish, 1988) assumed information to be localized in the optical flow field. In their studies, properties of the optical flow field were typically simulated on graphic displays. They found performance accuracy in the range of .2° to 9° depending on the location of the aim point, on whether or not the aim point was visible in the display, and on the method of judgment. ""

Cutting (1986; Cutting et al. 1992) assumed that information is localized within the retinal flow field. In his studies, properties of the retinal flow field were simulated on graphic displays. Cutting found accuracies in the range of approximately 2°. Many theoretical and methodological questions remain with respect to judgments of heading. One concern about all the research on the psychophysics of heading judgments is that they have relied exclusively on judgments made by passive observers. Because it is self-produced eye rotations that create the conflict between optical- and retinal-based theories of information, it will be important to study heading in which the observer is actively coupled with the displays.

A final study on human performance limitations on heading perception in low-altitude flight concerns still another meaning of the word *heading*. Heading is not totally synonymous with direction of travel, but also refers to the particular place in the environment toward which one is headed. It is possible for two places in the environment to be both directly in front of a moving aircraft, if one place is in front of the other. For example, a small hill might be in front of a large mountain. In a sense, as a pilot heads for the small hill, the pilot is also heading toward the large mountain. Because the small hill is nearer, the flow pattern for it is geometrically faster than the flow pattern corresponding to the nonoccluded part of the farther mountain background.

However, the local optical flow rates for both the near and far slopes might either both be below threshold or not be distinguishable from each other. If the lighting conditions are such that all other information to distinguish the near from the far slope is nonexistent or below threshold, then the smaller, nearer slope can perceptually fuse with or be "absorbed" into the larger but further slope. Such a scenario was analyzed by Previc (1989). He isolated motion information from digitized imagery for actual low-altitude missions over three different terrains and determined that the motion information was insufficient to detect the foreground ridge. Such a failure to detect a foreground ridge could result in a collision with the near hill—all the while, the pilot perceiving a clear heading to a far mountain slope.

3.4   Conclusion

This has been a brief and selective review of a few of the task and information variables that might contribute to an understanding of human performance in low altitude flight. A full analysis must include discussions of nonvisual sources of information (e.g., Riccio, this volume), of instrument displays as sources of information, and of the higher order invariants that result as a function of interactions across the multiple sources of information. Many of these issues have yet to be addressed within systematic research programs. Thus, this is a rich field for research with many questions yet to be asked.

There are two obvious applications for the research that has been presented in this chapter. First is training. From an ecological perspective the primary objective of training is the education of attention (e.g., E.J. Gibson, 1969). That is, the goal is to alert and sensitize the pilot to the information that is available. Obviously, this cannot be done until we first have an understanding of the information. Identifying the available information is necessary, but not sufficient. We must also understand how to most effectively communicate that information. For perceptual-motor skills, such as flying at low altitudes, it is likely that practice is going to be central to the communication process. The challenge will be to create opportunities for practice that highlight the invariants that specify critical control variables and that at the same time provide a safety net to encourage exploration and discovery (Flach et al., 1990).

For example, Warren (1988) explored a training procedure in which pilots were asked to fly a simulated display at zero altitude. At zero altitude the rate of change of splay is maximal and thus very salient. This is not an "ecologically valid" task and is only possible using a simulated display. However, the rationale was that practice under these conditions would draw attention to splay and would result in positive transfer to more natural tasks, in which splay was an important source of information. Graphical displays for training might also include supplemental "cues" to highlight important invariants. The aim point might be marked by a blinking "X" (Reardon, 1989; Reardon & Warren,

1989) or the splay lines might blink or change brightness or color. In using such artificial cues, the goal is to draw attention to the natural information, not to replace that information. For that reason, adaptive procedures in which these artificial cues are used only early in training or only when errors are particularly gross are recommended (see Lintern, 1980, or Lintern & Gopher, 1978, for discussions of adaptive training techniques.)

The second obvious application is display design. New electronic displays are gradually replacing the window as primary sources of information for flight control. The head-up display is an attempt to overlay instruments onto the flow field, and virtual imaging systems are being designed to completely replace the natural flow field with a computer-generated visual field. What should these new displays look like? We believe that understanding the information in the optical flow fields, that have richly supported control of self-motion up to now, will provide the foundation upon which to build new display configurations. We predict that successful new displays will be ones that integrate new sources of information within the information naturally provided by optical flow. Note that integrate does not mean overlay and does not require that optic flow be mimicked blindly. Creative new displays will preserve some aspects of optic flow, while eliminating others. Some invariants will be preserved, and new invariants, both local and global, will be created to reflect new levels of awareness about the environment.

Acknowledgments

John Flach was supported by a research grant from the Air Force Office of Scientific Research (F49620-92-J-0511) during preparation of this chapter. Support was also provided by the Armstrong Laboratory, Human Engineering Division, Wright-Patterson AFB, OH and the Psychology Department of Wright State University. Opinions expressed in this chapter are our own and do not represent an official United States Air Force position.

3.5   References

Berry, S. L. (1970). *A design for a visual landing simulator.* Unpublished master's thesis, University of Illinois, Savoy, IL.
Biggs, N. L. (1966). Directional guidance of motor vehicles: A preliminary survey and analysis. *Ergonomics, 9*, 193–202.

Calvert, E. S. (1954). Visual judgments in motion. *Journal of the Institute of Navigation*, London, *7*, 233–251.

Cutting, J. E. (1986). *Perception with an eye for motion.* Cambridge, MA: MIT Press.

Cutting, J. E., Springer, K., Braren, P. A., & Johnson, S. H. (1992). Wayfinding on foot from information in retinal, not optical, flow. *Journal of Experimental Psychology: General, 121*, 41–72.

Denton, G. G. (1980). The influence of visual pattern on perceived speed. *Perception, 9*, 393–402.

Flach, J. M. (1987). Consistent mapping, invariants, and the training of perceptual motor skills. In L.S. Mark, J.S. Warm, & R.L. Huston (Eds.), *Ergonomics and human factors: Recent research.* (pp. 131–137). New York: Springer-Verlag.

Flach, J. M., Hagen, B. A., & Larish, J. F. (1992). Active regulation of altitude as a function of optical texture. *Perception & Psychophysics, 51*, 557–568.

Flach, J. M., Lintern, G., & Larish, J. F. (1990). Perceptual motor skill: A theoretical framework. In R. Warren & A. H. Wertheim (Eds.), *Perception and control of self-motion* (pp. 327–355). Hillsdale, NJ: Lawrence Erlbaum Associates.

Flach, J. M., & Warren, R. (1995). Active psychophysics: The relation between mind and what matters. In J. M. Flach, P. A. Hancock, J. K. Caird, & K. J. Vicente (Eds.), *Global perspectives on the ecology of human-machine systems.* Hillsdale, NJ: Lawrence Erlbaum Associates.

Gibson, E. J. (1969). *Principles of perceptual learning and development.* New York: Appleton-Century-Crofts.

Gibson, J. J. (1947). *Motion picture testing and research* (Aviation Psychology Res. Rep., No. 7). Washington, DC: U.S. Government Printing Office. (Reprinted in D. Beardslee & M. Wertheimer (Eds.), *Readings in perception* (pp. 181–195) Princeton, NJ: D. van Nostrand, 1958.)

Gibson, J. J. (1950). *The perception of the visual world.* Boston: Houghton Mifflin.

Gibson, J. J. (1958). Visually controlled locomotion and visual orientation in animals. *British Journal of Psychology, 49*, 182–194.

Gibson, J. J. (1961). Ecological optics. *Vision Research, 1*, 253–262.

Gibson, J. J. (1966). *The senses considered as perceptual systems.* Boston: Houghton Mifflin.

Gibson, J. J. (1979). *The ecological approach to visual perception.* Boston: Houghton Mifflin.

Gibson, J. J., & Crooks, L. E. (1938). A theoretical field analysis of automobile-driving. *American Journal of Psychology, 51,* 453–471.

Gibson, J. J., Olum, P., & Rosenblatt, F. (1955). Parallax and perspective during aircraft landings. *American Journal of Psychology, 68,* 372–385.

Grosz, J., Rysdyk, R. T., Bootsma, R. J., Mulder, J, A., van der Vaart J C & van Wieringen, P. C. W. (1995). Perceptual support for timing of the flare in the landing of an aircraft. In J. M. Flach, P. A. Hancock, J. K. Caird, & K. J. Vicente (Eds.). *Local applications of the ecological approach to human-machine systems.* Hillsdale, NJ: Lawrence Erlbaum Associates.

Haber, R. N. (1987). Why low-flying fighter planes crash: Perceptual and attentional factors in collisions with the ground. *Human Factors, 29,* 519–532.

Hancock, P. A., & Chignell, M. H. (1995). On human factors. In J. M. Flach, P. A. Hancock, J. K. Caird, & K. J. Vicente (Eds.), *Global perspectives on human-machine systems.* Hillsdale, NJ: Lawrence Erlbaum Associates.

Hasbrook, A. H. (1975). The approach and landing: Cues and clues to a safe touchdown. *Business and Commercial Aviation, 32,* 39–43.

Johnson,W. W., Bennett, C. T., O'Donnell, K., & Phatak, A. V. (1988, June). *Optical variables useful in the active control of altitude.* Paper presented at the 23rd Annual Conference on Manual Control. Cambridge, MA.

Johnson, W. W., & Phatak, A. V. (1989). Optical variables and control strategy used in a visual hover task. In the *Proceedings of the 1989 IEEE Conference on Systems, Man & Cybernetics* (pp. 719–724). Cambridge, MA: IEEE Systems, Man, and Cybernetics Society.

Johnson, W. W., Tsang, P. S., Bennett, C. T., & Phatak, A. V. (1989). The visually guided control of simulated altitude. *Aviation, Space, and Environmental Medicine, 60,* 152–156.

Kelly, L., Flach, J. M., Garness, S., & Warren, R. (1993). Altitude control: Effects of texture and global optical flow. *Proceedings of the 7th International Symposium on Aviation Psychology* (pp. 292–295). Columbus, OH: Ohio State University, Department of Aviation.

Koenderink, J. (1986). Optic flow. *Vision research 26,* 161–180.

Koenderink, J. (1990). Some theoretical aspects of optic flow. In R. Warren & A. H. Wertheim (Eds.), *Perception and control of self-motion* (pp. 53–68) Hillsdale, NJ: Lawrence Erlbaum Associates.

Kraft, C. L. (1978). A psychophysical contribution to air safety:

In H. A. Pick, Jr. (Ed.), *Psychology: From research to practice* (pp. 363–385). New York: Plenum.

Langeweische, W. (1944). *Stick and rudder: An explanation of the art of flying.* New York: McGraw-Hill.

Larish, J. F., & Flach, J. M. (1990). Sources of optical information useful for perception of speed of rectilinear self-motion. *Journal of Experimental Psychology: Human Perception and Performance, 16,* 295–302.

Lee, D. N. (1976). A theory of visual control of braking based on information about time to collision. *Perception, 5,* 437–459.

Lee, D. N. (1980). The optic flow field: The foundation of vision. *Philosophical Transactions of the Royal Society of London, Series B, 290,* 169–179.

Lintern, G. (1980). Transfer of landing skill after training with supplementary visual cues. *Human Factors, 22,* 81–88.

Lintern, G., & Gopher, D. (1978). Adaptive training of perceptual-motor skills: Issues, results, and future directions. *International Journal of Man-Machine Studies, 10,* 521–551.

Lintern, G., & Liu, Y.-T. (1991). Explicit and implicit horizons for simulated landing approaches. *Human Factors, 33,* 401–417.

McNaughton, G. B. (1981, July). Notes on human factors mishaps. *Life Sciences—United States Air Force Safety Journal,* pp. 1–9.

Miller, M. (1983). *Low altitude training.* Training Manual for the 162d Fighter Weapons School, Arizona Air National Guard, Tucson, AZ.

Owen, D. H., & Warren, R. (1982, February). *Perceptually relevant metrics for the margin of safety: A consideration of global optical flow and density variables.* Paper presented at the conference on Vision as a factor in aircraft mishaps. USAF School of Aerospace Medicine, San Antonio, TX. (Also in D. H. Owen (Ed.). (1982) *Optical flow and texture variables useful in simulating self-motion* (I) (Interim Tech. Rep. for Grant AFOSR-81-0078, pp. E-1–E-15). Columbus: Ohio State University, Department of Psychology, Aviation Psychology Laboratory. (NTIS No. AD-A117 016).)

Owen, D. H., Warren, R., Jensen, R. S., Mangold, S. J., & Hettinger, L. J. (1981). Optical information for detecting loss in one's own forward speed. *Acta Psychologica, 48,* 203–213.

Owen, D. H., Wolpert, L., Hettinger, L. J., & Warren, R. (1984). Global optical metrics for self-motion perception. *Proceedings of the 1984 Image III Conference* (pp. 406–415). Phoenix, AZ: Air Force Human Resources Laboratory.

Human Resources Laboratory.

Owen, D. H., Wolpert, L., & Warren, R. (1983). Effects of optical flow acceleration, edge acceleration, and viewing time on the perception of egospeed acceleration. In D. H. Owen (Ed.), *Optical flow and texture variables useful in detecting decelerating and accelerating self-motion* (AFHRL-TP-84-4). Williams AFB, AZ: Air Force Human Resources Laboratory. (NTIS No. AD-A148 718)

Previc, F. H. (1989). Detection of optical flow patterns during low-altitude flight. *Proceedings of the 5th International Symposium on Aviation Psychology* (pp. 708–713). Columbus: The Ohio State University, Department of Aviation.

Powers, W. T. (1978). Quantitative analysis of purposive systems. Some spadework at the foundations of scientific psychology. *Psychological Review, 85,* 417–435.

Reardon, K. A. (1989). *Effects of emergent detail on a landing-judgment task.* Unpublished master's thesis, Wright State University, Dayton, OH.

Reardon, K. A., & Warren, R. (1989). Effects of emergent detail on descent-rate estimations in flight simulators. In *Proceedings of the 5th International Symposium on Aviation Psychology* (pp. 714–719). Columbus: The Ohio State University, Department of Aviation.

Riccio, G.E. (1995). Coordination of postural control and vehicular control: Implications for multimodal perception and simulation of self motion. In J. M. Flach, P. A. Hancock, J. K. Caird, & K. J. Vicente (Eds.). *Local applications of the ecological approach to human-machine sytems.* Hillsdale, NJ: Lawrence Erlbaum Associates.

Shaw, R. E. & Kinsella-Shaw, J. (1988). Ecological mechanics: A physical geometry for intentional constraints. *Human Movement Science, 7,* 155–200.

Smets, G. (1995). Designing for telepresence: The Delft virtual window system. In J. M. Flach, P. A. Hancock, J. K. Caird, & K. J. Vicente (Eds.). *Local applications of the ecological approach to human machine systems.* Hillsdale, NJ: Erlbaum.

Smith, K., & Hancock, P. A. (1993). Situation awareness is adaptive, externally directed consciousness. *Proceedings of the Conference on Situational Awareness.* Orlando, FL: University of Central Florida.

Tobias, S. B., & Owen, D. H. (1984). Global optical flow and texture variables useful in detecting decelerating self-motion. In D. H. Owen (Ed.), *Optical flow and texture variables useful in detecting*

Williams AFB, AZ: Air Force Human Resources Laboratory. (NTIS No. AD-A148 718)

Warren, R. (1976). The perception of egomotion. *Journal of Experimental Psychology: Human Perception and Performance, 2,* 448–456.

Warren, R. (1982). *Optical transformations during movement: Review of the optical concomitants of egospeed* (Final Tech. Rep. for Grant No. AFOSR-81-0108). Columbus: Ohio State University, Aviation Psychology Laboratory. (Also appears as Tech. Rep. AFOSR-TR-82-1028, Bolling Air Force Base, Washington, DC: Air Force Office of Scientific Research. (NTIS No. AD-A122 275).)

Warren, R. (1988). Visual perception in high-speed low-altitude flight. *Aviation, Space, and Environmental Medicine, 59* (11, Suppl.), A116–A124.

Warren, R. (1990). Preliminary questions for the study of egomotion. In R. Warren & A. H. Wertheim (Eds.). *Perception and control of self-motion* (pp. 3–32) Hillsdale, NJ: Lawrence Erlbaum Associates.

Warren, W. H., & Hannon, D. J. (1988). Direction of self-motion is perceived from optical flow. *Nature, 336,* 162–163.

Warren, W. H., & Hannon, D. J. (1990). Eye movements and optical flow. *Journal of the Optical Society of America, A7,* 160–169.

Warren, W. H., Mestre, D. R., Blackwell, A. W., & Morris, M. W. (1991). Perception of circular heading from optic flow. *Journal of Experimental Psychology: Human Perception and Performance, 17,* 28–43.

Warren, W. H., Morris, M. W., & Kalish, M. (1988). Perception of translational heading from optical flow. *Journal of Experimental Psychology: Human Perception and Performance, 14,* 644–660.

Wolpert, L. (1983). *The partitioning of self-scaled and texture-scaled optical information for detection of descent.* Unpublished master's thesis, The Ohio State University, Columbus.

Wolpert, L. (1987). *Field of view versus retinal region in the perception of self motion.* Unpublished doctoral dissertation, The Ohio State University, Columbus.

Wolpert, L. (1988). The active control of altitude over differing texture. *Proceedings of the 32nd Annual Meeting of the Human Factors Society* (pp. 15–19). Santa Monica, CA: Human Factors Society.

Wolpert, L., & Owen, D. H. (1985). Sources of optical information and their metrics for detecting loss in altitude. *Proceedings of the 3rd Symposium on Aviation Psychology* (pp. 475–481). Columbus, OH: The Ohio State University.

Wolpert, L., Owen, D. H., & Warren, R. (1983). Eyeheight-scaled versus ground-texture-unit scaled metrics for the detection of loss in altitude. *Proceedings of the 2nd Symposium on Aviation Psychology* (pp. 513–521). Columbus: The Ohio State University.

Zacharias, G. (1990). An estimation/control model of egomotion. In R. Warren & A. H. Wertheim (Eds.). *Perception and control of self-motion* (pp. 425–460). Hillsdale, NJ: Lawrence Erlbaum Associates.

Chapter 4

Perceptual Support for Timing of the Flare in the Landing of an Aircraft

Jacqueline Grosz[1], Rolf Th. Rysdyk[2], Reinoud J. Bootsma[3], J. (Bob) A. Mulder[2], J. (Hans) C. van der Vaart[2], Piet C. W. van Wieringen[1]

[1]*Faculty of Human Movement Sciences, Free University*
Amsterdam, The Netherlands

[2]*Faculty of Aerospace Engineering, Delft University of Technology,*
Delft, The Netherlands

[3]*UFR STAPS, University of Aix-Marseille II*
Marseille, France

4.0   Introduction

Of central concern in the ecological approach to perception (Gibson, 1966, 1979) is the perceptual support of activity. For perception to be useful for the regulation of activity, a *sine qua non* is that perception provides the perceiver/actor with reliable, and hence veridical, knowledge about the environment and ongoing activities. Many (e.g., Gregory, 1966, 1970; Helmholtz, 1867) have argued that due to the static, two-dimensional nature of the retinal image, the visual system simply cannot provide the required veridical information. Perhaps Gibson's most significant contribution to perceptual science was his insight that this argument rests on the wrong assumption: There is no a priori reason to assume that the visual system operates on static pictures in the form of retinal images! On the contrary, there is a lot of reason to assume that the visual system operates on the dynamic features of the optic array. According to Gibson, the optical information is not to be found in the stimulation by light, but in the structure in the light. Stimulation by itself is a necessary but not a sufficient requirement for perception to occur. One way to illustrate this is to picture one in a thick fog. Obviously, the receptors in the retina are continuously being stimulated by the incoming light, but one does not perceive anything for the very reason that there is no structure in the light reaching one's eyes.

In the world of living creatures, many surfaces co-exist, and each surface is densely covered with texture elements, patches of differing pigmentation, that reflect light differently from their neighbors. A densely structured optic array is thus available at every imaginable point in this "bath of light." Importantly, no matter where one positions a light-sensitive device, such as an eye, the structure of the optic array at that point is lawfully determined by the surroundings. Hence, the structure in the light lawfully relates the point of observation to the environment.

When an observer moves through a stable environment, the structure in the light at the point of observation will change continuously, and the manner in which it changes is fully determined by (a) the layout of the environment, and (b) the movements of the observer. Of course, objects in the environment may also be moving, and this too will lead to changes in the optic array. However, whereas motion of an object results in a *local* change in the optic array, movements of the observer generate a *global* change. The optic flow field, generated by changes in the optic array, thus provides the observer with information about the movements of the self, the layout of the environment and, possibly, the motion of objects in the environment.

In the present chapter we are concerned with the visual information sources that a pilot may use to help him or her land the aircraft. The medium through which the pilot moves is the air. Note that the optical information to specify air when it is clear and transparent is not at all obvious. The character of the visual world of the airplane pilot is therefore normally determined by the ground beneath him or her as well as the horizon in front of the pilot, not by the air through which he or she flies (Gibson, 1950). Of course, when there are patches of cloud nearby or faroff, the pilot can perceive the movement of the aircraft relative to these moving structures. But only when the ground is in sight can the pilot, other than by way of his or her instruments, visually pick up information about the way the plane is moving relative to the earth.

4.1   Optic Flow During Landing

The global changes in the optic array that are generated by the movements of the observer directly specify the direction in which the observer is moving. A forward movement, for instance, results in a *globally expanding flow field.* The singular point that forms the source of the radial expansion pattern (see Figure 4.1) specifies the point toward which one is heading (but see Warren, Mestre, Blackwell, & Morris, 1991, and Warren, Morris, & Kalish, 1988) for a qualification of this statement). A backward movement, on the other hand, results in a *globally contracting flow field,* and the sink of the radial contraction specifies the point from which one is coming. Gibson (1950, 1979) emphasized that the direction of heading is implicit everywhere in the flow field. Even when the focus of outflow itself is not in view (i.e., when the observer does not look in the direction in which he is moving), there is visual information available that specifies the direction of heading (see Figure 4.2).

Figure 4.1. *Optic flow as perceived by a pilot on level flight, looking ahead in the direction of locomotion. From The Perception of the Visual World, by J.J. Gibson, 1950, New York: Houghton Mifflin. Copyright © 1950 by Houghton Mifflin. Reprinted by permission.*

Figure 4.2. *Optic flow as perceived by a pilot on level flight, looking to the side. From The Perception of the Visual World, by J.J. Gibson, 1950, New York: Houghton Mifflin. Copyright © 1950 by Houghton Mifflin. Reprinted by permission.*

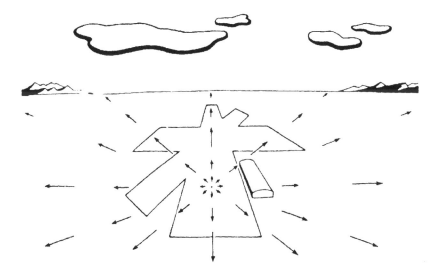

Figure 4.3. *Optic flow as perceived by a pilot approaching a runway. From The Perception of the Visual World, by J.J. Gibson, 1950, New York: Houghton Mifflin. Copyright © 1950 by Houghton Mifflin. Reprinted by permission.*

During the approach to the runway (see Figure 4.3), the focus of expansion is on the point on the runway toward which the glide slope is momentarily aimed. Presently, most airfields are equipped with an Instrumental Landing System (ILS) to help the pilot maintain the desired course during the approach phase of the landing procedure. The ILS brings the pilot to a certain height (normally 50 ft) above the ground at the beginning of the runway, most commonly via a 3° approach slope. If the pilot was to continue his or her descent in the manner depicted in Figure 4.3, he or she would crash into the runway with a considerable sink rate. In order to reduce the sink rate, at some time prior to touchdown the pilot slowly pulls back the control wheel, thereby adjusting the pitch angle of the aircraft (see Figure 4.4). Obviously, the timing of the onset and the control of the speed of execution of this maneuver, which is termed the *flare*, are of crucial importance to the quality of the landing. Finally, the landing gear of the aircraft makes contact with the runway, and thus, we touchdown. For a smooth landing, the sink rate at the moment of touchdown needs to be minimized, although not to the extent that the aircraft becomes airborn again. Immediately after touchdown the pilot starts braking and then, effectively, drives on the runway.

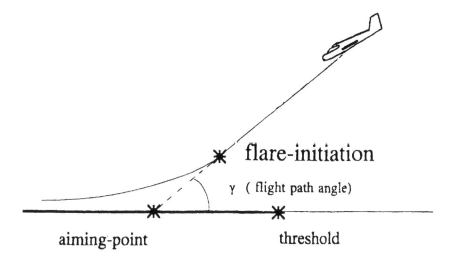

Figure 4.4. *Schematic view of an airplane approaching the runway.*

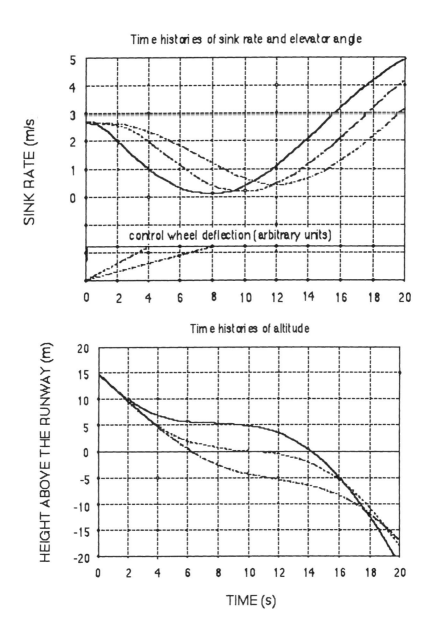

Figure 4.5. *Time histories of control wheel displacement and its subsequent effects on height and sink rate for a correct (dotted lines), too slack (dashed-dotted lines), and too fierce flare (solid lines).*

4.2   Dynamics of the Landing Maneuver

As mentioned earlier, during the last minutes before landing, the aircraft follows the ILS glide slope with constant flight speed. Depending on the slope of the glide path and the approach speed, the sink rate at this time in on the order of 2.5 to 3.5 m/ sec. A soft and safe landing requires a touchdown sink rate between 0.5 and 1.0 m/ sec^{-1}, indicating that a significant reduction in sink rate is to be achieved in the flare or roundout. By slowly moving the control wheel backward, the pilot pitches the nose of the aircraft up. The change in pitch angle causes a change in the direction of the airstream around the aircraft and generates additional lift force, thereby decelerating the downward motion component. Changes in horizontal flight speed are minimal during this maneuver.

In Figure 4.5, three different time histories of control wheel displacement and the subsequent effects on height and sink rate are presented. For all three maneuvers the initial height was set at 15 m and the initial sink rate at 2.75 m/ sec. If no control action was taken, the aircraft would hit the runway with unaltered sink rate after about 5.5 sec. When the control wheel is moved back in a ramp type fashion up to a certain maximum, as in Figure 4.5 (dotted lines), the flight path will be curved and after about 10 sec the desired reduction in sink rate is reached.

However, note that following the initial decrease, the sink rate will start to increase again if contact with the ground has not yet been made. This is due to a slow and periodic motion, the so-called *phugoid motion*, which is a characteristic common to all aircraft. Obviously, therefore, there is only a limited time window available in which a soft touchdown can be made, demonstrating the crucial importance of the time of initiation of the flare.

Also shown in Figure 4.5 are the time histories for the case of a landing, that is, for the given time of initiation of the flare, "too slack" (dashed-dotted lines), giving rise to a rather hard touchdown. Had the aircraft not made contact with the ground after about 6 sec, the minimal sink rate achieved would have been within the acceptable limits, and this particular maneuver could therefore also have been classified as "too late," because an earlier initiation of the flare would have given rise to a satisfactory landing. Finally, Figure 4.5 (solid lines) demonstrates that a "too fierce" execution of the flare (for a given time of initiation) also leads to a rough landing. In this case, minimal sink rate is reached (after about 8 sec) *before* the aircraft touches the runway and, as a result of the phugoid motion, the sink rate increases during the time before touchdown of up to more than 2 m/ sec. This latter maneuver could, of course, also be classified as "too early."

4.3   Time-to-Contact Information

In the foregoing we have seen that the direction in which the pilot is heading is specified by an invariant (the focus of expansion) in the globally expanding optic flow field. Using this as well as the information provided by the ILS instruments, the pilot can regulate the flight of the aircraft so as to correctly approach the runway. However, we have also seen that at some time prior to touchdown the flare has to be initiated (and executed properly) in order to make a soft landing. How does the pilot know when to do this, or put differently, what visual information might the pilot use to determine when to initiate the flare?

Lee (1974) argued that the optic flow field generated by movements of the point of observation also contains information about the distance to different structures in the environment. Interestingly, these distances are not available in absolute terms (i.e., not in meters), but are scaled to the observer's movement velocity. As distance scaled to velocity factually provides a metric in units time (m per m/ sec = sec), this amounts to the observer having access to the remaining times-to-contact with the different environmental structures, assuming the observer's movement velocity remains constant. This can also be demonstrated in the following way. The angle subtended by an environmental structure of size R at distance Z from the point of observation (see Figure 4.6) is given by

$$tan \frac{1}{2} \varphi = \frac{R}{2 \cdot Z} \tag{4.1}$$

For small φ, $tan \ (1/2) \ \varphi = (1/2) \ \varphi$, and hence

$$\varphi = \frac{R}{Z} \tag{4.2}$$

The rate of change of φ over time is therefore

$$\frac{d\varphi}{dt} = \frac{R \cdot dZ/dt}{Z^2} \tag{4.3}$$

The ratio of φ and $d\varphi/dt$ specifies the remaining time until contact, if dZ/dt is constant, as is evident from

$$\frac{\varphi}{d\varphi/dt} = \frac{R}{Z} \cdot \frac{Z^2}{R \cdot dZ/dt} = -\frac{Z}{dZ/dt} = time\text{-}to\text{-}contact \qquad (4.4)$$

(If the velocity with which the distance is closed is considered to be negative, the velocity with which the angle φ grows is, per definition, positive; hence, the minus sign.) Thus, whereas physical size (R), physical distance (Z), and physical velocity (dZ/dt) are visually ambiguous (see equations 4.2 and 4.3), the time remaining until contact (if speed of approach is constant) is directly available in the optic flow field, because it is completely specified in optical parameters in equation (4.4), as was first demonstrated by Lee (1974, 1976). The optic variable specifying this time relation between the observer and the environment ($\varphi/[d\varphi/dt]$) was coined τ (tau) by Lee. Thus, tau specifies the time remaining until the moving observer will make contact with the environmental structure under the condition that speed of approach remains constant. Of course, there are a number of situations in which the speed of approach is not constant (e.g., when the observer brakes). For reasons of conceptual clarity, Lee and Young (1985) suggested to term the first order time relationship specified by tau the $tau\text{-}margin$. The tau-margin, defined as the current remaining distance divided by current speed of approach, therefore only specifies the real remaining time-to-contact in the absence of accelerative forces.

In recent years, a number of experiments have been reported that corroborate the thesis that tau-margin information (as specified by τ) is used to guide behavior (Bootsma & Van Wieringen, 1988, 1990; Fitch & Turvey, 1978; Laurent, 1991; Laurent, Dinh Phung, & Ripoll, 1989; Lee, 1976, 1980; Lee, Lishman, & Thomson, 1982; Lee & Reddish, 1981; Savelsbergh, Whiting, & Bootsma, 1991; Todd, 1981; Van der Horst, & Brown, 1989; Warren, Young & Lee, 1986). Note that the logical consequence of adhering to tau-margin information is that accelerations (i.e., changes in velocity) are not taken into consideration. Strong support to corroborate this claim comes from an experiment by Lee, Young, Reddish, Lough, and Clayton (1983), in which subjects were required to jump up to hit balls dropped from varying heights. Because a dropped ball accelerates due to the gravitational force, τ does not specify the real time-to-collision with the point of observation in such a situation. In fact, the early difference between τ and time-to-collision is greater the greater the height from which the ball is dropped. Lee et al. (1983) demonstrated that subjects gear their actions to the tau-margin (specified by τ) and not to the real time-to-collision in such a situation.

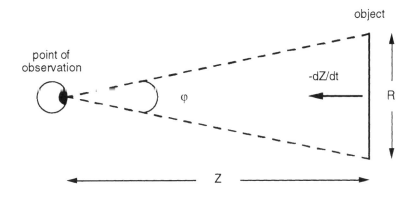

Figure 4.6. *Optic relationship between an observer and an object at current distance Z, approaching with velocity -dZ/dt.*

A situation that more closely resembles the kind of problem encountered in landing an aircraft is found when a car driver is closing in on a stationary obstacle (Lee, 1976; Yilmaz, 1991). The driver needs to know when and how hard to brake in order to avoid collision, but the information needed does not depend simply on the spatial proximity to the obstacle nor simply on the closing velocity. Somehow the driver must brake enough to allow him or her to stop before reaching the obstacle, but at the same time he or she wants to do this in a controlled manner (i.e., he does not want to "slam" the brakes every time he or she encounters an obstacle). What the driver needs is predictive information, telling how much time is left before he or she will reach the obstacle. With this type of information, continuously available via τ, the driver can then regulate his or her braking. Van der Horst and Brown (1989) experimentally evaluated this hypothesis by having subjects approach an obstacle with varying constant speeds, under the instruction to always start braking at the latest possible moment. As might be expected, the results revealed that the magnitude of the *tau*-margin at the moment of onset of braking was not constant, but increased with speed. The minimum *tau*-margin value reached during the braking process itself, however, was found to be independent of approach speed (about 1.1 sec). As this minimum *tau*-margin reached during braking provides an indication of how imminent a collision has been, these results can be taken to support the position that *tau*-margin information was used during the braking process to regulate the amount of braking.

Interestingly, a similar minimum *tau*-margin was found when the instruction to brake at the latest possible moment was withdrawn. Moreover, braking performance was substantially deteriorated when stroboscopic occlusion (hindering the possibility of using *tau*-margin information) was applied.

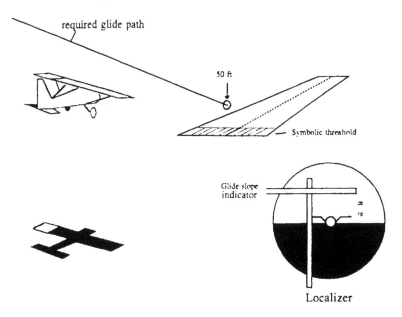

Figure 4.7. *Instrumental Landing System (ILS) and ILS indicator on the dashboard. The airplane is approaching the runway at the right of and below the required flight path, as indicated by the crossing needles of the ILS indicator.*

4.4   Timing the Onset of the Flare In Landing

As shown in Figure 4.4, the pilot also needs to "brake" (i.e., reduce sink rate) in order to make a soft contact with the runway. We argue that, similarly to braking in driving, the pilot uses predictive temporal information, as specified by τ, to determine the moment of onset and the speed of execution of the flare. Note, however, that whereas in driving, a too early or too fierce execution of the braking action can be easily corrected, landing an aircraft has the additional constraint that both a too early (or too fierce) and a too late (or too slack) execution of the flare will give rise to a hard contact with the runway.

We designed an experiment in which one component of the physical quantity specified by τ (i.e., the tau-margin) was indirectly manipulated. By varying the *slope* of the approach path to the runway (specified by the ILS) variations in *sink rate* could be obtained. Figure 4.7 provides an indication of the way in which the ILS instrument operates.

Pilots were asked to land a (simulated) aircraft under three different approach angle conditions ($\gamma = 2°$, $3°$, and $4.5°$). Because the aircraft needs to maintain a minimal airspeed during approach in order to prevent it from stalling, the sink rate is the highest for the $4.5°$ approach angle and the lowest for the $2°$ approach angle.

4.4.1   Method

Subjects: Three male subjects volunteered for participation in the experiment. All three were qualified jet transport pilots.

Task: Each pilot was asked to land a simulated aircraft a number of times under each of the three approach angle conditions, resulting in a total of 126 landings. The pilots were instructed verbally, without demonstration.

Apparatus: A computer-directed flight simulator was used. The simulator consisted of a 3 degrees-of-freedom motion system, a visual system, and a cockpit. The motion system and the visual system were controlled simultaneously by an Encore computer with a computational cycle time of 0.03 sec. The visual system provided a night view of the Amsterdam airport Schiphol, and the pilots were aiming for runway 06. The aircraft simulated was a Citation 1, model 500.

The visual display was generated by a beam penetration color Cathode Ray Tube, with an image resolution of about 1 arcmin. The visual field from the pilot's position was $40°$ in the horizontal plane and $20°$ in the vertical plane. The bandwidth of the motion system dynamics was around 6 Hz, which is better than for most existing simulators (see Baarspul, 1990). Computational timedelays were between 50 to 80 msec. Motion generation timedelays were negligible. The visual display generation lagged the motion generation by less then 10 to 20 msec. (Note that the most demanding criteria set by the FAA for so-called class II moving-base flight simulators allow this delay to be up to 150 msec.)

Procedure: Every single landing started with the aircraft set in a trimmed position on the ILS glide path, at 5 km distance (over ground) from the threshold of the runway, at a height corresponding to the ILS path selected. Data recording started at a distance of 1 km (over ground) from the runway.

Data Analysis: For each landing, a large number of variables were recorded every 0.03 sec, including height, ground distance to runway threshold, sink rate, air speed, pitch angle, elevator deflection, and elevator force. The elevator deflection records were selected for determining the moment of onset of the flare maneuver. Given that we were interested in the actual timing behavior of the pilots, this operationalization of the onset of the flare was preferred over one with reference to the flight path (such as, for example, a certain percentage of change of slope). Once this moment had been identified, the height, the distance to the runway threshold, and the sink rate were ascertained at this point in time together with the moment of touchdown and the sink rate at touchdown.

Table 4.1

Means and Standard Deviations (between parentheses) of Selected Variables for Each of the Approach Angles.

ILS	Height of Tf (m)	Sink Rate at Tf (m/s)	Tau-margin at Tf (s)	Distance to Threshold (m)	Sink rate at touch-down (m/s)	Time from Tf to touch-down (s)
2°	10.27 (3.07)	1.63 (0.52)	6.21 (1.75)	113.8 (94.4)	0.84 (0.26)	7.32 (2.33)
3°	13.09 (3.93)	2.31 (0.31)	5.70 (1.53)	-7.7 (77.7)	0.86 (0.52)	8.99 (2.38)
4°	17.18 (5.35)	3.82 (0.71)	4.52 (1.27)	-19.3 (68.1)	0.99 (0.57)	7.36 (2.42)

Note. A negative distance to threshold indicates that the flare was initiated before the airplane reached the threshold of the runway. T_f refers to the time at which the flare was initiated.

4.4.2   Results

First, the effectivity of using different ILS approach angles as an indirect manipulation of sink rate was evaluated. An analysis of variance

(ANOVA) on the sink rate at the time of initiation of the flare (T_f) revealed a strong effect of ILS approach angle [$\underline{F}(2,123) = 184.14$, $\underline{p} < 0.001$]. As shown in Table 4.1, sink rate indeed varied with ILS approach angle. Similar analyses of variance were performed on the dependent variables (a) height at T_f, (b) *tau*-margin at T_f, (c) duration of the flare, and (d) sink rate at touchdown. Significant effects of approach angle were found for height at T_f [$\underline{F}(2,123) = 30.15$, $\underline{p} < 0.001$)], *tau*-margin at T_f [$\underline{F}(2,123) = 14.19$, $\underline{p} < 0.001$], and duration of the flare [$\underline{F}(2,123) = 6.79$, $\underline{p} < 0.002$]. The sink rate at touchdown was found not to be affected by approach angle [$\underline{F}(2,123) = 1.28$, $\underline{p} > 0.2$]. Table 4.1 presents the mean results of all three pilots on these dependent variables.

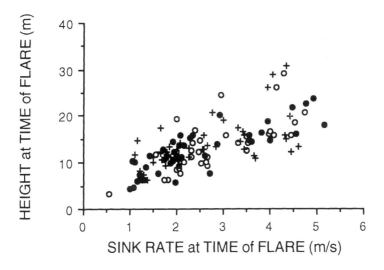

Figure 4.8. *Height at the moment of initiation of the flare as a function of momentary sink rate.*

The relation between sink rate and height of the aircraft at the moment of initiation of the flare is shown in Figure 4.8 for all landings of the three subjects. A linear regression analysis revealed a highly significant relation, $\underline{F}(1,124) = 125.56$, $\underline{p} < 0.001$, between sink rate and height[1] with the best fit line being described by Height (m) = 3.6 + 3.3

---

[1]Note that height has been calculated from the first point of the aircraft to contact the ground (i.e., the landing gear).

*Sink Rate (m/ sec). In line with the finding of an effect of approach angle in the ANOVA reported earlier, the intercept (3.6 m) of this best fit line was found to differ significantly from zero ($p < 0.01$). Hence, it must be concluded that the magnitude of the tau-margin at the moment of onset of the flare (T_f) was not constant over different sink rates.

4.4.3   Discussion

As the sink rate at the moment of touchdown was found to be similar for all three ILS approach angles, it must be concluded that the quality of the landing itself did not suffer from the approach angle manipulation. How was this performance achieved? Figure 4.8 illustrates two important findings from the present experiment. In the first place, it is clear that our pilots did not initiate the flare at a constant height above the runway, as the height at T_f was found to be significantly correlated with sink rate at T_f. At the same time, however, it is also clear that they did not initiate the flare at a constant *tau*-margin magnitude, specifying the time before they would make contact with the runway. In analyzing a car-braking task, Van der Horst and Brown (1989) also reported that the moment of onset of braking varied with the speed of approach. However, they found the *tau*-margin at the onset of braking to increase with increasing speed, whereas in the present experiment we found the tau-margin at the onset of the flare to decrease with increasing sink rate! Given that more "braking" (i.e., more reduction in sink rate) is needed (and in fact achieved, see Table 4.1) for the higher sink rates, we had not expected this result.

Given that (a) the *tau*-margin at T_f decreased with increasing sink rates, and (b) a constant sink rate was achieved at the moment of touchdown, the pilots must necessarily have performed a more forceful flare maneuver for the approaches with the higher sink rates. Put differently, they traded off the time remaining at the moment of onset of the flare for speed of execution. This strategy may also be held responsible for the fact that the actual times to touchdown were independent of the glide slope. The tradeoff results in differential effects on these times relative to the *tau*-margins at the onset of flare. Such a tradeoff between perceptual (*tau*-margin information) and executional (speed of execution) variables was also found by Bootsma and Van Wieringen (1990) for the manner in which table top tennis players achieve their accuracy when performing an attacking forehand drive. Such compensatory variability by itself, however, does not explain why the magnitude of the tau-margin at T_f varied systematically with sink rate.

Although until now we have concentrated solely on a sufficient reduction of sink rate as the requirement for a correct landing, there are, of course, more requirements that need to be met. For instance, not only must the pilot make a soft contact with the ground, he or she must also ensure that the plane lands at a position leaving enough runway to decelerate after touchdown. If the pilots would have initiated their flares at a constant value of the tau margin, irrespective of approach angle, they would still have had to adjust the speed of execution in order to attain a soft landing, because sink rates differ for the different ILS glide paths. Because the ILS sets them on a trajectory that is aiming toward a point at a height of 50 ft above the threshold, initiating the flare at a constant *tau*-margin would mean that touchdown would occur at different positions on the runway. For an approach with a low sink rate, this would mean that they already would have used up a significant portion of the runway before touchdown. For approaches with a high sink rate, on the other hand, landings on or even before the threshold of the runway could occur. Hence, we tentatively conclude that the pilots not only wanted to make a soft landing, but that they also wanted to make contact with the runway at a specific point. These two requirements naturally lead to the tradeoff between the moment of initiation of the flare and the speed of execution of the flare as found in the present experiment.

4.5   References

Baarspul, M. (1990). A review of flight simulation techniques. *Progress in Aerospace Science, 27,*1–120.

Bootsma, R. J., & Van Wieringen, P. C. W. (1988). Visual control of an attacking forehand drive in table tennis. In O. G. Meijer & K. Roth (Eds.), *Complex movement behaviour: "The" motor-action controversy* (pp. 189–199). Amsterdam: North-Holland.

Bootsma, R. J., & Van Wieringen, P. C. W. (1990). Timing an attacking forehand drive in table tennis. *Journal of Experimental Psychology: Human Perception and Performance, 16,* 21–29.

Fitch, H. L., & Turvey, M. T. (1978). On the control of activity: Some remarks from an ecological point of view. In D. Landers & R. Christina (Eds.), *Psychology of motor behavior and sport* (pp. 3–35). Champaign, IL: Human Kinetics Press.

Gibons, J.J. (1950). *The perception of the visual world.* New York: Houghton Mifflin.

Gibson, J. J. (1966). *The senses considered as perceptual systems.* Boston:

Houghton Mifflin.

Gibson, J. J. (1979). *The ecological approach to visual perception.* Boston: Houghton Mifflin.

Gregory, R. L. (1966). *Eye and brain: The psychology of seeing.* London: Weidenfeld and Nicolson.

Gregory, R. L. (1970). *The intelligent eye.* New York: McGraw-Hill.

Helmholtz, H. von. (1867). *Handbuch der physiologischen Optik.* Leipzig: Voss.

Laurent, M. (1991). Visual cues and processes involved in goal-directed locomotion. In A.E. Patla (Ed.), *Adaptability of human gait: Implications for the control of locomotion* (pp. 99-123). Amsterdam: North-Holland.

Laurent, M., Dinh Phung, R., & Ripoll, H. (1989). What visual information is used by riders in jumping. *Human Movement Science, 8,* 481–501.

Lee, D. N. (1974). Visual information during locomotion. In R. B. McLeod & H. Pick (Eds.), *Perception: Essays in honor of J. J. Gibson* (pp. 250–267). Ithaca, NY: Cornell University Press.

Lee, D. N. (1976). A theory of visual control of braking based on information about time-to-collision. *Perception, 5,* 437–459.

Lee, D. N. (1980). Visuo-motor coordination in space-time. In G. Stelmach & J. Requin (Eds.), *Tutorials in motor behavior* (pp. 281–293). Amsterdam: North-Holland.

Lee, D. N., Lishman, J. R., & Thomson, J. A. (1982). Visual regulation of gait in long jumping. *Journal of Experimental Psychology: Human Perception and Performance, 8,* 448–459.

Lee, D. N., & Reddish, D. E. (1981). Plummeting gannets: A paradigm of ecological optics. *Nature, 293,* 293–294.

Lee, D. N., & Young, D. S. (1985). Visual timing in interceptive actions. In D. J. Ingle, M. Jeannerod & D. N. Lee (Eds.), *Brain mechanisms and spatial vision* (pp. 1–30). Dordrecht, The Netherlands: Martinus Nijhoff.

Lee, D. N., Young, D. S., Reddish, D. E., Lough, S., & Clayton, T. M. H. (1983). Visual timing in hitting an accelerating ball. *Quarterly Journal of Experimental Psychology, 35A,* 333–346.

Owen, D. H. (1990). Perception and control of changes in self-motion: A functional approach to the study of information and skill. In R. Warren & A. H. Wertheim (Eds.), *Perception and control of self-motion* (pp. 289–322). Hillsdale, NJ: Lawrence Erlbaum Associates.

Savelsbergh, G. J. P., Whiting, H. T. A., & Bootsma, R. J. (1991).

'Grasping' tau!. *Journal of Experimental Psychology: Human Perception and Performance, 17,* 315-322.

Todd, J. T. (1981). Visual information about moving objects. *Journal of Experimental Psychology: Human Perception and Performance, 7,* 795–810.

Van der Horst, A. R. A., & Brown, G. R. (1989). *Time-to-collision and driver decision making in braking* (Rep. IZF 1989 C-23). Soesterberg: The Netherlands, TNO Institute for Perception.

Warren, W. H., Mestre, D. R., Blackwell, A. W., & Morris, M. W. (1991). Perception of circular heading from optical flow. *Journal of Experimental Psychology: Human Perception and Performance, 17,* 28–43.

Warren, W. H., Morris, M. W., & Kalish, M. (1988). Perception of translational heading from optic flow. *Journal of Experimental Psychology: Human Perception and Performance, 14,* 646-660.

Warren, W. H., Young, D. S., & Lee, D. N. (1986). Visual control of step length during running over irregular terrain. *Journal of Experimental Psychology: Human Perception and Performance, 12,* 259–266.

Yilmaz, E. H. (1991). *The use of visual expansion in braking regulation: An experimental study.* Unpublished manuscript, Brown University, Department of Cognitive Science, Providence, RI.

Chapter 5

Coordination of Postural Control and Vehicular Control: Implications for Multimodal Perception and Simulation of Self-Motion

Gary E. Riccio

Department of Kinesiology, Bioengineering Program, and Beckman Institute for Advanced Science and Technology
University of Illinois at Urbana Champaign
and
Human Engineering Division and Biodynamics and Bioengineering Division
Armstrong Laboratory, Wright Patterson AFB, OH

5.0   Prologue

5.0.1   Purpose

The purpose of this chapter is to reveal the aspects of perception and action that are essential for simulating self-motion in a virtual environment. It is assumed that a goal of the associated technology is to provide for compelling experiences of presence in virtual environments. The potential manifestations and applications of this emerging technology are sufficiently diverse that general discussions are difficult. Consequently, this chapter focuses on research that is relevant to flight simulation. Although the discussion is couched in the particulars of flight simulation, it is emphasized that the design principles and strategies are not limited to flight simulation. To promote generality, self motion in real and virtual environments is discussed from the perspectives of ecological psychology and control theory. The centrality of these complementary perspectives is due to the fact that they have generated a significant body of theoretical and empirical research in which the interaction between a human and an environment and the linkage between perception and action are fundamental. More extensive treatment of these perspectives, and their relevance to self-motion, are available elsewhere (Flach, 1990; Riccio, 1993b).

From ecological and control theoretic perspectives, one must start with identification of the task-relevant information (e.g., Riccio, 1993a, 1993b). Only then do issues concerning observability and controllability of a system's behavior become nonarbitrary, and only then can developments in simulator technology transcend the arbitrary constraints of convenient experimental paradigms. In other words, research and development in flight simulation should be based on an epistemology that is commensurate with the tasks performed in flight simulators and with those aspects of the human and environment that are meaningfully related to these tasks. The conceptual analysis presented in this chapter exemplifies such an epistemology (see also J. Gibson, 1979, pp. 7–43; Riccio, 1993b).

5.0.2   Caveats and Cautions

Extensive conceptual analyses, such as presented in this chapter, are unusual in published applications of control theory to perception and action. Control theoretic analyses usually proceed expeditiously from block diagram representations of a system to formal mathematical models of the system and its behavior. These models have definite epistemological implications, although the implications are rarely examined extensively. Such implications generally follow from mathematical assumptions (e.g., initial values, boundary values, parameterization, equation structure) that are heavily influenced by criteria such as solvability and computability. Different mathematical techniques have different assumptions and implications, and these differences can proliferate as models develop. If modeling techniques are driven primarily by domain-specific mathematical expedients, they are likely to diverge with respect to the phenomena that they can explain. It is in this sense that particular programs of research in control theory and ecological psychology may appear to be incompatible. Mathematical models must converge, however, if they are motivated by similar conceptual analyses of the systems and behavior they are intended to explain. Thus, the conceptual treatment of self-motion in this chapter should not be interpreted as an argument against mathematical modeling. Rather, it is a necessary precursor to any mathematical model of the perception and control of self-motion.

Topics such as self-motion and flight simulation are familiar in ecological psychology and in the study of human-machine systems. Most readers undoubtedly will have expectations about this chapter that are based on other treatments of self-motion and flight simulation. These expectations could be a problem for the reader insofar as this chapter presents a radically different style of analysis. The reader will be disappointed, for example, if a comprehensive summary of the literature is expected. As a conceptual analysis this chapter provides a guide to thinking about the phenomena of self-motion and flight simulation rather than a survey of the facts. A premise of this chapter is that we are largely factless in the scientific study of these phenomena. An increasing amount of detail has become available on a vanishingly small subset of self-motion and flight simulation. Much of this concerns mathematical modeling of vehicular control and visual perception of the objects and events depicted in a computer-generated display. Although these areas of investigation are important, they are not even close to being sufficient for an appreciation of real and simulated self-motion.

This chapter attempts to reorient the study of these phenomena so that subsequent analyses have sufficient scope. The conceptual reorientation sacrifices depth of analysis for breadth. Thus, readers who are familiar with the traditional approaches to real and simulated self-motion may find the conceptual reorientation somewhat unsatisfying. Familiarity with traditional approaches may actually be a problem because this chapter challenges some of the most pervasive assumptions in the study of self-motion. By relaxing these assumptions, the conceptual reorientation broadens the domain of inquiry and leads to some areas of scientific inquiry that have been relatively neglected in the study of self-motion. The intent of this chapter is to acquaint the reader with the diversity of research that is relevant to the perception and control of self-motion. A detailed discussion of such a diverse body of research is beyond the scope of this chapter. For these reasons, the reader should expect to be left with many more questions than answers. This chapter attempts to entice the research-oriented reader to consider a broader domain of inquiry. The design-oriented reader should expect to discover a significantly extended range of possibilites for the design of flight simulators or for any simulation of self-motion.

5.0.3   Pedagogical Foundation

Block diagrams are used in this chapter to graphically summarize categories of human-environment interactions that are essential in the perception and control of self-motion. In addition, such diagrams emphasize the coupling of perception and action. The mutuality of the human and the environment, and the coupling between perception and action, are fundamental principles in ecological psychology (J. Gibson, 1979). Block diagrams, on the other hand, are more familiar in control theory. The correspondence between ecological psychology and control theory is somewhat controversial. It has been argued, however, that there are deep conceptual connections between these two approaches (Flach, 1990; Riccio, 1993b). Both approaches begin by differentiating[1] between the essential and incidental characteristics of a system. In control theory, block diagrams document the conclusions of this initial phase of investigation. Block diagrams represent both the separable components of a system and the functional relations among these components. They describe the interactions between sensors, controllers, and controlled processes. These components are persistent characteristics of systems that can be identified over changes in systems. Thus, consideration of invariance and transformation underlies the development and use of block diagrams for the analysis of a system's behavior.

Block diagrams generally are constructed from arrows, branching points, circles, and rectangles. Arrows represent the "flow of control energy or information" from one component of the system to another (DiStephano, Stubberud, & Williams, 1967, p. 13; cf. Kugler & Turvey, 1987, pp. 7–9). Branching points indicate that energy or information

---

[1]Control-theoretic terminology is blended with ecological terminology to highlight similarities and differences between various programs of research in control theory and ecological psychology. One problem created by this blending of terminology is the term differentiation. It is never used in this chapter to refer to a particular mathematical operation. In other work, the mathematical usage is explicit when equations are included. In ecological psychology, differentiation is associated with discrimination and elaboration of detail through the process of perceptual learning and, more than any other concept, it emphasizes that perception unfolds over time and that living systems are in a continual process of development (see e.g., E. Gibson, 1991). To avoid confusion in this chapter, the term is always used in phrases such as "differentiation between" or "differentiation among" perceivables.

flows from one component of the system to several other components. Circles (summation points) generally represent simple concatenation operations such as addition or subtraction. The outputs of these operations are dimensionally similar to the inputs. Rectangles (blocks) represent cause–effect relations such as the transfer function for some subset of the system or the perception–action relations for some human activity. These dynamical[2] relations generally depend on the spatiotemporal characteristics of change in the system, and the outputs of these dynamical subsystems are often dimensionally different than the inputs. Block diagrams vary in detail because they reflect the level of detail in the corresponding analysis of a system. The amount of detail in a block diagram is determined by the invariants and transformations that are considered in the analysis of a system's behavior. Sections 5.1–5.2 provide a summary of the invariants and transformations and the elements of the associated block diagram representations that are considered essential to the perception and control of vehicular motion (see also Riccio, 1993b).

5.1   Vehicular Control

5.1.1   Vehicular Control Relative to the Surroundings

There can be more than one cause for an observed change in orientation or motion of oneself or the vehicle in which one is an occupant. This is illustrated in Figure 5.1 (Riccio, 1993b). The stimulation of the sensory organs ("input" to the "sensors") provides information about a change in state relative to the surroundings, irrespective of the cause of the change. Gibson (1979, p. 184) referred to this information as kinesthesis[3], and he was careful to distinguish it from feedback or

---

[2]The word "dynamical" is used in this chapter to refer to the relationship between motion and its causes, that is, characteristics that are considered in the science of "dynamics." Elsewhere, the word "dynamic" is sometimes used instead of "dynamical." This can be confusing, however, because "dynamic" may also refer to change irrespective of the cause of change, and it often has this meaning in psychology.

[3]J. Gibson wrote that "Visual kinesthesis specifies locomotion relative to the environment, whereas the other kinds of kinesthesis may or may not do so. The control of locomotion in the environment must therefore be visual." (1979, pp. 226-227). This should be interpreted as an argument for the necessity of vision and not for the sufficiency of vision in most situations. Furthermore, visual control of locomotion is not necessary in some special situations (Lee, 1990).

reafference about controlled movement (see also J. Gibson, 1964). Control actions ("outputs" from the "controller") are just one cause of change relative to the surroundings. Events in the surroundings can also result in change (e.g., motion) relative to the surroundings.

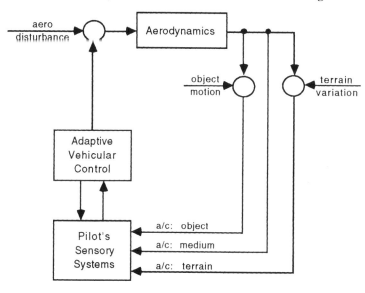

Figure 5.1. *Core components of block diagrams in this chapter (see also Riccio, 1993b). Bidirectional flow of information between sensors and adaptive controller is unconventional in control systems diagrams. It is used in all subsequent diagrams in this chapter as an expedient representation of a nested inner loop in which the controller adjusts the perceptual systems to optimize the pickup of information. The notation on the inputs to the sensory systems represents the fundamentally relational nature of perceivables. Events outside a system or subsystem are always perceived in relation to the system or subsystem. In this diagram, the system is the aircraft (represented by a/c), and the events are movements of the aircraft relative to some aspect of the meaningful environment. This diagram differentiates between three aspects of the meaningful environment: the terrain, objects (e.g., other aircraft), and the aerodynamic medium. The aerodynamic medium includes the medium of support (i.e., the air mass) and the inertial "reference frame."*

Vehicular motion is observed with respect to particular task-relevant aspects of the surroundings. Uncontrolled motion occurs whenever there are task-relevant changes or disturbances in the surroundings, and this relative motion need not involve a transmission of force from the surroundings to the vehicle (cf. J. Gibson, 1979, pp. 93–102). Three types of disturbance are represented in Figure 5.1: (a) a vehicle can be displaced relative to an object, the terrain, and the inertial environment[4] by some external force such as a wind gust; (b) a vehicle can be displaced relative to the terrain because of variation in terrain (e.g., hills and valleys); or (c) a vehicle can be displaced relative to an object because of motion of the object (e.g., movement of another vehicle).

5.1.2   Dynamical Variation and Vehicular Control

One can differentiate between disturbances if there is invariant structure in stimulation to specify persistence and change of objects and the terrain (J. Gibson, 1979, pp. 102–110). If a precipice is perceived as such, for example, the relative motion resulting from flight over the precipice is not confused with motion relative to the inertial environment (e.g., change in altitude due to wind shear). The availability of information in stimulation does not guarantee that the information can be picked up by an observer. Perceived variation in the level of the terrain before passage over the precipice could be different than the perceived change in above-ground-level during passage over the precipice. The difference could be incorrectly attributed to motion relative to the inertial environment, and there even could be a visceral experience associated with this event. Such misperceptions are mitigated by perception of motion relative to the inertial environment.

Perception of motion relative to the inertial environment is, by definition, equivalent to perception of change in the inertial environment. Perception of such motion is possible if there is invariant structure in stimulation that specifies change in the inertial environment. The inertial environment changes whenever a vehicle speeds up, slows down, or changes direction, that is, whenever there is change in the vehicular velocity vector. The inertial environment is part of the aerodynamic environment because it influences the response of the aircraft to control inputs and external forces. Factors that affect the inertial environment are represented on the input side of the aerodynamics in Figure 5.1. Aerodynamics constrain the effects of forces on the aircraft, although such constraints are not imposed on the

---

[4]The inertial environment includes any reference frame or aspect of a system's surroundings that is not accelerating with respect to the system.

effects of object motion or terrain variation.

The vestibular and somatosensory systems are stimulated whenever external forces cause variations in magnitude or direction of vehicular velocity (Stoffregen & Riccio, 1988). This is obviously not the case for variations in remote objects or the terrain. Thus, multimodal perception can be important for differentiation among disturbances. Riccio (1993b) argued that multimodal differentiation is necessary because (a) it is the only general way to detect persistence and change in the inertial environment, and (b) such information about the inertial environment is required to predict the effects of the pilot's control actions on the motion of the aircraft.

5.1.3 Effects of Vehicular Control on the Human Operator

Sustained acceleration in a vehicle changes the magnitude of force on occupants in the vehicle. These so-called "G forces" have both physiological and biomechanical effects on the occupants (see Brown, Cardullo, McMillan, Riccio, & Sinacori, 1991; Kron, Cardullo, & Young, 1980). Large G forces can lead to a progressive loss of peripheral vision ("gray out") and even a complete loss of vision and loss of consciousness ("black out"). Such events impose hard constraints or limits on perception and action by the pilot, and these limits partially determine the operational envelope of the pilot-aircraft system. Robust control of a system is facilitated if the states of a system relative to its limits are observable (cf. Riccio, 1993a; Riccio & Stoffregen, 1988, 1991). The physiological effects of increasing G forces provide information about the limits of the pilot-aircraft system with respect to its limits, and this information can support robust control of an aircraft. The informational role of these physiological effects is represented in Figure 5.2.

Although sustained "high G" manuevers are relatively rare, variations in the inertial environment involving lower magnitudes of force and acceleration are ubiquitous. Common sources of variation include transients due to sudden disturbances or sudden changes in direction and orientation. A source of more continuous variation is vibration. Vibration and transients have consequences for nonrigid organisms in general (Riccio & Stoffregen, 1988) and, in particular, for occupants of vehicles who are neither rigid nor rigidly attached to the vehicle (Bott & Lincoln, 1988, pp. 2076–2081; Griffin, 1975). Such disturbances generate motion of an occupant relative to controls, displays, and the seat in the cockpit. Relative motion stimulates

multiple perceptual systems, and there is information in this stimulation
that can facilitate vehicular control (Riccio, 1993b). Relative motion can
also degrade vehicular control, however, to the extent that it interferes
with perception and action in the cockpit. The information available in
nonrigidity and disturbances due to nonrigidity are represented in
Figure 5.2.

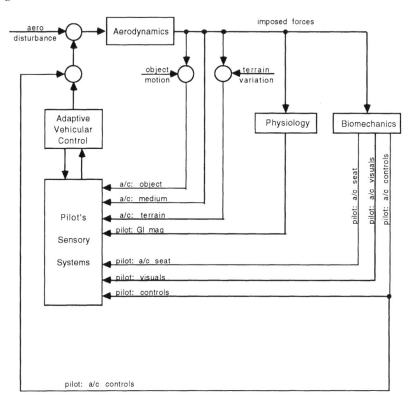

Figure 5.2. *Forces encountered during variation in the velocity vector affect
the cardiovascular system of the pilot. These physiological effects are due to
both microbiomechanical effects of G forces on tissues and fluids and
neurophysiological responses to the attendant changes in blood pressure. The
notation reflects that the events are perceived in relation to the pilot (e.g., the
size of the pilot's field of view). Forces encountered during variation in the
velocity vector can cause movement of the pilot relative to various aspects of the
cockpit environment. The notation reflects that the events are perceived in
relation to the pilot (e.g., looking at instruments, manipulating controls). The*

events are included in the vehicular control loop because they have affordances for vehicular control.

5.2 Coordination of Postural Control and Vehicular Control

5.2.1 Postural Control in the Cockpit

Pendulum dynamics, in general, and balance, in particular, are pervasive properties of human-environment interactions (Stoffregen & Riccio, 1988). Constraints that are specific to pendulum dynamics are imposed by gravitational and resistive inertial forces. The sum of these forces is sometimes referred to as the gravitoinertial force vector (see, e.g., Young, 1984, pp. 1046–1051). The magnitude of the gravitoinertial vector is nonzero whenever a body is in contact with a support surface, and the direction of this vector generally determines the direction of balance for a body in contact with a surface of support (Riccio & Stoffregen, 1990; Stoffregen & Riccio, 1988). When the orientation of the body deviates from the direction of balance, torque is produced by the nonalignment of gravitoinertial and support surface forces.

The dynamics of balance in the cockpit vary because of variation in the gravitoinertial vector within and across typical flight maneuvers (see, e.g., Brown et al., 1991). Linear acceleration and centripetal acceleration of the aircraft change both the direction and magnitude of the gravitoinertial vector. Thus, postural control in vehicles must be robust or it must adapt to variations in both the direction of balance and the consequences of imbalance. Adaptability is also important because postural control is influenced by situation specific factors other than torques on the body segments (Riccio, 1993a; Riccio & Stoffregen, 1988, 1991).

The visual system and the parts of the body involved in manual control are nested within the postural control system (J. Gibson, 1966, 1979; Reed, 1982). Postural orientation and configuration may be modified to facilitate perception and action. For example, the pilot may tilt the head away from the direction of balance to scan a "tilted" horizon. In addition, the pilot may move the head or torso to minimize movement with respect to the inside of the cockpit even when this increases movement with respect to the direction of balance. Such nonmechanical influences on postural control are analogous to effects that variations in remote objects and the terrain have on vehicular control. System diagrams for postural control should represent the observability of postural orientation and configuration with respect to balance, support surfaces (e.g., an aircraft seat), manipulanda (e.g., control stick in the cockpit), and objects of visual regard (e.g., cockpit instruments). Figure 5.3 is a simplified diagram that represents these essential aspects of postural control.

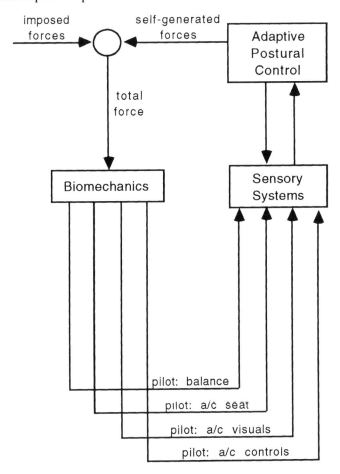

Figure 5.3. *Block diagram for postural control. This diagram emphasizes analogies between postural control and vehicular control. Note that information about the direction of balance is available in the activity of postural control.*

5.2.2 Coupled Control Systems

The importance of postural control in a vehicle indicates that the human operator is confronted with two concurrent control tasks. On the one hand, the operator needs to control the dynamical interaction between the vehicle and its medium or surface of support. On the other hand, the operator needs to control the dynamical interaction between his or her body and the interior of the vehicle. The two interactions can and should be represented as separate systems in the diagrams that represent the respective control problems. Each system exists in the context of the other in that the behavior of each system influences the other. The disturbance of each system by the other is one source of coupling between the systems. Another source of coupling derives from coordination between the controllers, that is, interdependence between the perception-action relations that are relevant to the respective control problems. The two sources of coupling are represented in Figure 5.4.

The coordination of postural and vehicular control suggests that these two control systems form a "superordinate" control system (Riccio, 1993b). The function that can be uniquely attributed to this superordinate system is the perception and control of self-motion. Self-motion is not adequately described by the component (or subset of components) of the system diagram that refers to vehicular control. This component does not differentiate the control of remotely piloted vehicles from the control of a vehicle in which the operator is an occupant. Thus, vehicular control considered in isolation can only lead to models of phenomena that are common to these two types of vehicles. The experience of constant velocity motion is such a phenomenon, albeit vanishingly rare. Because the control of self-motion involves variations in velocity and acceleration, the postural consequences of these variations must be considered in any model of the perception and control of self-motion.

A ubiquitous postural control activity provides "persistent excitation" of the human-environment system that reveals the changing dynamics of this interaction (Riccio & Stoffregen, 1991). In particular, postural control is sensitive to variations in the magnitude and direction of the gravitoinertial force vector. Postural control activity is analogous to "dither" of the aircraft's controls in that it can provide information about changes in the inertial or aerodynamic environment (Riccio, 1993b). The obtaining of such stimulation allows the pilot to "feel" the changing dynamics of the aircraft. A possible advantage of the

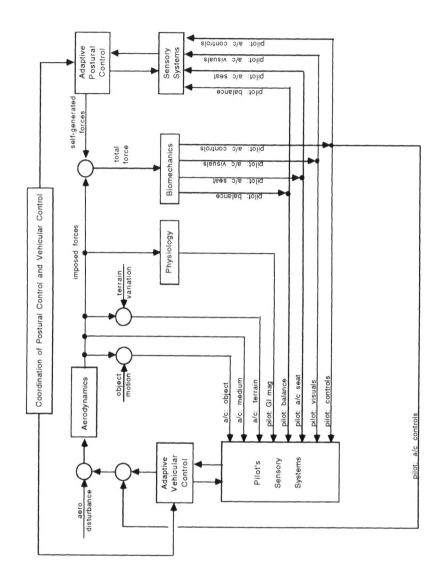

Figure 5.4. *Coupling of postural and vehicular control systems. Control systems and the associated diagrams are linked in three ways: (a) Variation in vehicular velocity imposes forces on the postural control system. (b) Postural control and vehicular control are coordinated in that the actions of each system reflect the constraints on both systems. (c) Movement of the pilot relative to various aspects of the cockpit environment has affordances for both postural control and vehicular control. Note that the control of posture in the cockpit provides information about the direction of balance that otherwise is not available in the vehicular control system (cf. Figure 5.2).*

information in postural dither is that it does not interfere with vehicular control actions to the extent that dither on the control stick can. Another advantage is that this information is always available unless the pilot is completely restrained in the cockpit. One reason to emphasize this source of information is that it reveals another way in which control systems can be coupled. The postural control system can be considered as an "informative dynamical component" of the sensors used for vehicular control (Riccio, 1993b); that is, the vehicular control loop is closed through the postural control system of the human operator. Such "active observation" may be a general form of coupling in nested systems (J. Gibson, 1979; Riccio, 1993b; Shaw, Kadar, Sim, & Repperger, 1992). It could be generally important in the control of self-motion because postural control is relevant in all forms of pedal and vehicular locomotion. This is consistent with the importance attributed to the basic orienting system in an ecological approach to perception (J. Gibson, 1966, 1979; Riccio, 1993; Riccio & Stoffregen, 1988, 1991).

5.3   Simulated Vehicular Control

The sine qua non of flight simulation is generally considered to be the capacity to induce perception of self-motion through an environment without moving the observer. This capacity becomes useful if the observer is allowed to control the simulated self-motion, if the observer-actor can achieve goals. Most goal-directed motion through the environment requires perception of objects and surfaces that are distant from the observer. Visual perception is thus crucial for goal-directed motion, and there is no question that visual display systems are necessary in flight simulation. Furthermore, there is general agreement that further developments in visual display systems are important because recognition of familiar objects and layouts increases the range of flight tasks that can be performed in the simulator. For example, the detail on a tanker aircraft is important in the approach and docking phases of in-flight refueling; the depth of a ravine or the presence of telephone wires is important in low-level flight. The issue in flight simulation over which there is the greatest controversy, and which has the greatest design consequences, is whether there are any situations in which visual display systems are not sufficient (e.g., Cardullo & Sinacori, 1988; Hosman & van der Vaart, 1981; Lintern, 1987). Consequently, there is some disagreement about whether nonvisual displays are necessary for high fidelity simulations.

There is a lack of general agreement about the criteria for fidelity of flight simulators. This makes it difficult to resolve controversies about the sufficiency of particular displays. Fidelity criteria fall into two general categories (Martin, McMillan, Warren, & Riccio, 1986; Riccio & Hettinger, 1991; R. Warren & Riccio, 1985): (a) Subjective experience in the simulator and the aircraft should be similar (experiential fidelity). Ideally, the simulation should not be perceived as such (i.e., as a simulation), but rather as motion of the pilot in an environment with recognizable objects. (b) Flight control skills acquired in the simulator and those acquired in the aircraft should be similar (action fidelity). Ideally, training in the simulator should transfer to the aircraft. Progress in flight simulation has been limited by a poor understanding of experiential fidelity and action fidelity, and this can be attributed to neglect of many of the issues reviewed in this chapter (although, see Brown et al., 1991; Riccio, 1993b).

The controversy about the sufficiency of visual displays in flight simulation has theoretical as well as practical relevance. The general theoretical issues concern multimodal perception of self-motion and control of the nested systems involved in orientation and self-motion (Riccio, 1993b). The real and simulated events for which these issues are most important involve variation in orientation, variation in the velocity vector for self-motion, and the consequences of these variations for perception and action. The consequences of variation in orientation or velocity derive from the forceful interaction of the aircraft or simulator with the pilot's body. Vehicular motion, and the forces that they impose on the pilot's body, stimulate multiple perceptual systems.

It is a common assumption in many areas of research, including those concerned with flight simulation, that such multimodal stimulation is either redundant or conflicting (see Stoffregen & Riccio, 1988, 1991). This assumption is inappropriate, however, given that nonredundancies are both common and informative for a nonrigid body (Riccio & Stoffregen, 1988, 1991). Multimodal stimulation is more accurately described as complementary, and this complementarity has nontrivial implications for the perception and control of self-motion. Although redundant stimulation may be considered unnecessary in many situations, complementary stimulation would be necessary if it provided task relevant information not available to individual perceptual systems. Rigidity of the body and redundancy or conflict of multimodal stimulation are two pervasive assumptions that are rejected in this chapter. This leads to a completely different view of action fidelity and experiential fidelity in flight simulators.

5.4   Vehicular Control Relative to the Simulated Surroundings

One set of goals for locomotion includes chasing, avoiding, or moving with another object (see Warren, 1990). Such events can be simulated simply by providing information about the relative location of a moving object. Visual displays that provide this minimal information are common in target-following or pursuit-tracking experiments that investigate the spatiotemporal characteristics of manual control (see, e.g., Poulton, 1974). It can be difficult, however, to differentiate controlled and uncontrolled movements when only the relative location of the target is visible. For example, vehicular motion and motion of another object can each result in a change in the relative location of the object. Differentiating between these two causes of relative motion makes possible the identification of vehicular dynamics (i.e., effects of one's control actions) and the identification of any regularities in the motion of the other object. Knowledge of the dynamics of the controlled system or the dynamics of disturbances can improve control performance. Thus, displays that support differentiation between vehicular motion and object motion can improve target following in a simulated vehicle.

The results of Hosman and van der Vaart (1981), for example, demonstrated this point. They found significant improvements in target-following performance when either peripheral visual displays or whole-body motion were added to a central visual display (18% and 33% reduction in tracking error, respectively). Peripheral displays and whole body motion provided information about vehicular motion relative to the stationary environment, whereas the central display provided information about relative location of the target. This effect was probably not due to differences across conditions in the perceivability of vehicular dynamics because vehicular dynamics did not vary. It presumably was due to differences in the pilot's ability to predict movements of the target. In fact, the analysis of the control behavior revealed that pilots anticipated movements of the target when

[5]This is indicated by "phase lead" in the Bode plots presented in Hosman & van der Vaart (1981, p 283). The operator's control actions were phase advanced compared to motions of the target at corresponding frequencies, particularly at low frequencies.

information about vehicular motion was available (Hosman & van der Vaart, 1981, p. 283)[5].

The results of Hosman and van der Vaart (1981) indicated that human operators do not simply respond to an instantaneous deviation from a desired state. Humans perceive regularities in a controlled system and in the surroundings of a system as events unfold over time. Knowledge of these regularities can improve control of a system such as a simulated vehicle. Simulator displays that support the differentiation between controlled and uncontrolled movements facilitate the perception and exploitation of regularities in the simulated vehicle and the surroundings. As noted earlier, however, uncontrolled motion can result from forces in the surroundings that move the vehicle as well as events in the surroundings, such as object motion, that do not affect the vehicle. The former is the case in disturbance regulation or compensatory control experiments (see, e.g., Poulton, 1974). In such experiments, information about vehicular motion relative to the stationary environment does not necessarily support differentiation between controlled and uncontrolled movements. This should make it difficult to perceive and exploit regularities in a disturbance.

Hosman and van der Vaart (1981) also investigated disturbance regulation with central visual displays, peripheral displays, and whole body motion. Their results did not show any evidence of anticipatory control behavior in a disturbance regulation task. Performance was significantly improved when whole-body motion was added to the central and peripheral displays (see also Junker & Replogle, 1975), but the data on control behavior indicate that this was due to increased sensitivity to higher rates of vehicular motion (Hosman & van der Vaart, 1981, p. 282)[6].

The results from disturbance regulation and target-following experiments emphasize an important general point about flight simulation. Additional displays can improve performance either by increasing the number of events that can be separately perceived (e.g., differentiation between object motion and vehicular motion) or by increasing sensitivity to less specific events (e.g., increased sensitivity to relative motion between object and vehicle). The task or goal of locomotion determines the importance of differentiation and sensitivity.

---

[6]This is indicated by enhanced gain, with whole-body motion, at the higher frequencies in the Bode plots presented by Hosman & Van der Vaart (1981, p. 282). Subjects were more effective in compensating for high-frequency disturbances in the whole-body motion condition.

Because natural tasks are generally nested, displays may have to serve both of these functions. Chasing or avoiding another object, for example, requires sensitivity to relative motion, and increased sensitivity to relative motion can improve performance in such tasks. It is also important, however, to differentiate among the causes of relative motion. Such differentiation supports perception and exploitation of regularities in the various disturbances, and it supports the observability and controllability of vehicular motion relative to the stationary (or inertial) environment. Vehicular motion relative to the inertial environment has consequences for control because there are boundary conditions on controllability. For example, it is important to perceive the linear or centripetal acceleration of a vehicle relative to the limits on thrust or turning radius in order to manage efficiently the energy transfers that occur during locomotion. Simulations that do not provide the information required for such activities are fundamentally deficient.

5.5 Dynamical Variability in Simulators and Vehicular Control

5.5.1 Perception of the Aerodynamic Environment

Perception of vehicular motion relative to the inertial environment is not important when acceleration is either small, nonexistent, or irrelevant to the task. This is the case in many experiments on simulated vehicular motion, but is relatively rare for locomotion in general because slowing down, speeding up, and changing direction are almost always necessary in the attainment of one's goals. Thus, perception of vehicular motion relative to the inertial environment, and the fidelity with which such motion is simulated, determines which tasks or situations can be reasonably recreated in the simulator. The extent to which an occupant of a flight simulator can move, or be moved, with respect to the inertial environment is extremely limited. In fact, this is the essential difference between a simulator and a real vehicle.

In current fixed-base simulators, the perception of acceleration is heavily, if not exclusively, dependent on the visual perception of linear and centripetal acceleration (i.e., visual kinesthesis) and perception of the forces implicit in visually displayed acceleration. Unfortunately, there are few experimental data on visual sensitivity to accelerative self-motion (Todd, 1982; W. Warren, Mestre, Blackwell & Morris, 1991). The experiental data suggest that such sensitivity is extremely limited, and this conclusion is supported by anecdotal evidence[7]. Perception of acceleration in a fixed-base simulator may be limited because vehicular acceleration is fundamentally a multimodal phenomenon (Riccio, 1993b). This implies that multimodal perception is relevant even when simulations or experiments do not involve actual movement of the observer relative to the inertial environment. Multimodal stimulation in current fixed base simulators is specific to remote control (teleoperation) of a vehicle. It is not specific to self-motion.

Vestibular and somatosensory systems are involved in the perception of self-motion, or absence thereof, even when simulators or experimental manipulations are limited to visual displays. This view contrasts with an assumption that information about motion is unavailable to the vestibular and somatosensory systems when there is no motion. Such an assumption is implicit in most research on postural control and simulated self-motion. Two common manifestations of this assumption about the vestibular and somatosensory systems are that (a) the magnitude of motion is related to the informativeness of the attendant stimulation, and (b) a modality is irrelevant to experimental results unless the corresponding stimulation is either experimentally manipulated or indirectly affected by such manipulations. A revealing exception is an experiment on the multimodal perception of rotation and orientation (DiZio & Lackner, 1986). Stationary observers looked at a rotating disk under conditions that generally induce an experience of self-motion (i.e., perceived rotation of the body). The disk rotated at constant velocity in a body-referenced roll axis, whereas the disk and the observer's head were oriented in different ways with respect to "vertical" (Figure 5.5).

This experiment is interesting in the present context because the investigators examined effects on the perception of orientation as well as the perception of self-motion. Perception of orientation in the various experimental conditions was constrained by the information available to

---

[7]Such evidence is often available in the informal, and sometimes unsolicited, comments of skilled participants in an investigation (cf. Riccio, 1993a). This was the case in the development of a simulation for an experiment on terrain following (Middendorf, Fiorita, & McMillan, 1991). Prior to the experiment, a pilot was brought in to evaluate the fidelity of the experimental simulator. The pilot indicated that the controllability of acceleration and gravitoinertial force ("G scheduling") was not sufficient when, in fact, the variations in "G" generated by the pilot were unrealistically large (M. Middendorf, personal communication). The most parsimonious interpretation of this psycho-physical inconsistency is that the pilot was relatively insensitive to the visually displayed acceleration.

the vestibular and somatosensory systems that indicated an absence of variation in orientation with respect to gravity. In general, the orientation of the disk was perceived to be in the horizontal plane, even when the disk was, in fact, tilted 30° relative to the horizontal plane. This perceived or misperceived orientation was consistent with the visual perception of a constant velocity roll motion and the vestibular or somatosensory perception of invariant orientation with respect togravity. Both kinesthesis and proprioception[8] are relevant to phenomenology in this experiment (cf. Riccio, 1993a, 1993b). Similar intermodal constraints should be relevant in the simulation of self-motion.

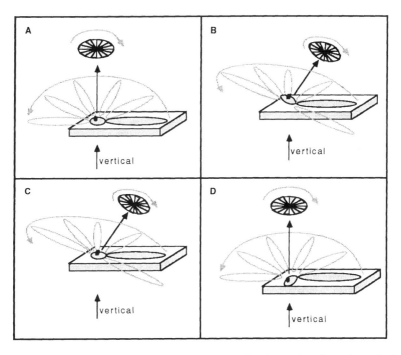

Figure 5.5. *Four experimental conditions for investigating the relation between perceived orientation and perceived self-motion. (Adapted from "Perceived Orientation, Motion, and Configuration of the Body During Viewing of an Off-Vertical Rotating Surface," by P.A. DiZio & J.R. Lackner, 1986, Perception & Psychophysics, 39, p. 42. Copyright 1986 by Psychonomics. Reprinted by permission.) This schematic representation is an oblique view with respect to a horizontal plane. The orientation of a circular disk and of the subject's head was manipulated across conditions while the torso always was*

oriented in the horizontal plane. Constant velocity rotation of such a disk generally induces an experience of self-rotation when viewed by a restrained (or recumbant) subject in an otherwise dark room. The subject's task was to describe the orientation of the disk, the head, and the torso relative to vertical and to indicate whether self rotation was experienced. The trajectories represented in each panel would be experienced if the subject perceived the correct orientation of the disk relative to vertical and also experienced self-rotation in a plane parallel to the disk (note that this implies changing orientation of the subject with respect to vertical in conditions B and C). In 20 out of 24 trials in conditions B and C, subjects misperceived the disk to be oriented in the horizontal plane. They also perceived self-rotation and invariant orientation of the head and torso with respect to vertical in these trials. Presumably because of this, they also misperceived the orientation of the eyes relative to the head, the head relative to vertical, and/or the torso relative to vertical.

Consider a curved trajectory of self-motion. The centrifugal forces on the body, during centripetal acceleration of the body, change the gravitoinertial force vector and the direction of balance relative to the surroundings. Orientation to the inertial environment is independent of orientation to the visible or tangible surroundings. During a coordinated turn in an aircraft, for example, the roll orientation of the aircraft changes as heading changes, and it remains approximately aligned with the changing direction of balance (Figure 5.6). Rotation and changes in orientation of the aircraft relative to the visible

---

[8]The use of the terms, kinesthesis and proprioception, is highly variable and somewhat confusing across the various communities in which perception is studied (see, Owen, 1990). The following definitions are used in this chapter. Kinesthesis refers to perception of change in location of the body, as a whole, relative to some aspect of the environment (i.e., self motion or "egomotion"). Visual kinesthesis, for example, refers to visual perception of self motion relative to remote objects or surfaces. Proprioception refers to perception of the position and motion of various body parts in relation to one another (i.e., postural configuration) or in relation to the environment (e.g., location of the eyes relative to an object of regard, location of the hand relative to manipulanda, or orientation relative to the direction of balance). Note that kinesthesis refers to perception of states that are relevant to locomotion, and proprioception refers to perception of states that are relevant to postural control. Kinesthesis and proprioception are related because locomotion and postural control are related, not because of redundancy or lack of specificity in the definitions for these terms.

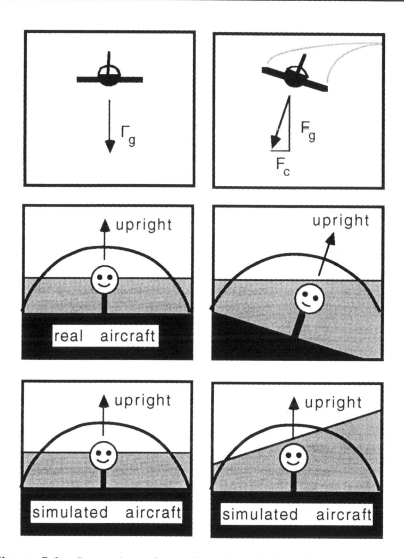

Figure 5.6. *Comparison of straight and level flight (left panels) with a coordinated turn (right panels). Fg is the force of gravity and Fc is the centrifugal force due to centripetal acceleration. The vector sum of Fg and Fc is the gravitoinertial force vector. The direction of balance (or upright) is contraparallel to the gravitoinertial vector. Upper panels depict the relative orientation of aircraft, the terrain, and upright. Lower panels depict these orientations for a fixed base simulator.*

surroundings can be recreated in a fixed-base simulator, but rotation and changes in orientation relative to the inertial environment are absent. Stimulation of the visual system and the otolith organs in the simulator is similar to such stimulation in the aircraft if one maintains the same orientation with respect to the direction of balance (e.g., if one remains upright), whereas stimulation of the semicircular canals is different. In this case, stimulation of the semicircular canals allows the simulator to be perceptually differentiated from the aircraft. In particular, sensitivity to the absence of variation in canal simulation allows the simulator to be perceived as such. More generally, such perceptual differentiation is based on intermodal relations among the perceptual systems (e.g., between otolith and canal stimulation, between visual and canal stimulation) because stimulation of the individual systems depends on the postural control strategies used in the cockpit. Postural control and movement of the body in a simulator is not necessarily the same as in a real aircraft.

Multimodal patterns of stimulation are different when driving a car along a curved path on level terrain. During a level turn, the direction of balance with respect to the terrain changes as heading changes, whereas the orientation of the car is (by definition) invariant with respect to the terrain (Figure 5.7). Stimulation of the visual system and the semicircular canals in the simulator is different than such stimulation in the terrestrial vehicle if one remains upright, whereas stimulation of the otolith organs is similar. In this case, visual stimulation or canal stimulation allows the simulator to be perceptually differentiated from the real vehicle. Visual stimulation and canal stimulation in the simulator is similar to that in the real vehicle, however, if one maintains orientation of the body with respect to the terrain (and the vehicle). This postural control strategy leads to different patterns of otolith stimulation in the simulator and the real vehicle, and sensitivity to the absence of variation in otolith stimulation allows the simulator to be perceived as such. Thus, in both aircraft and terrestrial vehicles, information may be available in the absence of what is generally considered to be the "adequate stimulus" for a perceptual system. As in the experiment of DiZio and Lackner (1986), such null information can influence what kind of self-motion is perceived.

Presumably, a simulated event can be perceived as self-motion only if one fails to differentiate the simulator from the vehicle that is simulated. The sensitivity of nonvisual systems to the presence or absence of variation is important in this differentiation. Nonvisual sensitivity is dependent on frequency and amplitude of body movement

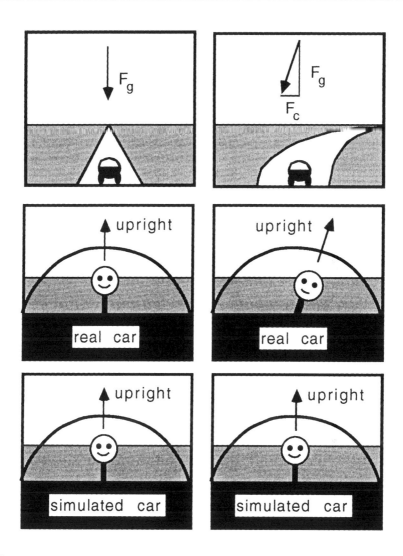

Figure 5.7. *Comparison of straight and level motion (left panels) with a level turn (right panels). Fg is the force of gravity and Fc is the centrifugal force due to centripetal acceleration. The vector sum of Fg and Fc is the gravitoinertial force vector. The direction of balance (or upright) is contraparallel to the gravitoinertial vector. Upper panels depict the relative orientation of the car, the terrain, and upright. Lower panels depict these orientations for a fixed base simulator.*

with respect to the inertial environment. This frequency and amplitude dependence is different for the otoliths and the semicircular canals (see Ormsby & Young, 1977; Zacharias & Young, 1981). Thus, the experience of self-motion in a simulator should be dependent on the relative magnitude of linear and angular motion of the observer relative to the inertial environment, and on the frequency and amplitude of such motion, as well as on the frequency and amplitude of motion relative to the visually displayed environment. Because the relation between the inertial environment and the visible environment is different for coordinated turns and level turns, the relative importance of visual and nonvisual sensitivity may be different in the simulations of these events. Moreover, postural control should be considered in the simulation of curvilinear self-motion because movement of the body in the cockpit influences the stimulation of the various motion-sensitive perceptual systems[9]. The role of postural control in the simulation of self-motion is addressed in more detail later.

5.5.2   Exploration of Aerodynamics

Simulation of vehicular motion can be useful even if it is not perceived as self-motion. In principle, the human operator can learn the dynamics of vehicular control in a simulator. In learning to control the simulator, the human operator learns the effect that actions on the controls have on movement of the simulated vehicle. If these dynamical relations are simulated with high fidelity, control skills acquired in the simulator could be useful in the real vehicle. The dynamics of aircraft control are complex because they are dependent on a number of aerodynamic parameters in which variation is common (McRuer et al., 1973, pp. 36–60). The precision and accuracy with which aerodynamics are modeled in a flight simulator is dependent on tasks that are performed

---

[9]One should not assume that postural control necessarily increases the complexity of the multimodal patterns of stimulation that specify self motion. It is possible that purposeful movement of sensory organs can simplify patterns of stimulation through the creation and exploitation of symmetries or high-order properties that facilitate the pick up of information (cf. J. Gibson, 1979, pp. 209–219). Virtually all the segments of the body, from the head to the feet, can be considered as "accessory structures" for the sensory organs of the head. Most of the body can be, and perhaps should be, considered as part of the visual system or the vestibular system in the context of the obtaining of stimulation and the pick up of information by an active perceiver (Riccio, 1993b).

in the simulator and on various technological limitations (see, e.g., Rolfe & Staples, 1986). Variation in aerodynamic parameters is negligible in a "small signal" disturbance regulation task, for example, and a linear model about the associated aerodynamic operating points would be sufficient for such a task (see, e.g., Martin et al., 1986). Manuevers that involve significant variation in G, altitude, orientation, and air speed (see, e.g., Brown et al., 1991), however, require nonlinear or piecewise-linear aerodynamic models to adequately simulate the associated variation in maneuverability of the aircraft. This is a potential problem because complex aerodynamic models can lead to computational transport delays that degrade the temporal fidelity of the simulation. Thus, design tradeoffs in the development of a high-fidelity simulation are common, but they are difficult because little is known about human perception and control of such complex dynamical systems. Data on transfer of training between simulators with different transport delays can help support such design tradeoffs (McMillan, Riccio, & Lusk, in preparation).

There are tradeoffs in the development of visual displays for flight simulators because complex visual displays also can lead to computational transport delays. It may be necessary to reduce display complexity in order to increase temporal fidelity of the simulation. Reductions in the spatial detail of a visual display could reduce sensitivity to patterns of optic flow that provide information about aircraft motion (e.g., linear or centripetal acceleration). In any case, limitations on the visual perception of aircraft motion necessarily reduce sensitivity to aircraft dynamics in a fixed base simulator. These limitations are less important when there are other ways to perceive aerodynamics. In an actual aircraft, aerodynamics are more or less observable in the movability[10] of the controls (Riccio, 1993b). Aerodynamics could be perceived through dither on a control stick, for example, if constraints on these movements are similar to constraints on the movement of the aircraft. Stick dynamics and aerodynamics are similar to the extent that some aerodynamic parameters that affect maneuverability of the airframe also affect the movability (or

---

[10]"Moveability" refers to the ease of difficulty of moving some object (e.g., the control stick in an aircraft cockpit). This property is often expressed in terms of force-displacement relations. The relation between force and displacement is often dependent on the rate of application of force, and this complex multiple relation is generally described mathematically as a mechanical impedance.

impedance) of the control surfaces and, indirectly, impedance of the control stick. Other similarities may exist because of perceptual factors that have been implicit in the evolution of aircraft design and, specifically, in the dynamical connections between the control stick and the control surfaces.

The purpose of control loading systems in flight simulators is to mimic as closely as possible the impedance (i.e., the "feel") of the controls in the actual aircraft. It is recognized, in the design of control loaders, that simulation of the stick impedance is important at frequencies that are at least an order of magnitude higher than the frequencies required to simulate aircraft control (Rolfe & Staples, 1986, pp. 77–81). The stiffness and damping of the stick at these high frequencies and low amplitudes could be explored with minimal effect on movement of the aircraft, and this exploration may provide information that is relevant to aircraft control at lower frequencies and higher amplitudes. In this sense, dither on the control stick could be viewed as a nested inner loop or informative dynamical component in the human vehicle control system. This view is similar to recent work in teleoperation (Repperger, 1991) in which manipulator dynamics are designed according to observability criteria rather than impedance matching criteria. Such an approach to teleoperation makes explicit what may have been implicit in the design of aircraft.

5.6   Effects of Simulated Vehicular Control on the Human Operator

5.6.1   Physiological Effects

Although variations in G (i.e. gravitoinertial force) are important in flight, they are not represented with much fidelity in common flight simulators. Physiological effects, in particular, are conspicuously absent in flight simulators with high-fidelity visual displays. The simulation of physiological effects is important insofar as these effects constrain perception and action in the cockpit. Presumably, simulation of these effects could have some training benefit since perception and action in the cockpit are a part of piloting skill. More generally, the operational envelope of the pilot-aircraft system is influenced by physiological limitations of the pilot, and representation of physiological boundaries in a simulator can be considered as part of the problem of accurate "aero" modeling. Knowledge of these boundaries is considered to be important for both pilots and designers.

Human centrifuges are an important tool for training and scientific examination of physiological effects and limits. Such devices are impractical for most aspects of flight simulation because it is difficult and expensive to include high fidelity visual displays and because there are undesirable effects such as relatively large coriolis forces on occupants when they move in the cockpit. More exotic devices could be used to recreate or simulate physiological effects in otherwise conventional flight simulators. Such devices have been treated extensively in other reports (Brown et al., 1991; Kron et al., 1980). Examples include lower body negative pressure (LBNP) devices and variable transmission visors. LBNP devices reduce the impedance of blood vessels in the lower parts of the body and, consequently, reduce blood flow toward the head. LBNP could recreate the lower blood pressure in the head during moderate positive-G maneuvers. Variable transmission visors, on the other hand, would simulate the effects of lower blood pressure in the head. Such devices could simulate gray-out by progressively reducing the field of view.

5.6.2 Biomechanical Effects

Variations in gravitoinertial force are more common at magnitudes that do not have noticeable physiological effects. Devices such as G seats and G suits are currently used to simulate the somatosensory effects of relatively low-magnitude inertial variations (see, e.g., Rolfe & Staples, 1986). These devices employ inflatable bladders that recreate some of the tactile pressure patterns that result when the aircraft pushes and pulls against the pilot. G seats can be enhanced with hydraulically or pneumatically driven actuators that increase the specificity of somatosensory stimulation (see, e.g., Brown et al., 1991; Kron et al., 1980). The seat pan, back rest, and seat restraints can be driven independently in such dynamic seats and, to some extent, movement in rotational and translational axes can be represented independently. The potentially high bandwidth of dynamic seats can be useful in the simulation of transients and vibration. At the very least, such devices allow a pilot to become acquainted with some of the challenges that vehicular motion (i.e., whole-body motion) pose for perception and action in the cockpit. Beyond this, laboratory experiments have demonstrated that skilled tracking performance, manual control behavior, and perception of the magnitude of simulated vehicular motion in a dynamic seat simulator can be similar to that in a whole body motion device (see, e.g., Flach, Riccio, McMillan, & Warren, 1986; Martin et al., 1986).

The whole-body motion device in the experiments previously mentioned (a roll-axis tracking simulator or RATS) moved only in the roll axis, and the axis of rotation was through the seat pan. The purpose of these experiments was to make the dynamic seat as similar as possible to the RATS with respect to various measures of perception and performance. It was found that the dynamic seat could be made more similar to the RATS if the dynamics of these two devices were different. Unlike in the RATS, the motion of the dynamic seat was not exactly the same as the motion of the simulated vehicle (e.g., motion relative to the horizon in the visual display). In particular, matching the dynamic seat to the RATS required a drive algorithm for which the dynamic seat position was proportional to a visually simulated vehicular position (i.e., roll angle) at low frequencies, and for which the seat velocity was proportional to vehicular position at high frequencies. The general interpretation of this matching algorithm was that it provided a level of sensitivity to high frequency motion in the dynamic seat that is naturally provided by the vestibular system in the RATS.

Although drive algorithms can successfully lead to the development of similar manual control skills in the dynamic seat and the RATS, transfer of training between such devices has been disappointing. Furthermore, the lack of transfer is inexplicable within common models of human operator behavior such as the crossover model or the optimal control model (see, e.g., Flach, 1990). To understand why, one must examine the assumptions and limitations of these models and the associated experimental techniques. All the psychophysical and manual control techniques in these experiments were based on the operator's sensitivity to variation irrespective of what the operator perceived as varying. This was consistent with models of vehicular control behavior that assumed vehicular states were the only task-relevant parameters. In other words, the models considered the role of kinesthesis but not proprioception in vehicular control. It follows that the experiments and models are limited to those situations in which movement and control of the body in the cockpit is irrelevant to perception and control of vehicular motion. The lack of transfer between the dynamic seat and the RATS suggests that movement of the body in the cockpit is relevant to vehicular control in one or both of the devices. That is, both kinesthesis and proprioception seem to be relevant to control of these devices.

Movement in the cockpit also provides an alternative explanation of the whole-body matching algorithm for the dynamic seat. In the dynamic seat simulator, the operator perceives motion of the body with respect to the seat and perceives the body to be stationary with respect to the inertial environment; that is, the dynamic seat is perceived as such. In the RATS, the operator perceives motion of the body with respect to the seat and motion of the body with respect to the inertial environment; that is, this whole-body motion device is perceived as such. Motion with respect to the inertial environment is a salient difference between the dynamic seat and the RATS, and this difference has consequences for postural control in the cockpit. The devices also differ in the motion of the body with respect to the seat pan. The difference is primarily at low frequencies of motion (i.e., rotational acceleration) because the body tends to move with the RATS at lower frequencies and accelerations. Thus, somatosensory stimulation is attenuated at low frequencies in the RATS. Somatosensory stimulation becomes increasingly similar in the dynamic seat and the RATS as frequency of motion increases because the body increasingly resists motion at higher frequencies and accelerations in the RATS. This suggests that a matching algorithm would have to attenuate motion in the dynamic seat at low frequencies but not at high frequencies. In fact, the matching algorithm does exactly this (Flach et al., 1986; Martin et al., 1986). Note that the vestibular interpretation does not address, and seems to be inconsistent with, the attenuation of dynamic seat motion in a whole-body matching algorithm.

The motion of the dynamic seat would have to be attenuated even further to match the somatosensory stimulation in an aircraft. Roll-axis motion in an aircraft is usually accompanied by changes in the lift vector and the gravitoinertial vector. In coordinated turns, for example, change in roll angle leads to minimal change in orientation with respect to the gravitoinertial vector. (This is due largely to the aerodynamic characteristics of the airframe.) Somatosensory stimulation may be important for the perception and control of large amplitude transients such as a snap roll (see, e.g., Brown et al., 1991) or precision tasks such as landing during downdrafts (Strughold, 1950, p. 995), but the variation in the pilot's orientation with respect to the seat is generally much less than the variation in the aircraft's orientation. This has obvious implications for dynamic-seat drive algorithms in flight simulators. For example, drive algorithms for the dynamic seat that match performance in a device such as the RATS may have unrealistically large excursions compared to real aircraft. Although such augmented information can improve tracking performance, pilots may indicate that the experience is different than in a real aircraft (Cress, McMillan, & Gilkey, 1989, p. 100).

Such phenomenological reports by skilled participants (e.g., pilots) can play an important role in the investigation of human-environment interactions (cf. Riccio, 1993a), and in the present context, they suggest that the nominal vehicular control task (e.g., compensatory tracking) is only part of the nested task structure when one interacts with the environment while moving. Pilots must be sensitive to movement of the body in the aircraft as well as movement of the aircraft.

5.7   Coordination of Postural Control and Simulated Vehicular Control

5.7.1   Postural Control in Simulated Vehicles

Phenomenological comparisons between simulators and real vehicles are consistent with data on performance matching and transfer of training between simulators. The nonrigidity of the human operator is difficult to ignore from either of these perspectives. The salient fact of nonrigidity suggests that control of self-motion involves coordination of nested systems. Human operators can learn to control a moving seat in ways that are superficially similar to control of a whole-body motion device, but different skills are acquired in doing so. The perception–action relations are different in the two devices. Control of the moving seat requires sensitivity to the changing pattern of contact between the seat and the body and the ability to couple manual control action to these patterns. Control of whole-body motion is different because of movement in the cockpit. Neither vestibular nor somatosensory simulation alone is specific to movement of the device. The human operator must be sensitive to intermodal relations between vestibular and somatosensory stimulation in order to perceive motion of the device (cf. Stoffregen & Riccio, 1988). Moreover, a whole-body motion device presents a different challenge to postural control than does a moving seat.

In both devices, the human operator must stabilize the hand-arm system with respect to the control stick and stabilize the visual system with respect to the visible surroundings. Such interactions with the cockpit environment impose general constraints on postural control, but the way in which perception and action constrain posture in a whole-body motion device is different than in a dynamic seat simulator. Movements of the RATS, for example, affect the position and motion of the body with respect to the control stick, the visual display, the seat, and the direction of balance. Postural adjustments must be coordinated with movements of the device in order to stabilize the perception and

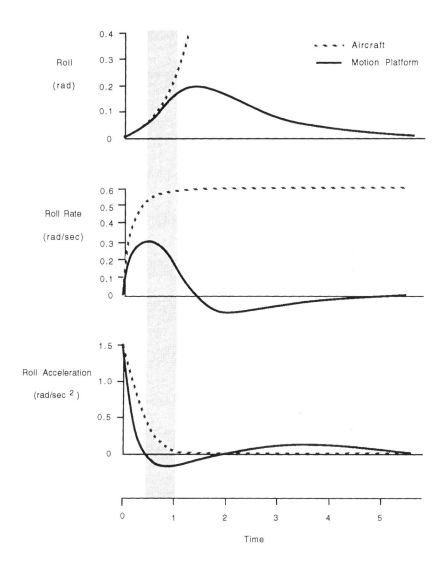

Figure 5.8. *Roll response of an aircraft and a simulator to an aileron step input. (Adapted from "Three Generations of Motion Cues on Six Degree of Freedom Motion System (Tech. Rep. LR 248), by M Baarspul, 1977. Delft, The Netherlands: Delft University of Technology. Copyright by Delft University of Technology. Reprinted by permission.) The visual displays in a simulator can represent the aircraft trajectory with reasonably high fidelity. The motion platform follows a different trajectory due to the effects of washout filters (the*

effect of a linear washout filter is shown in this figure). The washout filter
returns the roll state of the platform to a level orientation so as to simulate the
experience of orientation during a coordinated turn (cf. Figure 5.6). Note that
in doing so there is a period of time (shaded) over which the motion platform
moves in a direction opposite to the state of the simulated aircraft (presumably
represented in the visual displays of the simulator).

action systems with respect to the cockpit. The meaning of vestibular
and somatosensory stimulation in a whole-body motion device is
different than in a dynamic seat simulator. The information in
stimulation is different, and it has different affordances for the
coordination of postural control and vehicular control, in the two
devices (cf. J. Gibson, 1979; Riccio, 1993a; Riccio & Stoffregen, 1988,
1991).

5.7.2   Perception of Postural States in Simulated Vehicles

Postural control in a simulator provides information about the direction
of balance in the simulator, but the direction of balance cannot be
manipulated in most simulators. This is a problem because the direction
of balance changes frequently during goal-directed self-motion. Thus,
perception of the direction of balance in the simulator imposes
limitations on the fidelity with which self-motion can be simulated.
Sensitivity to the direction of balance is due primarily to a balancing of
the head when other body parts are passively stabilized, and such
sensitivity increases with tilt from the direction of balance (see Riccio,
Martin, & Stoffregan, 1992). Noticeable deviations from the direction of
balance occur in a flight simulator when the rotational component of a
coordinated turn is reproduced with a motion platform, but tilt in an
aircraft is minimal because the gravitoinertial vector remains roughly
perpendicular to the wings as the aircraft rolls. This creates a tradeoff in
flight simulator design. Platform rotation increases the fidelity of the
simulation with respect to vehicular motion, but it decreases the fidelity
of the simulation with respect to vehicular orientation. The tradeoff has
resulted in specialized drive algorithms that partially decouple the
perception of rotation and orientation (see, e.g., Rolfe & Staples, 1986,
pp. 166–120).
       High-pass washout filters minimize sustained tilt with respect to the
direction of balance but allow high-frequency variations in rotation and
transient changes in orientation to be preserved in the motion platform
(Figure 5.8). Further decoupling of rotation from transient changes in

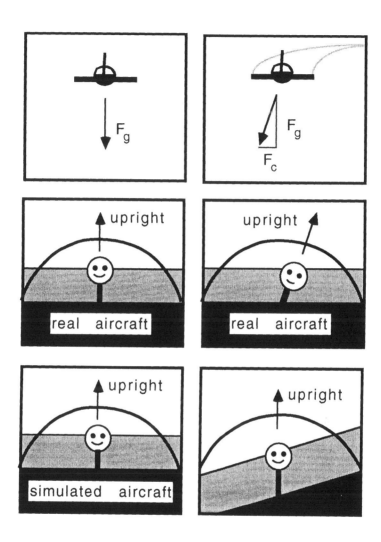

Figure 5.9. *Comparison of straight and level flight (left panels) with a level turn (right panels). Fg is the force of gravity and Fc is the centrifugal force due to centripetal acceleration. The vector sum of Fg and Fc is the gravitoinertial force vector. The direction of balance (or upright) is contraparallel to the gravitoinertial vector. Upper panels depict the relative orientation of aircraft, the terrain, and upright. Lower panels depict these orientations for a motion-base simulator.*

orientation can be accomplished by coordinating rotation and translational acceleration of the motion base. Pitching upward, for example, can be accompanied by a transient backward acceleration so that changes in orientation with respect to the gravitoinertial vector are minimized. Rotation and translation can also be coordinated to simulate sustained accelerations. Tilting the motion platform with respect to gravity, for example, simulates changes in the gravitoinertial vector with respect to the aircraft during linear or centripetal acceleration of the aircraft (Figure 5.9). Through the use of such algorithms, the forces on the pilot's body in a simulator can closely approximate some of the forces in a real aircraft (see, e.g., Parrish, Diedonne, Bowles, & Martin, 1975). This has implications for both experiential fidelity and action fidelity in the simulator.

Coordination of postural control and vehicular control should be explicitly considered in the evaluation of experiential and action fidelity in a simulator. Figure 5.10 represents these nested control systems in a motion-base flight simulator. The skill of coordinating the nested postural and vehicular systems develops in the context of the forceful interactions between these systems. Motion devices in simulators provide the potential for development of more appropriate skills of coordination. The realization of this potential requires a level of analysis for postural control in simulators that is commensurate with extant analyses of vehicular control. A starting point is to examine the multimodal and multicriterion control of posture in a simulator. Consider, for example, an event involving the simulation of rotation with a motion-base simulator. Low-frequency rotation of the simulated vehicle, represented in the visual display, would be attenuated by washout filters in the motion base. There would be a low correlation between visually displayed rotation and actual rotation of the pilot and the simulator at low frequencies. Furthermore, there could actually be brief periods of time during which there is a negative correlation between visually displayed rotation and actual rotation (Figure 5.8). This would be the case if high-frequency components of rotation, preserved in the motion base and the visual display, were opposite in sign and lower in amplitude than the low-frequency components of rotation preserved only in the visual display. This could present a problem for postural control given the multiple criteria that include stabilizing posture with respect to the visible surroundings and the inertial environment. Although this problem may not be insurmountable for the postural control system, it could lead to motion sickness and to the development of postural control skills in the

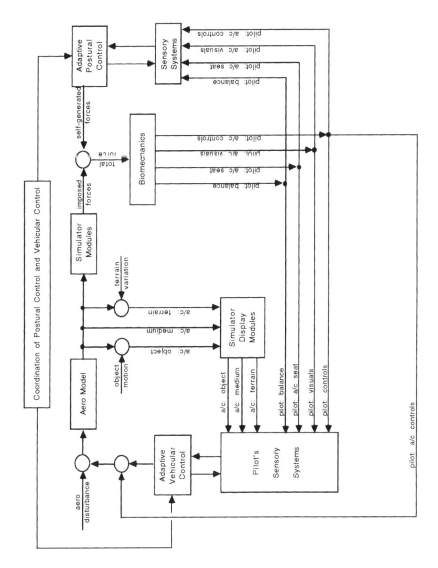

Figure 5.10. *Coupling of postural and vehicular control in a motion-base simulator. Forces imposed on the postural control system by simulated vehicular motion require mechanical devices (e.g., motion platforms) that push and pull on the pilot's body. Nonvisual display systems (e.g., control loaders) required for perception of the simulated aerodynamic environment are represented in the same box as visual display systems required for perception of simulated vehicular motion relative to visible surroundings. Simulation of physiological effects is not represented in this diagram. Note that postural control in the cockpit provides information that is relevant to self-motion, even if it is not explicitly considered in the design and evaluation of the simulator.*

Pickup of this information allows the simulator to be perceived as such, and it is inconsistent with the perception of self-motion. This information also supports the coordination of postural control and vehicular control. Postural control, and its coordination with vehicular control, can be different in the simulator and an actual vehicle.

simulator that are different than those in an aircraft. The potential problems of motion-base simulators with respect to coordination of postural and vehicular control are not necessarily an argument in favor of fixed-base simulators. Fixed-base simulators present different problems for coordination of postural and vehicular control. In fixed-base flight simulators, deviations from the direction of balance could result from the use of habitual strategies for coordinating postural and vehicular control. This effect is represented in Figures 5.11 and 5.12. During a simulated change in orientation, for example, one may lean relative to the cockpit of the simulator to maintain stability with respect to the visually displayed horizon. This has been observed in videotapes of occupants in a fixed-base flight simulator (Riccio, 1991). The exploratory observations indicate a tendency to align the interocular axis with the visually displayed horizon. This primarily involves head tilt, but small amounts of lean in the torso have also been observed. Presumably, alignment with the horizon facilitates conjugate scanning eye movements. Small transient movements of the head and torso have also been observed during sudden turns or changes in orientation of the simulated vehicle. Such postural adjustments in a simulator could be destabilizing because the forces on the pilot are different than in actual self-motion. Adaptation to postural constraints of the simulator could lead to the development of postural control skills in the simulator that are different than those in the aircraft. These differences may be greater than the differences between postural control skills in motion-base simulators and real aircraft because the forces on the pilot can be represented with higher fidelity in motion-base simulators, although this is not necessarily the case.

5.8   Simulated Vehicular Motion and Postural Perturbations

5.8.1   Differentiation of Postural Events

The effects of visual displays on postural control in flight simulators

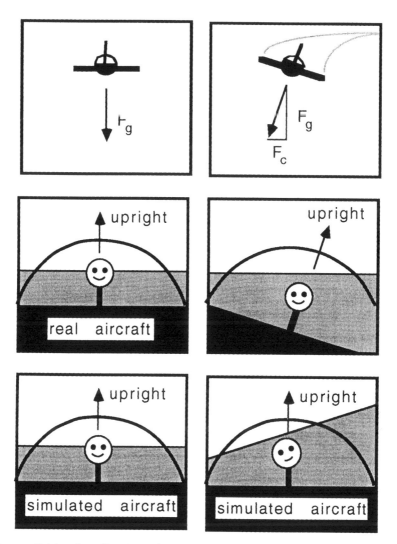

Figure 5.11. *Coordination of the head and torso during straight and level flight (left panels) and during a coordinated turn (right panels). Fg is the force of gravity and Fc is the centrifugal force due to centripetal acceleration. The vector sum of Fg and Fc is the gravitoinertial force vector. The direction of balance (or upright) is contraparallel to the gravitoinertial vector. Upper panels depict the relative orientation of aircraft, the terrain, the head, the torso, and upright. Lower panels depict these orientations for a fixed-base simulator.*

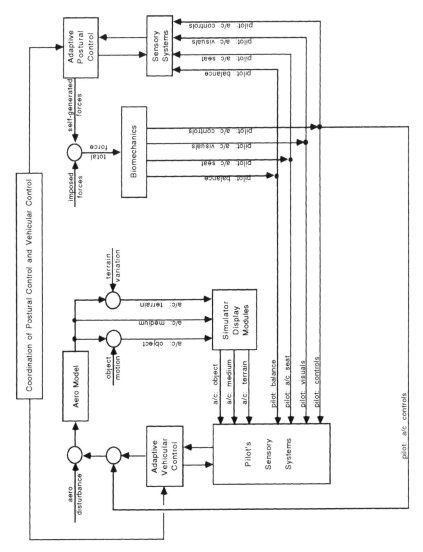

Figure 5.12. *Coupling of postural control and vehicular control in a fixed-base simulator. There are no forces imposed on the postural system by simulated vehicular motion. As in a motion-base simulator (see Figure 5.10), postural control in the cockpit provides information that is relevant to self-motion. Pickup of this information allows the simulator to be perceived as such, and it is inconsistent with the perception of self-motion. This information also supports the coordination of postural control and vehicular control. Postural control, and its coordination with vehicular control, in the simulator can be different than in an actual vehicle.*

indicate that such displays are relevant to both proprioception and kinesthesis. The effects are analogous to the well-known effects of motion in the visible surroundings on the control of stance (Dichgans & Brandt, 1978; Lee & Lishman, 1975; Lestienne, Soechting, & Berthoz, 1977; Lishman & Lee, 1973). Such effects would tend to destabilize the body with respect to the inertial environment, and this instability could interfere with manual control in the cockpit. Furthermore, visual displays that are likely to affect postural control could degrade control performance in a flight simulator. Although there are no experiments that directly test this hypothesis, some results are suggestive. For example, one study revealed that disturbance-regulation performance in a flight simulator could be worse with a wide field-of-view display than with a narrow field-of-view display (McMillan, Cress, & Middendorf, 1990). This effect could be due to the fact that larger displays can have a greater effect on both postural control and the perception of self-motion (see, e.g., Brandt, Wist, & Dichgans, 1975; Stoffregen, 1985). The effect of visible surroundings on postural control is useful to the extent that it stabilizes the visual system with respect to the surroundings. Presumably, the destabilizing effects of large displays would be mitigated by stable visible surroundings inside the cockpit. Visibility inside the cockpit surroundings was minimized in the experiment by McMillan et al. (1990). Thus, the effects of large displays on manual control may not apply to other simulators. In any case, experiments that minimize visibility of inside-the-cockpit surroundings should consider postural destabilization as a potential explanation for the effects of display manipulations on manual control performance.

Postural control can be influenced by many factors other than the size of a visual display. Compelling presentations of motion in depth, such as provided by flow fields, can influence both postural control and the perception of self-motion, even when displays are limited to central vision (see, e.g., Andersen & Braunstein, 1985; Andersen & Dyre, 1989). This suggests an alternative explanation for some of the effects optic flow has on the control of simulated vehicular motion. Consider an experiment on altitude maintenance in the presence of a disturbance (R. Warren & Riccio, 1985). In this experiment, performance was better with a display that contained only a horizon and a longitudinal "roadway" than with a display that contained a flow field of random texture elements on the terrain. There was also little transfer of training between these displays. Several explanations for the differences between displays were considered (see also Riccio & Cress, 1986; Wolpert, 1990), but postural confounds were neglected. It is reasonable to predict that postural destabilization would be greater for the flow field display than for the roadway display, and that the effect would be greater for optical disturbances of low magnitude (cf. Stoffregen, 1986) and low frequency (cf. Dichgans & Brandt, 1978; Lestienne et al., 1977). These predictions are consistent with the altitude dependence and frequency dependence of the manual-control data from the experiment of Warren and Riccio (see Riccio & Cress, 1986, p. 128; Zacharias, Warren, & Riccio, 1986).

Disturbances can occur in any of the three translational and three rotational axes of motion. Disturbances differ with respect to their affordances for the nested systems that subserve perception and action. It has been suggested that optical perturbations affect postural control because of the role of the postural system in maintaining stability of the visual system in relation to the visible surroundings (Riccio, 1993b). Thus, various axes of optical perturbation should be evaluated with respect to their effect on visual stability and the ability of the postural systems and other action systems to compensate for the optical perturbations. Optical disturbances in the vertical axis, for example, cannot be compensated by postural sway, whereas conjugate eye movements can be sufficient for this purpose. Such optical disturbances should not affect postural control. Lateral postural sway can compensate for optical disturbances in the lateral axis and, to some extent, in the yaw axis; however, conjugate eye movements reduce the need for postural compensation. Tilting of the torso and especially the head can compensate for optical disturbances in the roll axis, and such adjustments may facilitate conjugate eye movements for scanning optical structure by making the interocular axis parallel or perpendicular to this structure (cf. Boeder, 1961). The effects of motion in depth are more complicated. Fore-aft sway, as well as accommodation and convergence, would be useful for optical disturbances in depth and, to some extent, in pitch if the disturbances were due to motion of actual surfaces in depth. Such compensatory movements would be inappropriate for common visual displays in which the actual optical distances do not vary (i.e., the visual display is in a fixed location). Simulation can create problems for the visual system and its coordination with the postural system, to the extent that the apparent distances of surfaces in a visual display do not correspond to the actual optical distances.

The effect of any optical disturbance on postural control should be evaluated in the context of the nested lens, pupil, eyeball, head, and body system (cf. J. Gibson, 1979). Putative effects of "retinal location" or "flow structure" on postural sway (e.g., Stoffregen, 1985, 1986), for example, are confounded with the availability and effectiveness of other action systems (e.g., lens, pupil, eyeball, and head) for stabilizing the visual system with respect to the visible surroundings (e.g., focusing and tracking). Posture can be decoupled from optical disturbances because of insensitivity to the disturbance or because other action systems compensate for the disturbance. The effects of visible flow (central vs peripheral) and looking direction (toward the focus of expansion or to the side) should be different in real and virtual environments because in the latter (e.g., in a flight simulator) the actual locations of the visible surfaces do not change when the depicted locations change. Thus, data on the postural effects of motion relative to the visible surroundings in real environments may not apply to virtual environments.

5.8.2   Simulator Sickness

When other action systems are unavailable or ineffective, it may be difficult for the postural system to satisfy multiple criteria such as stability relative to balance, stability relative to the visible surroundings, and stability relative to the tangible surroundings (Riccio, 1993a; Riccio & Stoffregen, 1988). Limitations on the multicriterion control of posture could lead to imbalance and instability in situations that impose significant demands on vision and posture, such as in a flight simulator, and prolonged instability may cause motion sickness (Riccio & Stoffregen, 1991). This suggests that prolonged exposure to optical perturbations could lead to motion sickness when there are limitations on the control of posture and other action systems with respect to multiple criteria. In fact, visually induced motion sickness was observed in flight simulators and other virtual environments (see e.g., Kennedy, Hettinger, & Lilienthal, 1990). As with mechanical disturbances, optical disturbances at frequencies in the postural control range were found to be particularly nausogenic (Hettinger, Berbaum, & Kennedy, 1990). This is consistent with the postural instability theory of motion sickness.

The postural affordances of optical disturbances are different from those of mechanical disturbances (Riccio, 1993b). Optical disturbances should be most destabilizing and, thus, most nausogenic in roll axis. Data are not yet available, however, for the differential effect of axis of optical disturbance on visually induced motion sickness. Finally, it should be noted that simulator sickness includes symptoms such as eye strain, headache, blurred vision, and difficulty focusing that are not common in other forms of motion sickness (see Kennedy et al., 1990). Such symptoms may be due to disruptions in the eye-head system that are similar to the disruptions in postural control. Disruptions of the eye-head system may be due to the optical peculiarities of visual displays. Presumably, postural instability is responsible for the symptoms that are common to all forms of motion sickness (Riccio & Stoffregen, 1991).

5.9   Dynamical Variability in Simulators and Postural Control

5.9.1   Constraints During Curved Trajectories

A central theme in this chapter is that variation in the magnitude and direction of the velocity vector must be considered in the analysis of vehicular control. Variation in heading, in particular, is an important part of all forms of locomotion. Such variations generate centripetal accelerations and centrifugal forces that have consequences for postural control. It may be desirable to lean into a turn, for example, to minimize the tilt of various body parts with respect to the direction of balance. Postural control must adapt to the torques produced by variations in heading, even when tilt is not minimized. In general, although there are a variety of ways in which posture and locomotion can be coordinated, some form of coordination is required to maintain the stability of these activities. Coordination can be more or less reactive and more or less proactive. One can compensate for a perceived postural perturbation produced by the torques on the body segments during a change in heading, and one can anticipate the postural consequences of a controlled change in heading. Anticipatory activity depends on one's experience with the constraints on posture imposed by a particular type of locomotion.

Constraints on posture are particularly challenging in aircraft because of the relatively large variations in the direction and magnitude of the gravitoinertial force vector. Pilots thus experience relatively large variations in the direction of balance and in the consequences of tilt or imbalance. Because orientation of the aircraft tends to covary with the direction of balance, as in a coordinated turn, variation in the direction of balance may have relatively little effect on postural control skills in the cockpit. Variation in the consequences of imbalance, however, could be problematic. The forces on the pilot during common trajectories of aircraft motion, and the postural consequences of these forces, are similar to those experienced during "vertical axis vibration" (Riccio, 1993b). The nausogenic property of vibration in the vertical axis suggests that such events can lead to postural instability. Presumably, pilots adapt to these constraints by developing postural control skills that are unique to an aircraft. One aspect of this adaptation might be establishing a perception–action linkage between controlled changes in heading and the strength of compensatory postural actions.

The constraints on postural control in an aircraft are different than those in flight simulators. There are no variations in the forces on the pilot's body in a fixed-base flight simulator. Even in motion base simulators, variations in the magnitude of the gravitoinertial force vector are generally quite small. Thus, variations in the consequences of imbalance in flight simulators are minimal or entirely absent. Consider how this might affect postural stability in a fixed-base simulator for a pilot who has learned to coordinate postural control with vehicular control. (This coupling is represented in Figure 5.12.) The pilot might produce a particular muscular torque to counterbalance the anticipated consequence of imbalance during a controlled change in heading; however, the linkage would be inappropriate in the simulator. The gain on the postural control system would be too large in the simulator. It would not be surprising if this led to unstable oscillations in the postural control system, just as it would in many control systems (see, e.g., Stark, 1968). This kind of instability would not be expected, at least not to the same degree, for a nonpilot in a flight simulator. A nonpilot would not have learned the pattern of coordination in the aircraft that is inappropriate in the simulator. Similarly, one would not expect as much postural instability by a passenger in a flight simulator. The passenger would tend not to anticipate the control actions of the pilot and would tend not to use postural control skills that are destabilizing in the simulator. This suggests that nonpilots and passengers in flight simulators should be less susceptible to simulator sickness. This implication of a postural instability theory of motion sickness is consistent with the evidence on susceptibility to simulator sickness (see Kennedy et al., 1990).

5.9.2   Visual Control of Posture in Context

The destabilizing effects of inappropriately high gains on postural control in a simulator could be minimized if body segments are aligned with the direction of balance. This would lead to postural orientations and configurations that are similar in the aircraft and the simulator, and it would constitute a robust control strategy for posture in aircraft and simulators. Although such a strategy could reduce postural instability after transfer from an aircraft to a simulator, or vice versa, it could create other problems. For example, visual perception of the surroundings outside the cockpit could be impaired if the head is not tilted with the surroundings during a coordinated turn.

Little is known about the effects of head tilt on visual perception, but it is sufficiently important that occupants of vehicles consistently tolerate the imbalance produced by such head tilt during changes in heading. This is prima facie evidence that, within a certain range, head tilt can have some desirable effects that outweigh the undesirable effects of imbalance. It follows that visually induced head tilt is likely to be a factor that contributes to simulator sickness. Tilting the head stabilizes the visual system with respect to the outside the cockpit scene, but the muscular torques required to produce and maintain these tilts are smaller in a simulator than in an aircraft. Thus, the presence of visible variation in the roll axis together with the absence of inertial variation in the vertical axis is likely to elicit a postural control strategy that is inappropriate in a simulator. These events should be especially destabilizing for frequencies of variation in the postural control range. It is not surprising, then, that such low frequency variation in the simulated trajectory of vehicular motion induces simulator sickness (Hettinger et al., 1990).

5.9.3   Coordination in the Simulator and its Aftereffects

Experience in a simulator should allow one to develop postural control skills that are appropriate to the simulator. At the same time, adaptation to the simulator could create problems outside the simulator. Anecdotes about disorientation after prolonged exposure to vehicle simulators are pervasive (see, e.g., Kennedy et al., 1990), but there are few data on such aftereffects. One study indicates that aftereffects on postural control are measurable and that the effects can be somewhat specific (Hamilton, Kantor, & Magee, 1989). The length of time that subjects could maintain stance in an unstable posture was evaluated, in this study, before and after flying a standard left turn toward a runway in the simulator. Duration of stance was reduced (stance was less stable) only with the eyes open and only in the subjects who experienced balance-related symptoms of simulator sickness. There were no significant differences in stability for the subjects who did not experience such symptoms, and there were no differences for any of the subjects when their eyes were closed.

The link between simulator sickness and postural aftereffects is important. Presumably, subjects who do not experience simulator sickness do not experience the postural instability that necessitates a change in postural control strategies. From the present perspective, postural aftereffects would be expected only if there were changes in postural control in the simulator. The fact of specificity of postural aftereffects is also consistent with the present perspective; however, predictions about the nature of the specificity requires careful consideration of the patterns of simulation, the constraints on postural control, and the postural control strategies in the simulator.

The simulator used in the study by Hamilton et al. (1989) rotated continuously around a vertical axis through the pilot (i.e., yaw motion). The consequences of imbalance in such a rotating system are different than in a nonrotating system. In particular, the relation between displacements and torques are different. Consider, for example, the consequences of imbalance when the pilot leans forward to activate a switch (Hamilton et al., 1989, p. 247). A given displacement of a body segment in the pitch axis requires additional muscular torque to compensate for the centrifugal forces on that body segment in the rotating coordinate system. In addition, movement in the pitch axis gives rise to coriolis force in the roll axis that requires compensatory torque to stabilize the body segment.

Adaptation to centrifugal and coriolis forces in the simulator could create problems for postural control outside the simulator if compensatory muscular torques are linked to perceived displacement of the various body segments. Instability in a particular axis could result because the gain of the pilot's response in that axis would be too high. Additional instability would result from a coordinated response in pitch and roll to a perceived displacement in pitch. Although adaptive in the simulator, such a pattern of coordination could lead to instability in the roll axis outside the simulator. It is reasonable to expect aftereffects of this kind when the eyes are open because postural sway is affected by displacements relative to the visible surroundings.

Adaptation to centrifugal and coriolis forces in the simulator would not necessarily create problems outside the simulator if compensatory actions are linked to perceived force on the associated body segment in each axis of motion. Such resistance to aftereffects would be more likely with the eyes closed because the role of the vestibular system and force feedback in postural control is more important when displacement relative to the visible surroundings is irrelevant.

5.10   Coupling of Real and Simulated System Components

5.10.1 Felt Presence and Interaction with Virtual Environments

The coupling between postural control and vehicular control is different in simulators and real vehicles and is different in different simulators. The major reason for this is the effect of vehicular disturbances on postural control (compare Figures 5.4, 5.10, and 5.12). Devices can be categorized with respect to this coupling: (a) Real vehicles generate transient and sustained motion of the occupants and are a source of mechanical disturbances on postural control. (b) Fixed-base simulators do not move the occupants, and there are no mechanical disturbances on posture control that are similar to those in a real vehicle. (c) Simulators that contain locally moving components, such as dynamic seats, do not move the whole body, although some mechanical disturbances on postural control can be similar to those in a real vehicle. (d) Motion-base simulators generate transient, but not sustained, motions of the whole body and the transient mechanical disturbances on postural control can be similar to those in a real vehicle. (e) Continuously rotating devices, such as centrifuges, create both transient and sustained whole-body motions, but sustained rotation gives rise to mechanical disturbances on postural control that are different than in a real vehicle.

There should be qualitative differences among these categories with respect to the postural control skills acquired in the associated devices. The examples discussed in this chapter reveal that the coordination of postural control and vehicular control required in each category is fundamentally different. In principle, similar vehicular control performance and manual control behavior can be obtained in these devices, but transfer of training between devices may be poor because of differences in postural control in the various devices. Coordination of postural and vehicular control is important in all real or simulated vehicles. This is the case even when there are no mechanical disturbances of posture. In a fixed-base simulator, for example, vehicular control influences the relative orientation of the displayed surroundings outside the cockpit which, in turn, can influence the orientation of the head and configuration of the body in the cockpit. The latter is presumably due to a coupling between postural control and control of the eye-head system (i.e., the biomechanical components of the visual system). This emphasizes that the coupling between postural control and vehicular control is not unique in perception and action.

Coupling of control systems is pervasive and occurs at many levels of analysis. Little is known about these flexible and adaptive couplings, but they are an essential aspect of human environment interactions (Riccio, 1993a; Riccio & Stoffregen, 1988) and must be taken into account in the analysis and design of virtual environments. The multiplicity of coupled control systems within the human body is evident in the high degree of coordination and coherence in human movement. The direct coupling of some of these systems with the environment establishes a multiplicity of indirect couplings between the human and the environment. This web of connections with the environment is presumably the major determinant of felt presence in an environment. Felt presence is necessarily reduced when connections with the environment are broken.

Connection with an environment and felt presence in an environment are minimal when mechanical coupling with the environment is absent. One may feel connected with a device such as a fixed-base simulator. One may even feel a sense of "oneness" with the device when high levels of proficiency are achieved. The feeling of felt presence in the simulated environment, however, is not compelling when simulations are limited to visual displays. One may feel a strong sense of vection in such devices (if variation in velocity is small), but the phenomenology of vection is different than the phenomenology of actual self-motion (Riccio, in press). The multiplicity of couplings that proliferate from a fundamental mechanical coupling, and perception of the associated affordances, may explain why.

5.10.2   Informative Interactions with Virtual Environments

Motion bases and devices that push or pull on the body can simulate some aspects of the mechanical coupling between a real vehicle and the body of the human operator. In particular, the biomechanical consequences of transient changes in the direction and magnitude of gravitoinertial vector, and sustained changes in the direction of this vector, can be simulated with a reasonable degree of fidelity. The biomechanical consequences of sustained changes in the magnitude of the gravitoinertial vector, however, cannot be simulated with any fidelity in extant devices. Gravitoinertial magnitude can be manipulated in centrifuges and related devices, but the biomechanical consequences are different than in a real vehicle. The consequences of imbalance are different because the attendant coriolis forces in rotating devices necessitate different patterns of postural compensation. Imbalance, and its consequences for perception and action, are important and ubiquitous perceivables.

The balancing and control of unrestrained body segments is a source of information about the inertial environment that is always available in simulators. Simulators do not differ in the extent to which they include information about the inertial environment because such information is always available in the constraints that the inertial environment imposes on postural control. Simulators differ in the nature of the constraints they impose on postural control. The information available in postural control is the mapping between the constraints on movement and the environment in which one moves. The constraints are specific to the environment, and information is specification. The pickup of this proprioceptive information is responsible for a feeling of presence in that environment. The achievement of felt presence in a simulated or virtual environment depends heavily on the extent to which the simulator can recreate the constraints that a real environment imposes on movement.

The pickup of information in postural control limits the perception of visually displayed variations in the velocity vector of self-motion, and it provides a basis for differentiating the simulator from that which is simulated. Proprioception constrains kinesthesis. Visual perception of variation in vehicular velocity that is accompanied by nonvisual perception of invariance in the consequences of imbalance, for example, necessitates that the displayed motion be perceived as something other than self-motion. The control of self-motion is not the superordinate function of the coupling between postural control and vehicular control in such a simulation. The superordinate function could be described as teleoperation of a vehicle. Teleoperation of a vehicle is proprioceptively and experientially different than control of self-motion. Experiences in fixed-base simulators generally resemble the former more than the latter.

Differences between teleoperation and self-motion are not limited to phenomenology. The sources of information available for energy management are also different. In a fixed-base flight simulator, for example, information about G force and the associated potential energy is available only in the kinematics of the displayed trajectory of aircraft motion. In real aircraft, however, curvilinear trajectories are fundamentally multimodal phenomena. Nonvisual systems provide sensitivity to the constraints that variations in the velocity vector impose on postural control. Such sensitivity is useful because these constraints on postural control are related to G forces on the aircraft. The vehicular control loop can be closed through the postural control system of the human operator. The absence of posture as an informative dynamical component in simulated vehicular control limits the precision and accuracy of energy management in the simulator. It is important to emphasize that such limitations on energy management would be important in any simulation of self-motion, not just simulated flight, because energy management is important in all forms of locomotion.

5.11 Conclusions

5.11.1 Felt Presence

Perception science, movement science, and mathematical modeling all have contributed to developments in flight simulation. Technological developments have been limited, however, by dominant scientific assumptions and paradigms. Ecological psychology provides an alternative approach to the phenomenon of self-motion and to any technology that attempts to simulate movement of an observer–actor in a virtual environment (Hettinger, & Riccio, 1993). Ecologically essential issues for such technology are presented in this chapter, with special emphasis given to those issues that have been previously neglected. In particular, the broad analysis of self-motion emphasizes the nonrigidity of a pilot (or human operator in any vehicle) and the multicriterion task structure that is necessitated by the fact of nonrigidity. From this perspective, the task of vehicular control cannot be considered alone or independently of the task of postural control in the vehicle. The nesting of tasks, and the associated nesting of control systems, is partially due to the mechanical coupling between vehicular control and postural control.

Vehicular control normally influences the dynamics of balance in the cockpit, and the absence of such influences in a simulator seriously limits felt presence in the virtual environment. The absence of mechanical coupling between vehicular control and postural control means that the human operator is not present in the simulated environment. This information is fundamentally multimodal because information is available to the visual system about variation in the vehicular velocity vector, and information is available to the nonvisual system about the presence or absence of variation in the dynamics of balance in the cockpit. Information about the dynamics of balance is always available in subtle patterns of nonvisual stimulation obtained through ubiquitous postural control activity. The absence of covariation (i.e., coupling) between vehicular velocity and the dynamics of balance, for example, is revealed in multimodal patterns of stimulation even in a simulator that contains only a visual display. Technological improvements that promote felt presence and movement in a virtual environment require explicit consideration and control of these multimodal patterns.

5.11.2 Dynamics of Balance

This chapter constitutes a strong set of arguments for multimodal approaches to virtual environments and, in particular, for nonvisual devices in flight simulators. The arguments do not imply that existing motion-base systems are superior to existing fixed-base simulators with respect to experiential or action fidelity. The relative efficacy of these devices depends on how well they simulate or recreate constraints on postural control in real vehicles. The constraints on postural control in a motion-base simulator depend heavily on the mathematical algorithms that are used to link states of the simulated vehicle to motions of the platform and on the technological implementation of the algorithms. A motion-base simulator could have poor fidelity because of inappropriate algorithms, inappropriate implementation, or insufficient calibration of the moving platform. If properly designed and maintained, motion platforms can recreate many of the forces that real aircraft impose on the pilot's body. The conspicuous exception is the magnitude of the gravitoinertial force vector (see Parrish et al., 1975). An ecological approach indicates that this is important because of the effect that gravitoinertial magnitude has on the consequences of imbalance.

Although variation in gravitoinertial magnitude can be achieved in centrifuges, coriolis forces create consequences for imbalance that are different from those encountered during common trajectories of aircraft motion. A different type of device is needed for this essential category of information. One possibility was suggested by the experiment of Riccio et al. (1992). In this experiment, the perception of upright (or vertical) was determined primarily by the experimentally controlled dynamics of balance for the body as a whole (cf. Riccio & Stoffregen, 1990; Stoffregen & Riccio, 1988). A similar concept could be adapted to a fixed or moving cockpit by connecting torque motors to the pilot's helmet (cf. Brown et al., 1991; Kron et al., 1980). Both the direction of balance and the consequences of imbalance could be manipulated through the orientation dependence of torque on the head. Such a helmet loading system could simulate the perceivable effects of variation in the direction and magnitude of the gravitoinertial vector. The results of Riccio et al. (1992) suggest that the helmet loader would promote experiential fidelity with respect to this important category of information. Furthermore, it could promote action fidelity by improving postural control in the cockpit, minimizing simulator sickness and reducing postural aftereffects.

5.11.3 Nonrigidity of Human Operator

Variation in the gravitoinertial force vector is caused by variation in the velocity vector of self-motion. Variation due to controlled changes in heading or altitude generally occurs over long time scales, whereas variation due to vibration and transients can occur over relatively short time scales. It is unnecessary, and perhaps unwise, to use a helmet loader to recreate the effects that gravitoinertial variations have on the direction of balance and the consequences of imbalance when these variations are faster than the corresponding postural control activities. These high-frequency variations in force tend to move the various body segments, and it is best to consider them as sources of postural perturbations. Driving a helmet loader at high frequencies could lead to injury. Such high-frequency forcing also would tend to create patterns of head–torso motion that would be different than in an aircraft in which forcing is applied through the seat. Such high-frequency effects could be attenuated, whereas low-frequency variation in balance constraints could be preserved with a low-pass filter in the helmet loading system. An undesirable effect of a low-pass filter on a helmet loader would be the removal of information about high-frequency vehicular motion that otherwise would be available in patterns of head–torso disturbance. This information would be preserved in the high-frequency fidelity of a motion platform. Helmet loaders and motion platforms would be complementary; together they could provide for high-fidelity simulation of transient and sustained changes in the inertial environment.

High-frequency motion of a nonrigid pilot could be included in a fixed-base simulator by visually simulating and manipulating the inside cockpit surroundings. Instruments and edges of the windscreen, for example, could be included in the visual display and could be moved independently of the visually displayed scene outside the cockpit. The relative motion of inside-the-cockpit structure and outside the cockpit structure, and the accretion and deletion of the latter by the former, would correspond to a changing point of observation in the cockpit; that is, it could simulate the movement of the pilot in the cockpit due to transients and vibration. In this way, simulated inside-the-cockpit structure would be analogous to a dynamic seat. Motion of these simulator components should be based on the fact that the pilot is not rigidly attached to the cockpit. Motion of these two components would be different to the extent that the head and the upper part of the torso move differently than the lower part of torso.

Together, inside-the-cockpit structure and a dynamic seat could simulate the nonrigidity of the pilot. This may not require a quantitatively precise biomechanical model to determine the relative motion of the two displays. Qualitative accuracy could be achieved simply by driving the inside-the-cockpit structure (i.e., simulated head motion) so that it lags the dynamic seat (i.e., simulated motion of lower torso). The simulation of nonrigidity is represented in Figure 5.13. Visual simulation of nonrigidity would provide an alternative to motion platform systems that actually move the pilot in the cockpit. Nevertheless, some nonvisual devices such as dynamic seats and helmet loaders are necessary for simulation of self-motion.

5.11.4 A Different Approach

This chapter presents a unique conceptual analysis of real and simulated self-motion. As such, it is an alternative to the more traditional mathematical and engineering analyses. It may be complementary to some of the more traditional treatments. Irrespective of such complementarity, however, the present analysis should lead to a more accurate "picture" of self-motion. The conceptual analysis emphasizes the most important issues in the perception and control of self-motion. These issues have considerable generality. The various sections in the chapter, and many of the details presented in the context of real and simulated flight, can be applied to all forms of real and simulated locomotion. In any case, it should be clear that the present perspective motivates and emphasizes different types of displays than do traditional approaches to flight simulation and virtual environments. Helmet loaders and near-field optical structure, in particular, may be useful for simulating movement of an observer–actor in any virtual environment.

The difference in emphasis is important because it indicates that the present perspective is not tantamount to an argument for increased complexity or physical (or pictorial) fidelity in flight simulation and virtual environments. To the contrary, it suggests that displays need only be approximately correct if the essential categories of information are represented in the simulation. Indeed, if such categories of

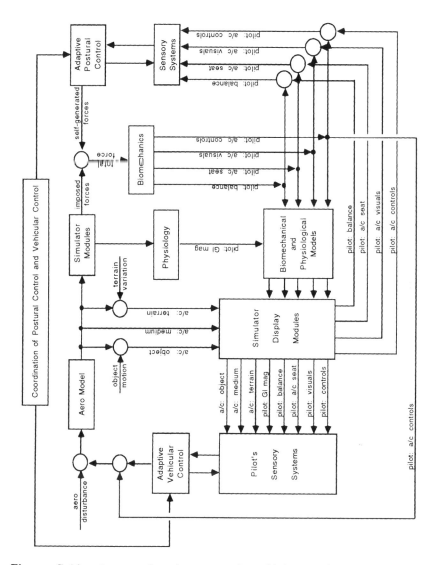

Figure 5.13. *A comprehensive approach to flight simulation. This is an elaboration of Figure 5.10. Physiological effects of high G maneuvers can be simulated with specialized devices that either recreate the physiological effects (e.g., lower blood pressure in the head) or simulate the consequences of these effects (e.g., reduce the field of view). Biomechanical effects of variation in gravitoinertial vector can be simulated with mechanical devices that manipulate the dynamics of balance in the cockpit. Other biomechanical effects, such as nonrigid motion of the pilot due to vehicular vibration and transients, can be simulated with a dynamic seat or with a visual display that includes a simulation of inside-the-cockpit surroundings. Note that the mechanical or*

175

optical simulation of biomechanical effects influences postural control in the
cockpit as well as control of the simulated vehicular. Note also that simulation
of physiological and biomechanical effects require mathematical models that
would be analogous to the aeromodel in a flight simulator.

information can be preserved, it may be desirable to reduce the
complexity of displays to increase temporal fidelity (Martin et al., 1986;
McMillan et al., in preparation) or to facilitate perceptual learning
(Warren & Riccio, 1985).

Acknowledgments

This paper was written during a year in residence at Wright Patterson
Air Force Base, Ohio, in a division of a national laboratory (the
Armstrong Laboratory) that was formerly the Armstrong Aerospace
Medical Research Laboratory. Support for this year in residence was
provided by the Air Force Office of Scientific Research University
Resident Research Program (AFOSR URRP, FY92). The ideas presented
in this paper grew out of several projects in which I was involved while
employed at the Wright Patterson AFB (1984–1988). One project
involved the analysis of drive algorithms for a dynamic seat that would
allow the associated pattern of somatosensory stimulation to substitute
for the vestibular stimulation that is present during whole body
movement (AAMRL contract F33615-85-C-0541). Another involved the
development of OCM based models for learning in flight simulators
(AAMRL contract F33615-86-C-0532). The most relevant project involved
the analysis of new approaches to motion cuing in flight simulators
(AAMRL contract F33615-85-C-0541). I am indebted to Grant McMillan
(the dynamic seat project), Greg Zacharias (the optimal control project),
Frank Cardullo and John Sinacori (the motion cuing project) for their
insights and creativity regarding multimodal simulation of natural
environments. I am also indebted to Rik Warren for bringing me to
Wright Patterson AFB and for supporting me on the AFOSR URRP. I
would also like to thank Victoria Tepe and John Flach for comments on
the manuscript.

5.12   References

Andersen, G. J., & Dyre, B. P. (1989). Spatial orientation from optic flow
in the central visual field. *Perception and Psychophysics, 45,*
453–458.

Andersen, G. J., & Braunstein, M. L. (1985). Induced self-motion in central vision. *Journal of Experimental Psychology: Perception and Performance, 11,* 122–132.

Baarspul, M. (1977). *The generation of motion cues on a six degree of freedom motion system* (Tech. Rep. LR 248). Delft, The Netherlands: Delft University of Technology.

Roeder, P. (1961). The cooperation of extraocular muscles. *American Journal of Ophthalmology, 51,* 469–481.

Boff, K.R., & Lincoln, J.E. (Eds.). (1988). *Engineering data compendium: Human perception and performance.* Wright-Patterson Air Force Base, OH: Harry G. Armstrong Aerospace Medical Research Laboratory.

Brandt, T., Wist, E. R., & Dichgans. J. (1975). Foreground and background in dynamic spatial orientation. *Perception and Psychophysics, 17,* 497–503.

Brown, Y. J, Cardullo, F. M., McMillan, G. R., Riccio, G. E., & Sinacori, J. B. (1991). *New approaches to motion cuing in flight simulators* (AL TR 1991 0139). Wright Patterson AFB, OH: Armstrong Laboratory.

Cardullo, F. M., & Sinacori, J. B. (1988). *A broader view of in cockpit motion and force cuing.* AIAA Paper No. 88-4585. American Institute of Aeronautics and Astronautics.

Cress, J. D., McMillan, G. R., & Gilkey, M. J. (1989). The dynamic seat as an angular motion cuing device: Control of roll and pitch vs. the control of altitude and heading. *Proceedings of the Flight Simulation Technologies Conference* (pp. 94–100). New York: American Institute of Aeronautics and Astronautics.

Dichgans, J., & Brandt, T. (1978). Visual-vestibular interaction: Effects on self-motion perception and postural control. In R. Held, H. Leibowitz, & H. Teuber (Eds.), *Handbook of sensory physiology* (Vol. 8, pp. 755–804). New York: Springer-Verlag.

Distephano, J.J., Stubberud, A.R., & Williams, I.J. (1967). *Theory and problems of feedback and control systems.* New York: McGraw Hill.

DiZio, P. A., & Lackner, J. R. (1986). Perceived orientation, motion, and configuration of the body during viewing of an off-vertical, rotating surface. *Perception and Psychophysics, 39,* 39–46.

Flach, J. (1990). Control with an eye for perception: Precursors to an active psychophysics. *Ecological Psychology, 2,* 83–111.

Flach, J., Riccio, G., McMillan, G., & Warren, R. (1986). Psychophysical methods for equating performance between alternative motion simulators. *Ergonomics, 29,* 1423–1438.

Gibson, E. J. (1991). *An odyssey in learning and perception.* Cambridge, MA: MIT Press.

Gibson, J.J. (1964). The uses of proprioception and propriospecific information. In: Reed, E., & Jones, R. (Eds.) (1982). *Reasons for realism: Selected essays of James J. Gibson* (pp. 164–170). Hillsdale, N.J.: Lawrence Erlbaum Associates.

Gibson, J. J. (1966). *The senses considered as perceptual systems.* Boston: Houghton Mifflin.

Gibson, J. J. (1979). *The ecological approach to visual perception.* Boston: Houghton Mifflin.

Griffin, M.J. (1975). Levels of whole-body vibration affecting human vision. *Aviation, Space and Environment Medicine, 46,* 1033–1040.

Hamilton, K. M., Kantor, L., & Magee, L. E. (1989). Limitations of postural equilibrium tests for examining simulator sickness. *Aviation, Space, and Environmental Medicine, 60,* 246–251.

Hettinger, L., & Kennedy, R. (1990). Vection and simulator sickness. *Military Psychology, 2,* 171–181.

Hettinger, L., & Riccio, G.E. (1993). Visually induced motion sickness in virtual environments. *Presence: Telerobotics and virtual environments, 1,* 306-310.

Hosman, R., & van der Vaart, J. (1981). Effects of visual and vestibular motion perception on control task performance. *Acta Psychologica, 48,* 271–287.

Junker, A. M., & Replogle, C. R. (1975). Motion effects on the human operator in a roll axis tracking task. *Aviation, Space, and Environmental Medicine, 46,* 819–822.

Kennedy, R. S., Hettinger, L. J., & Lilienthal (1990). Simulator sickness. In G. H. Crampton (Ed.), *Motion and space sickness* (pp. 318–341). Boca Raton, FL: CRC Press.

Kron, G. J., Cardullo, F. M., & Young, L. R. (1980). *Study and design of high G augmentation devices for flight simulators* (TR F 33615 77 C 0055). Binghamton, NY: Singer Link Flight Simulation Division.

Kugler, P.N., & Turvey, M.T. (1987). *Information, natural law, and the self-assembly of rhythmic movement.* Hillsdale, N.J.: Lawrence Erlbaum Associates.

Lee, D. N. (1990). Getting around with light and sound. In R. Warren & A. H. Wertheim (Eds.), *Perception and control of self-motion* (pp. 327-356). Hillsdale, NJ: Lawrence Erlbaum Associates.

Lee, D. N., & Lishman, J. R. (1975). Visual proprioceptive control of stance. *Journal of Human Movement Studies, 1,* 87–95.

Lestienne, F. G., Soechting, J., & Berthoz, A. (1977). Postural readjustments induced by linear motion of visual scenes. *Experimental Brain Research, 28*, 363–384.

Lintern, G. (1987). Flight simulation motion systems revisited. *Human Factors Society Bulletin, 30*(12), 1–3.

Lishman, J. R., & Lee, D. N. (1973). The autonomy of visual kinesthesis. *Perception, 2*, 287–294.

Martin, E. A., McMillan, G. R., Warren, R., & Riccio, G. E. (1986). A program to investigate requirements for effective flight simulator displays. In International Conference on Advances in Flight Simulation (23 pp.). London: Royal Aeronautical Society.

McMillan, G. R., Cress, J., & Middendorf, M. (1990). Dynamic seat cuing with wide versus narrow field of view displays. In *Proceedings of the AIAA Flight Simulation Technologies Conference* (pp.). Washington, DC: American Institute of Aeronautics and Astronautics.

McMillan, G., Riccio, G., & Lusk, S. (in preparation). *Time delay and asynchrony in flight simulators: Human performance, control behaviour, transfer of training, and handling qualities.* Wright-Patterson AFB,OH: U.S. Department of Defense, Crew Systems Ergonomics Information Analysis Center (CSERIAC).

McReur, D.T., & Jex, H.R. (1967). A review of quasi-linear pilot models. *IEEE Transactions on Human Factors in Electronics, HFE-8*, 231–249.

Middendorf, M. S., Fiorita, A. L., & McMillan, G. R. (1991). The effects of simulator transport delay on performance, workload, and control activity during low level flight. *Proceedings of the AIAA Flight Simulation Technologies Conference.* Washington, DC: American Institute of Aeronautics and Astronautics.

Owen, D. H. (1990). Lexicon of terms for the perception and control of self-motion and orientation. In R. Warren & A. H. Wertheim (Eds.), *Perception and control of self-motion* (pp. 33–50). Hillsdale, NJ: Lawrence Erlbaum Associates.

Ormsby, C., & Young, L. (1977). Integration of semicircular canal an otolith information for multisensory orientation stimuli. *Mathematical Biosciences, 34*, 1–21.

Parrish, R. V., Diedonne, J. E., Bowles, R. L., & Martin, D. J. (1975). Coordinated adaptive washout for motion simulators. *Journal of Aircraft, 12*, 44–50.

Poulton, E. C. (1974). *Tracking skill and manual control.* New York: Academic Press.

Reed, E., & Jones, R. (Eds.). (1982). *Reasons for realism: Selected essays of*

James J. Gibson. Hillsdale, N.J.: Lawrence Erlbaum Associates.

Repperger, D. W. (1991). Active force reflection devices in teleoperation. *IEEE Control Systems, 36,* 52–56.

Riccio, G.E. (1991). [Videographic analyses of postural control in real and simulated vehicles]. *Unpublished raw data.*

Riccio, G. E. (1993a). Information in movement variability about the qualitative dynamics of posture and orientation. In K. M. Newell & D. M. Corcos (Eds.), *Variability and motor control* (pp. 315–337). Champaign, IL: Human Kinetics.

Riccio, G.E. (1993b). *Multimodol perception and multicriterion control of nested systems: Self motion in real and virtual environments.* (UIUC-BI-HPP-93-02). University of Illinois at Urbana-Champaign: Beckman Institute for Advanced Science & Technology.

Riccio, G.E. (in press). Perception of motion with respect to multiple criteria. *Behavioral & Brain Sciences.*

Riccio, G. E., & Cress, J. (1986). Frequency response of the visual system to simulated changes in altitude and its relationship to active control. *Proceedings of the 22nd Annual Conference on Manual Control* (pp. 117–134). Wright-Patterson AFB, OH: Aeronautical Systems Division.

Riccio, G. E., & Hettinger, L. (1991). Experiential and action fidelity in flight simulators. In *Proceedings of the Aviation Psychology Conference.* Columbus, OH: The Ohio State University.

Riccio, G. E., Martin, E. J., & Stoffregen, T. A. (1992). The role of balance dynamics in the active perception of orientation. *Journal of Experimental Psychology: Human Perception and Performance, 18,* 624–644.

Riccio, G. E., & Stoffregen, T. A. (1988). Affordances as constraints on the control of stance. *Human Movement Science, 7,* 265–300.

Riccio, G. E., & Stoffregen, T. A. (1990). Gravitoinertial force versus the direction of balance in the perception and control of orientation. *Psychological Review, 97,* 135–137.

Riccio, G. E., & Stoffregen, T. A. (1991). An ecological theory of motion sickness and postural instability. *Ecological Psychology, 3,* 195–240.

Rolfe, J. M., & Staples, K. J. (1986). *Flight simulation.* Cambridge, UK: Cambridge University Press.

Shaw, R.E., Kadar, E., Sim, M., & Repperger, D.W. (1992). The intentional spring: A strategy for modeling systems that learn to perform intentional acts. *Motor Behaviour, 24,* 3–28.

Stark, L. W. (1968). *Neurological control systems: Studies in bioengineering.*

New York: Plenum.

Stoffregen, T. A. (1985). Flow structure versus retinal location in the optical control of stance. *Journal of Experimental Psychology: Human Perception and Performance, 11,* 554–565.

Stoffregen, T. A. (1986). The role of optical velocity in the control of stance. *Perception and Psychophysics, 39,* 355–360.

Stoffregen, T. A. & Riccio, G. E. (1988). An ecological theory of orientation and the vestibular system. *Psychological Review, 95,* 3–14.

Stoffregen, T. A., & Riccio, G. E. (1991). A critique of the sensory conflict theory of motion sickness. *Ecological Psychology, 3,* 159–194.

Strughold, H. (1950). The mechanoreceptors of the skin and muscles under flying conditions. In Department of the Air Force (Ed.), *German aviation medicine: World War II* (Vol. II, pp. 994–999). Washington, DC: U.S. Government Printing Office.

Todd, J. (1981). Visual information about moving objects. *Journal of Experimental Psychology: Human Perception and Performance, 7,* 795-810.

Warren, R. (1990). Preliminary questions for the study of egomotion. In R. Warren & A. H. Wertheim (Eds.), *Perception and control of self-motion* (pp. 579–619). Hillsdale, NJ: Lawrence Erlbaum Associates.

Warren, R., & Riccio, G. (1985). Visual cue dominance hierarchies: implications for simulator design. *Transactions of the Society for Automotive Engineering, 6,* 937–951.

Warren, W., Mestre, D., Blackwell, A., & Morris, M. (1991). Perception of curvilinear heading from optical flow. *Journal of Experimental Psychology: Human Perception and Performance, 17,* 28–43.

Wolpert, L. (1990). Field-of-view information for self-motion perception. In: R. Warren & A.H. Wertheim (Eds.), *Perception and control of self-motion,* (pp. 101–126). Hillsdale, N.J.: Lawrence Erlbaum.

Young, L. R. (1984). Perception of the body in space: mechanisms. In I. Smith (Ed.), *Handbook of physiology--The nervous system* (Vol. 3, pp. 1023–1066). New York: Academic Press.

Zacharias, G., Warren, R., & Riccio, G. (1986). Modeling the pilot's use of flight simulator visual cues in a terrain following task. *Proceedings of the 22nd Annual Conference on Manual Control.* Wright Patterson AFB, OH: Aeronautical Systems Division.

Zacharias, G., & Young, L. (1981). Influence of combined visual and vestibular cues on human perception and control of horizontal rotation. *Experimental Brain Research, 41,* 141–171.

Chapter 6

Designing for Telepresence: The Delft Virtual Window System

Gerda J.F. Smets

Delft University of Technology
Department of Industrial Design Engineering

6.0   Introduction: A True Story

Some time ago a director of a plant came to see us about a technological problem. He had a transport company and owned several oil tankers. Once every few years those tankers had to be cleaned of the sediment that remained. Although this was a very unhealthy job, because of the gases in the hold, it had to be done by people, because no other satisfactory solution was available. He had considered a lot of alternatives, inspired by robotics. All those solutions followed the same track. They relied on an autonomous robot that was brought into the hold and cleaned the ship. However, no robot proved to be flexible enough to fulfill this task. Its cleaning performance may have fulfilled the most demanding standards, but it failed immediately if the task deviated in the slightest from the specific constraints for which the robot was built. The man had come to ask if we knew of other solutions.

We answered that, fortunately, it is not necessary to depend entirely on a fully competent and autonomous robot to solve this problem. Instead of looking for a robotics solution, it might be worth considering a teleoperating solution. In this case a teleoperator is used, which is not a robot in a strict sense, because the machine is controlled by a human's perceptual and motor skills, rather than by the central processing unit of a computer.

The following definitions of teleoperation and telepresence were adapted from Sheridan (1989a). A *teleoperator* includes at the minimum artificial sensors, arms and hands, a vehicle to carry these, and communication channels to and from the human operator. The term *teleoperation* refers to direct and continuous human control of the teleoperator. Teleoperation problems can be distinguished in four broad categories: (a) telesensing, including vision, resolved force, touch, kinesthesis, proprioception, and proximity; (b) teleactuating, combining motor actuation with sensing and decision making; (c) computer aiding in human supervision of a teleoperator; and (d) meta-analysis of the human-computer teleoperator task interaction. This chapter is limited to telesensing, and more specifically to depth television, from a meta-analytical point of view. *Telepresence*, then, is the ideal of sensing sufficient information about the teleoperator and task environment and communicating this to the human operator in a sufficiently natural way, so that the operator feels physically present at the remote site.

Once told of these kinds of machines, the man replied that he knew of them. The problem with them, he felt, is that they are built as humanlike as possible, which means that a large amount of information exchange between the operator and the workstation is required.

The transmission requirements for control signals are relatively modest, but the sensing signals typically depend on stereoscopic images and employ two broadband color television channels. What visual information, he asked, is necessary to reach a teleperformance that matches a direct manipulation performance and (i.e., create a sense of remote presence)? And what visual information is redundant and might be skipped?

We could see his point, after considering Figure 6.1. It is a prototype of a telepresence system. The operator wears a sort of harness and looks at two miniature helmet-mounted television screens that are directly in front of the operator's eyes. The (nondepicted) workstation itself has two television cameras linked to the screens and two microphones linked to the operator's headphones. The controls of the teleoperator consist of an exoskeleton framework that follows the operator's movements and commands the motions of the station's effectors with a master–slave correspondence. The sense of remote presence in such an operator station is quite strong, and there is virtually no need for the operator to think about how to translate his or her movements into motions at the remote work location. The remote environment can be real, but may be computer-generated as well, which might be useful for computer-aided design (CAD).

So we told the man we agreed with him that the major challenges now lie at the human-machine interface. We also gave our opinion that the burden of communication from the sensors to the visual display can be eased by taking advantage of the ecological approach to visual perception, developed by Gibson (1966; 1979). In this approach perception and action are seen as tightly interlocked and mutually constraining. This theory is unique in the optical variables it chooses as visual input and that, indeed, have been shown to provide important information for control of action.

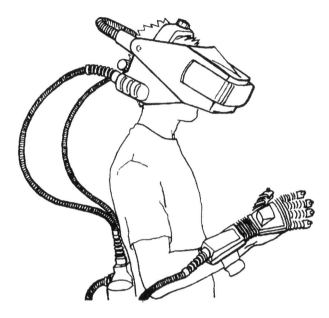

Figure 6.1. *A prototype of a teleoperator system, creating telepresence. Telepresence is commonly claimed to be important for direct manual telemanipulation. The (nondepicted) workstation itself has two television cameras to send visual information to the head-mounted LCD screens and two microphones to send auditory information to the operator's headphones.*

To summarize, the question of how we come to see a world containing objects that have stable sizes and shapes and positions is of technological relevance for the design of workstations that can carry out physical tasks safely in hostile environments (e.g., the hold of an oil tanker or a nuclear hot cell). However, this question is also of high theoretical interest: The psychology of perception has been dominated by it. We want to show how the answers to this question offered by perception psychology, and especially the answer given by the ecological theory of perception, might be of help in optimizing the design of teleoperators. We limit ourselves to the visual aspects of it: television. The whole world has been exposed to television, and the technology is inexpensive and highly developed. However, there remain some problems that, although not critical in ordinary television programming, are critical for teleoperation. Foremost among these is depth perception, which continues to be a major reason why direct manipulation performance is not matched by telemanipulation

(Sheridan, 1989a). In analogy to color television we speak of depth television when referring to television allowing for telemanipulation.

In the second section a history of applications of teleoperation—extension of human sensing and manipulating capability by coupling to (remote) artificial sensors and actuators—is given in order to situate the Delft Virtual Window system. This review is an illustration of the fact that what is considered of major importance in depth perception determines the research about it and, hence, the applications. Traditionally, the accent was on binocular disparity and, hence, most teleoperators are stereoscopic systems. But what if we consider an alternative theory? Depth perception can be explained in several theoretical frameworks. We have two eyes for it, but we can also move our head to get different view angles. These questions are considered in the third section. This theoretical background leads to an alternative working principle for television allowing for telemanipulation and enhanced telepresence. The fourth section presents an overview of this Delft Virtual Window system, referring to definitions of concepts from robotics and perception psychology that are crucial in order to explain the working principle of the system, as well as the system's performance measurements. The fifth section discusses telepresence applications using the system. We conclude by arguing that implementation in its turn leads to new challenges for fundamental research on perception.

6.1   Short History of Teleoperation and Telepresence

From well before the 16th century there were teleoperators in the form of simple arm extensions, for example, fire tongs. Early in the 19th century there were crude teleoperators for earth moving, construction, and related tasks. By the 1940s, prosthetic limb fitters had developed arm hooks activated by leather thongs tied to other parts of the wearer's body. In about 1945, the first modern master–slave teleoperators were developed. These were mechanical pantograph mechanisms by which radioactive materials in a "hot cell" could be manipulated by an operator outside the cell. Electrical servomechanisms soon replaced the direct mechanical type, and closed-circuit television was introduced, so that now the operator could be an arbitrary distance away. In the late 1950s, there was interest in applying this new servomechanism technology to human limb prostheses. Probably the first successful development was that of Kobrinskii (1960) in Moscow, a lower arm prosthesis driven by minute myoelectric signals picked up from the muscles in the stump or upper arm. This was followed rapidly by similar developments in the United States and Europe, including teleoperators attached to the wheelchairs of quadriplegics that could be commanded by the tongue or other remaining motor signals. From the early 1960s, telemanipulators and video cameras were being attached to submarines by the U.S., U.S.S.R., and French navies and used experimentally. Offshore mineral extraction and cable-laying firms soon became interested in this technology to replace human divers, especially as oil and gas drilling operations got deeper. When, also in these times, the race to the moon began, a particularly vexing problem was posed by transmission time delay between earth and moon. By 1970, the Western interest in teleoperation had turned to the undersea, for there was great economic demand for offshore oil. By 1980, industrial robots had wrist-force sensing and primitive computer vision[1]. The striking experience of remote presence came in reach with the development of head-mounted systems, accompanied with gloves that gave teletouch. To create graphic and force-feedback displays to enable the human operator to feel he or she is "there," when "there" exists nowhere other than in the computer, poses a particularly interesting challenge in many applications, especially in computer aided design (CAD).

This short review of relevant historical developments shows that teleoperation is primarily used for manipulation in hazardous or inaccessible environments (e.g., outer space, undersea, or nuclear "hot laboratories") or to compensate for motor deficiencies (e.g., for the motorically handicapped) or, more generally, when forces have to be exerted that are far too large for a human operator. Until now, teleoperation has not been used to overcome visual (or, more general, perceptual) deficiencies. Through the recent attainment of the sense of telepresence in telemanipulation, however, we are convinced that this is a promising, as yet virgin, territory for applications. Two applications illustrate how we faced this challenge. We demonstrate the advantage of a simple human-supervised teleoperation for visual tasks, using the Delft Virtual Window working principle. Until now most teleoperation systems use stereoscopic visual input. This can be explained considering perceptual theory.

6.2 Perceptual Theory and Telepresence: Two Approaches

We begin by contrasting two positions, which we call traditional and ecological, although there is much diversity of opinion within each viewpoint. The traditional approach accentuates the role of binocular

---

[1]It became clear that teleoperation for working in space, undersea, or other hazardous environments was to follow a course different from that of industrial robots. Industrial robots justify themselves by repetitively performing mostly predictable tasks in controlled environments with speed and accuracy. In human teleoperation there is little or no repetition of the same identical task. Although an industrial robot can go unmonitored by a person for a long period of time, in direct teleoperation a person must remain in the control loop continuously.

disparity. Although the ecological approach indicates the importance of a dynamic approach to perception, the difference between motion and movement as to the perceptual input is neglected. We describe how this came about and what the consequences are for telemanipulation and telepresence.

By traditional approach we mean psychologists and computational theorists of the tradition who describe the visual input in terms of point intensities of light at the retina. By ecological theory of perception we mean those theorists who argue that the input for vision can be described in terms of the structure of light. We regard the two approaches as complementary.

6.2.1   The Traditional Approach

Often the eye is thought of as a camera acting to focus light onto a mosaic of retinal receptors. At any instance of time, therefore, one can conceive of the pattern of excitation of retinal receptors as a *picture* curved around the back of an eyeball (Bruce & Green, 1990, p. 142). Though curved, the image is two-dimensional, and yet our perception is of a three-dimensional world. How might depth be recovered? Without invoking learning or memory, pictorial cues and binocular disparity are the most popular answers, with motion parallax being a good third. The movement of the observer is thought to be irrelevant for visual perception. In fact, experiments about visual perception, even those about parallax effects, are often designed so that the subject cannot move, with his or her head being fixed by a chin rest, for example.

As a consequence, the adherents of this approach create telepresence by reproducing a high-resolution stereoscopic image of the teleenvironment. As far as motion is concerned, it is restricted to presenting parallax shifts on the stereo display as a consequence of performances in the teleenvironment. Movement of the human operator is important to control the mobility of the effectors of the workstation, but it is not considered important for visual perception as such. The human operator provides largely symbolic commands (concatenations of typed symbols or specialized key presses) to the computer. However, some fraction of these commands still are analogical (body control movements isomorphic to the space-time-force continuum of the physical task) because "they are difficult for the operator to put into symbols" (Sheridan, 1989a, p. 488), yet this analogy is not considered important to enhance teleperformance nor telepresence.

A stereo display typically employs two broadband color television channels. The transmission requirements for those are high. The potential for overloading the communications system between workstation and operator is reduced by, among other solutions, transmitting a high resolution monochrome image along one television channel and a low resolution color image along another and combining the images at the operator's station in a color stereo image (Pepper, Cole, & Spain, 1983).

A rather unpleasant human factor these teleoperator designers must contend with is simulator sickness, a euphemism for heavy nausea, if there is a conflict between the operator's perception of motion (shifts on the stereo display of his head-mounted system) and his or her own movements (which is strengthened by the stereo display often producing pseudoparallax and pseudoconvergence).

6.2.2   The Ecological Approach

In the traditional approach, shape and depth perception are expressible in essentially static and pictorial terms. However, they can also be defined, in accordance with Gibson, as the detection of formless invariants over time. For a moving observer, the retinal projection undergoes a continuous serial transformation on the retina, due to movement parallax, that is unique for this movement pattern in this environment. We believe that movement parallax is sufficient to create (tele)presence by rigidly linking the observer's proprioceptive experiences to his exteroception, providing for exteroproprioceptive information.

Yet, the role of movement and movement parallax takes a remarkably minor place in research on depth perception. Even Bruce and Green, who devote explicit attention to the perception–action loop, do not differentiate between parallax shifts caused by moving objects (motion) or a moving observer (movement). A few authors, however, reported interesting results about how movement constrains depth perception.

When reading about those investigations, considered later, bear in mind that we do not claim that the ecological theory of perception is the right one and the traditional approach is wrong. Depth perception can be explained in several theoretical frameworks. We have two eyes, but we can also move our head to get different view angles. The point we want to make is that both differ, however, in what is considered of major importance. Therefore their research and, hence, their applications differ. For example, Figure 6.2 (top) is a picture of a bird high in the sky. It is only a crude image and hard to recognize. How can recognition be improved? An adherent of an indirect, bottom-up approach would first try his or her best to enhance the spatial resolution of this static picture (Figure 6.2, bottom). An adherent of the Gibsonian approach will not bother about the quality of this image. He or she will try to enhance time resolution keeping spatial resolution constant. This moving pattern can be simulated by riffling through the frames at the bottom right of the pages of this chapter, containing successive low resolution views of the

Figure 6.2 *Two pictures of a bird and a simulation of a moving bird on the top right of this chapter's pages. The top picture cannot be recognized easily. However, a moving pattern with the same low spatial resolution can be recognized as easily as the bottom high-resolution picture. This moving pattern can be simulated by riffling through the frames at the bottom right of the pages of this chapter, containing successive views of the moving bird.*

moving bird.

Ullman (1979) and Lee (1980) are examples of both research approaches respectively. Their difference lies in the nature of the information that forms the starting point for their analysis of motion perception. For Ullman, it is an image (from which he recovers edges and then establishes correspondences between successive images over time), for Lee, it is flow.

As to application, there are abundant examples of how the psychology of perception[2] has consequences for designing our environment (e.g., Norman, 1988). Yet, it seems that, until now, engineers have not fully detected the potential of the ecological theory of perception for industrial design engineering, as the editors point out in the preface of this book.

6.3   Definitions and System Overview of the Delft Virtual Window System

---

[2]Apart from perception psychology and human factors research, there are two other subdisciplines of psychology explicitly concerned with the human-environment interaction: experimental esthetics (founded by Fechner in 1876, the year when his *Vorschule der Aesthetik* was published) and environmental psychology (emerging during the 1960s, with Proshansky, Ittleson, & Rivlin's [1970] book as a first survey of the field). It is astonishing that neither psychologists nor engineers are familiar with the results of experimental esthetics (studying the preferences for and the expressiveness of perceptual stimuli) or environmental psychology (studying the interaction between social behavior and environmental variables), because both subdisciplines deal with the study of affordances although, indeed, the methodology of many of their experiments can be criticized.

6.3.1   Definitions

When talking of *depth television* or *virtual window system*, we are talking of television allowing for telemanipulation and providing telepresence. The television screen forms a virtual window in which what lies behind it forms a rigid whole with the world before it, just like a window in daily life.

The present work is embedded in the search to account for the perceptual consequences of active movement generated by an observer as opposed to merely passive motions presented to a passive onlooker. Therefore we define the concepts of motion and movement along with the related concepts of movement parallax and motion parallax[3]. For an understanding of those differences, the concept of reciprocity between the individual and environment is fundamental. Perceiving is an act *of* an individual *in* an environment, not an activity in the nervous system (Gibson, 1979, pp. 239–240; Mace, 1977; Owen, 1990), which gives *exproprioceptive information*: about the position of the body or parts of it relative to the environment[4].

Displacements of an observation point[5] can be generated by the observer or can be imposed[6] on the observer. Following Cutting (1986), we reserve *motion* for imposed displacements, resulting in a purely exteroceptive experience, and *movement* for observer-generated displacements, occurring when they behave or perform with any of the motor systems of their body, giving potential information for proprioception and exteroception as well as exteroprioception.

---

[3]Because it has not always been considered relevant to make a sharp distinction between movement and motion, lexical discussions about those concepts are rather complicated (see, Owen, 1990).

[4]The visual system has a dual role, being at once both *exteroceptive* (giving information about extrinsic events in the environment) and *proprioceptive* (giving information about the actions and reactions of the individual; Gibson, 1966) Following Lee (1976), we also mention *exproprioceptive* information about the position of the body or parts of it relative to the environment.

[5]Gibson (1966) defined the *ambient optic array* as the arrangement of variations of the light intensities (structured light) available at an observation point, a point in space in which an observer can sample light. For each observation point, a unique optic array exists. The *optic flow* is the optic array at a moving observation point.

[6]Imposed displacements occur when members of the body are moved, when the head is accelerated or turned, when the whole individual is passively transported and the eyes are stimulated by motion perspective, or when the observation point is passively moved as when looking at motion pictures.

principle have been obtained for Europe and the United States (Smets, Stratmann, & Overbeeke, 1988, 1990). The original version of the system used a simple video camera and an observer watching with one eye only. Other versions used stereo and computer-generated images, but already the monocular version provides a compelling and reliable spatial impression contrary to what most textbooks on perception and the majority of existing spatial display systems (reviewed in Overbeeke & Stratmann, 1988) suggest, in both of which space perception is often equated with the use of two eyes.

Figure 6.3: *The Delft Virtual Window system operates by sensing the observer's head position and moving the camera in the remote site accordingly. The screen then does not act as a screen anymore, but as a window, where what lies behind it forms a rigid whole (and therefore a virtual reality) with what is before it.*

Exproprioception is lost when the visual input is considered without the observer's movements, as is the case with ordinary television programming. In such a case the visual input cannot be scaled relative to the observer's body movements, which means that the perception of a rigid world, allowing for adequate spatial behavior and, hence, telepresence, is impossible. Shape perception, however, as we all know, remains possible.

Experiments with the Delft Virtual Window system (Overbeeke, Smets, & Stratmann, 1987; Overbeeke & Stratman, 1988) indicated a perceptual advantage of the active observer, whose head movements steer the camera, over the passive onlooker, whose movements are not coupled to the display output (Figure 6.4). Both active and passive observers received identical output on their monitor screens, yet their perceptual input differed. The former received movement parallax

In accordance with the distinction between motion and movement, we differentiate between *motion parallax* and *movement parallax*. This difference is comparable to Sedgwick's (1986) distinction between relative and absolute motion parallax. Parallax refers to the fact that when two objects are at different distances from a moving observation point, the objects seem to shift relative to each other. We speak of movement parallax when the movement of the observation point is self-generated, and of motion parallax when this motion is imposed. Suppose that one is walking in the countryside and looking at the landscape. Let us also suppose that the gaze direction is perpendicular to the moving direction, that the direction of movement is from right to left, and that one is gazing at a fixation point. Under these conditions, all of the objects closer to the observer than the fixation point would appear to move in a direction opposite to his or her movement. On the other hand, objects that are further away will appear to move in the same direction the observer is moving. Not only the direction, but also the the speed of movement varies with the object's proximity to his or her fixation point and movement speed and direction. This is movement parallax. Motion parallax contains the part of it used by a cinematographer. He or she uses the apparent movements around the fixation point as registered by camera movements, without them being linked to the spectators' movements.

With movement parallax, potential information for the absolute distance of the object from the observer would be available if the observer was able to register both the translatory component of the eye's movement relative to the object and the visual direction of the object. With motion parallax, there is no potential information for absolute distance (the distance between ourselves and the perceived object). Both movement and motion parallax offer potential information for shape perception, a distance order within a limited cluster of coherent features (defined by Koenderink, 1990).

With the help of those definitions, an overview of the Delft Virtual Window system is given, with reference to performance measurements.

6.3.2   System Overview and Performance Measurements of the Delft Virtual Window System

Exproprioceptive information is provided for if perception and movement are tightly interlocked as is the case in movement parallax. This offers potential information for the perception of absolute distances. Therefore, we contend that depth television can be created by continuously updating the display output to match the observer's movements in front of the screen. This is accomplished by sensing the observer's head position and moving the camera in the remote site accordingly (Figure 6.3). It worked so well that patents on the working information, the latter disposed of motion parallax. The performance required consisted of aligning wedges. Telemanipulations of the active operator almost matched direct manipulation performance: Errors were consistently small, and their variance did not differ significantly from the variance of errors in real-life condition (F = 1.24, n.s., with 10 subjects in each condition). The teleperformance of the passive onlooker, on the contrary, was not consistent, although the average error size remained small. The variance of his error size was significantly greater than that of the active observer (F = 3.25, p < 0.01, 10 subjects in each condition) and than that obtained in the real-life setup (F = 2.25, p < 0.01, 10 subjects in each condition). Data indicate that the active condition, in which movement parallax information was provided on the display, even allowed for things apparently leaping out of the screen. In this case, perceived depth was somewhat less compelling, however, than in the real-life setup. There was a systematic underestimation in aligning a wedge virtually leaping out of the screen with a real wedge mounted in front of the display, yet the variance remained relatively small. Movement parallax allowed for telemanipulation; motion parallax did not.

Figure 6.4. *Performance measurements of the Delft Virtual Window system: a sketch of the experimental design. Two observers receive the same display output, yet the head movements of the left observer steer the camera. His or her visual input contains movement parallax information, whereas the observer on the right disposes of motion parallax information. Although the displays are the same, the perceptual input differs. This proves to have a significant effect on spatial performances. The performance of the left observer is almost as good as*

*that of an observer in a comparable real-life setup, whereas the performance of
the right observer is not.*

Of course, there remains a lot of work to be done concerning the
performance of the Delft Virtual Window system, for example, to define
its constraints (in time delay, for example). Elsewhere we did some
experiments as to the spatial resolution that is sufficient to allow for
telemanipulation using this working principle (Overbeeke & Smets,
1990; Smets & Overbeeke, 1992).

6.4   Motion and Movement Differ as to their Input in the Perceptual System

The adherents of the ecological approach devote much attention to the
study of optic flow. In this case, the perceptual input of motion of the
whole environment relative to the observer is mostly considered as
equal to the input of movement of the observer relative to the
environment. Experimental data, however, do not support this
equivalence as to spatial performance.

6.4.1   About the Reduction of Movement Parallax to Motion Parallax

Movement parallax causing space perception was first described by von
Helmholtz (Gibson, Gibson, Smith, & Flock, 1959). He postulated that a
difference in angular velocity between the projections of objects on the
retina is an indication of the perception of distances between objects and
between the observer and an object. Typical experiments were designed
so that the subject could not move, thus reducing movement parallax to
motion parallax. This had been done partly for reasons of convenience—
it was sometimes easier to simulate movement of the display than to
allow the head to move—and partly to eliminate questions about the
role of proprioceptive information concerning head movements
(Sedgwick, 1986, p. 45).

At the same time a distinction was drawn between the role of
movement parallax in absolute and relative distance estimation, with the
emphasis being placed on estimation of relative distances. Using
Howard–Dolman types of setups, it was shown that motion parallax
provides information about the relative distances of points or surfaces
around the subject, but it cannot specify absolute distances or surfaces.
A human or an animal may be able to solve this problem by scaling its
visual input through head or body movement of a fixed velocity (Bruce
& Green, 1990, pp. 259–260), thus using movement parallax instead of
motion parallax. One such animal is the locust which makes side-to-side
swaying movements with its head before jumping from one surface to

another. Wallace (1959) showed that the force of a jump is controlled by distance information obtained during the swaying movement; when he moved the target surface in time with the locust's head, but in the opposite direction, the insects consistently jumped short of the target.

Similar head movements are seen in other animals, gerbils, for instance (Ellard, Goodale, & Timney, 1984), and have been proven to be used for comparable behavior. A well-known example of a stereotyped head movement is the head bobbing of many bird species, including doves, pigeons, and chickens. During walking, the head moves backward and forward relative to the body, and there is a brief "hold" phase in each cycle in which the head is almost stationary to the surroundings, followed by a "thrust" phase in which the head moves forward more quickly than the body. It is known that head bobbing in doves and pigeons is controlled by optic flow produced by walking (Davies & Green, 1988; Friedman, 1975; Frost, 1978;), but its possible significance for vision is not fully understood. The rapid forward movement of the head during the thrust phase will increase the velocities of texture patches over the retina and so "amplify" relative motion between them. A function of head bobbing may therefore be to aid the detection of small objects on the ground (Bruce & Green, 1990, pp. 259–260).

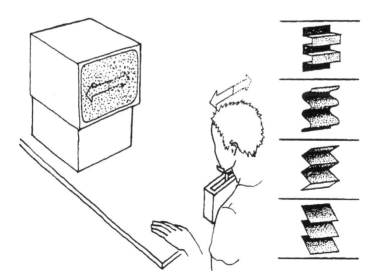

Figure 6.5. *Experimental setup of the Rogers and Graham (1979) experiment.*

Head movements by human beings might be useful to their visual perception as well. Rogers and Graham (1979; Graham & Rogers, 1982) built a setup in which movements of the subject's head caused shifts on a screen (Figure 6.5), corresponding to the projection of different wave patterns (square wave, sine wave, triangular wave, or sawtooth wave) on the screen. When the subject kept his or her head still the subject saw a random dot pattern. However, if the subject moved he or she had a convincing and unambiguous shape impression of the different patterns, seen from different angles. The phenomenon only occurred when the pattern on the screen was shifted according to the movements of the observer.

Perceived depth was more pronounced than in an analogous experiment in which motion parallax was used. In this experiment the observer's head remained fixed, and the oscilloscope moved back and forth with internal depth in the display specified by motion gradients tied to the movement of the oscilloscope. This demonstrates the effectiveness of movement parallax alone in producing vivid perceptions of surface layout in depth and suggests that previous failures to find such pronounced effects have been due to limitations in experimental displays used to simulate movement parallax rather than to limitations in the visual system's ability to make use of it (cf. Sedgwick, 1986, pp. 21–46).

6.4.2   About the Importance of the Task

Whether motion is sufficient for adequate spatial behavior or movement is required depends on the visual task. For some perceptual tasks, shifts in the optic array can be reduced to shifts on the retina. This occurs in those tasks in which two dimensional form perception is at stake without shape nor depth perception being necessary. For this kind of task, the movement of the observer is irrelevant. Think of the recognition of a bird high in the sky (as in Figure 6.1), or the recognition of a landmark deep down when one looks out of an airplane, or a graphic task. This type of task can easily be simulated on a computer display, with standard animation software and videoprocessor hardware. The screen is perceived as a flat screen.

Yet, most perceptual behavior, all manipulation tasks for instance, implies shifts *in* the optic array[7]. Different parts in the optic array shift relative to each other as well as to the observer. Those kinds of tasks are concerned with shape and depth perception or, in other words, with the perception of a rigid world. For them the visual input cannot be

---

[7]Flow on the retina or in the image does not necessarily copy flow in the array (Torrey, 1985). How the failure to distinguish flow in the optic array from flow on the retina can result in confusion is clearly stated by Owen (1990, pp. 44–47).

simulated in the same way as for the former tasks because the exploratory movements of the observer causing the serial transformation in optic flow are essential. Simulating a virtual world in which the visual input forms a rigid whole with the perception–action system of the observer, which allows for telemanipulation, requires the movements of the observer to be coupled to the shifts in his or her optic array. This can be realized on an ordinary television screen if the video camera in the teleenvironment is steered by the head movements of the observer. This creates a monocular virtual window display. There is no need for stereo images. As long as movement (head movements of the operator) and motion (shifts on the display) are consistent, there is no simulator sickness.

This second type of task calls for experiments with exploring actors supplementing those with passive observers, leading to active psychophysics, supplementing the traditional passive psychophysical methods (Warren, 1990). A third implication when studying the second kind of task is that we think it might be necessary to extend the models of optic flow so as to include the movements of the observer. This is explained next.

6.4.3   About Shape Through Motion Algorithms

Once the difference in spatial performance as a consequence of movement as compared to motion has been established, the next question is to ask about the algorithms and the physiological bases that might underlie or support this difference.

6.4.3.1   Optic Flow Models

The concepts of optic array and optic flow are widely embraced by the scientific community, both by traditional and ecological theorists. They have been the subject of several formalized mathematical models (e.g., Koenderink, 1986, 1990; Koenderink & Van Doorn, 1976, 1981; Longuet-Higgins, 1984; Longuet-Higgins & Pradzny, 1980) and played a part in stimulating research on physiological mechanisms involved in computing optical flow. Recent results of both physiological and computational research have led to the development of models that describe the separation of the perceptual input into object motion and observer movement (e.g., Andersen, 1990; Wertheim, 1990).

Two types of optic flow model, or two distinguishable stages in the analysis of optic flow, can be found in the literature, corresponding to the two types of tasks mentioned previously. One is mainly used in computer vision context and considers the optic array as a two-dimensional map of light intensities. Most work on this model concerns the analysis of sequences of video images, trying to recover object outlines by processing the image (e.g., Marr, 1982; Ullman, 1979), and identifying displacements of "landmarks." The other mathematical model takes these landmarks as given and considers the optic array as a two-dimensional set of directions to landmarks and the optic flow as changes in those directions. The model is used to recover shape and position of objects from these changes in direction (e.g., Clocksin, 1980; Koenderink, 1986; Koenderink & van Doorn, 1975; Pradzny, 1980).

Optic flow research efforts have to do with the role of the simplest variable of optic flow—optical velocity—and with that of the simplest variable of the layout of the environment—distance. The role of both is assumed to be very important in computing optic flow properties. The idea behind this is that, optic flow being a transformation of the optic array, the first thing we have to do is to define a velocity field—an array of measurements of image velocity in each small region of the image— which can in turn be used as input to other processes computing higher level properties of image motion. And then motion can be used to compute distances to define the layout of the environment[8]. Yet, the statement that "you know the layout of the environment if you know the distances to the landmarks" is at the same time a truism (for physicists) and completely misleading (when studying the perceptual process) (Koenderink, 1990).

Although we know that velocity and distance information is detected by the visual system, it does not seem to be necessary for optic flow properties to be computed from a velocity field (Bruce & Green, 1990; Koenderink, 1986), and it is probably never the case that we detect shape (a distance order within a limited cluster of coherent features) on the basis of distances (Koenderink,1990). Yet, the latter is almost invariably what many "shape from motion algorithms" from robotics and artificial intelligence (AI) do! They take a number of measurements, grind awhile, and come up with a set of distances for the landmarks that may eventually be combined again to higher order shape descriptors (Koenderink,1990, pp. 61–62).What the visual system seems to be able to do in many cases is to bypass distance estimates completely and instead detect higher order descriptors (like tilt, slant, curvature, etc.) directly (Koenderink, 1990). Lee (1980), for example, demonstrated that the timing approach to objects and surfaces can be obtained independently of an animal's velocity and distance by a more complex parameter of the optic flow field. This parameter, which Lee called *tau*, is the ratio of the time *t* of the distance of any point from the center of an expanding optical pattern to its velocity away from the center. Time-to-contact is likely to be a useful piece of information to animals, for example, to a diving gannet, which has to streamline its wings as it strikes the water. It

---

[8]Andersen (1990) claimed that the goal of a third stage would be the separation of two components: one determining object motion, the other observer motion.

is crucial to the bird's hunting success and survival that its streamlining is accurately timed, and Lee and Reddish (1981) provided evidence that this was achieved by detecting *tau*[9].

One important issue is that, in transferring the term ambient optic flow into mathematically formalized models, the concept has failed to retain much of what is new and crucial in Gibson's definition. They appear to be insufficiently developed to treat the visual information available to an active perceiver.

The same pattern of optic flow results from an observer moving while the environment is stationary, as well as from the observer being stationary while the environment moves in the opposite direction. These models are perfectly suited for tasks implying two dimensional form perception, in which the movement of the observer is unimportant. They are not suited for spatial tasks.

6.4.3.2   Motion–Movement Ambiguity of Optic Flow Models

For the models of optic flow to be suited for spatial tasks the motion–movement ambiguity has to be solved[10]. To resolve motion–movement ambiguity (and scale ambiguity as well), it might be necessary to extend optic flow models to include actively controlled movements, and thus, accelerations into the models of optic flow. This points to a kinetic concept of optic flow rather than a merely kinematic one. In kinematics spatial variables, such as distances, velocities, and accelerations, are always relative, that is, they are always taken to some reference point. In studying the kinematics of two objects, we can picture either of them at rest and attribute the motion to the other. In a kinetic analysis, however, it is not arbitrary which particle is accelerating, even though distance and velocity are still relative. Coupled with accelerations, mass and inertia enter into the description. The accelerations objects undergo are independent of the frame of reference. An illustration of the difference between kinematics and kinetics is given by the Copernican revolution (although it was not the concern at the time). Earth-centered (Ptolemean) and sun-centered (Copernican) frames of reference are kinematically equivalent, but only the sun-centered system facilitates the explanation of movements through gravity.

---

[9]Lee and Reddish (1981) also described a *tau* strategy for a diving gannet, its velocity increasing throughout its dive as it accelerated under gravity.

[10]There are other ambiguities as well. An example is scale ambiguity. The usual solution presented to the scale ambiguity problem is to point at the intrinsic scaling through eye height (Lee, 1980; Sedgwick, 1980; Warren, 1984).

For kinematic models it does not matter how the displacements of the observation point relative to the environment are created. A kinetic model adds extra constraints, such as the laws of mass and force that restrict possible movements of material bodies from impossible ones. Although the observation point is not a material body, its movements are those of the observer's head, which is. Apart from restricting patterns of motion we can see in our environment, a kinetic model also opens the door to multimodal specification, for example, when proprioceptive information complements optic flow information.

The usual approach to movement patterns such as head bobbing in doves or the undulation components of the human gait is to regard them as nuisance factors, noise on the "ideal" linear gait pattern that have to be filtered out. In the movie industry a whole range of patented expensive and intricate devices are used to stabilize a camera when the cameraman walks (Steadycam devices). Removing the undulations is no sinecure, but anyone who has seen bad homemade films knows that the gait pattern of the cameraman is disturbing to the viewers. But what would annoy the passive in the cinema does not disturb the active observer, who himself generates the flow pattern. The periodic undulation component of head movements in gait may carry information allowing for scaling the visual information in terms of action capabilities (Smets, Overbeeke, & Stappers, 1991; Stappers, 1992).

In summary: Optic flow models, impressive as they are, are only beginning to capture the richness of visual information available to the exploring observer. One way in which the models can be extended is by including the peculiarities of the movements of the observer that generate the optic flow, such as the periodic undulation components of the human gait. If this proves to affect spatial behavior, it can be used to redesign virtual reality systems.

Both issues mentioned point to the fact that the importance of choosing an optic flow model should not be underestimated. It is crucial to our understanding of what constitutes the effective stimulation for the perception–action loop (and points to what needs to be conveyed for low-bandwidth television).

For a complete understanding of the functional meaning of movement parallax, we need to understand both the "how" (the modeling) and the "why" (in a functional sense). Although the "how" has received much attention, the "why" has been relatively neglected: The building of a solid active psychophysics still has to be done. Attention to this "why" cannot only open the door to adjustments of optic flow modeling and virtual reality systems, but also to the design of new perceptual aids.

6.5 Applications: New Visual Aids Using the Delft Virtual Window System[11]

A short history of applications of teleoperation was given in order to situate the applications being developed in our laboratory. We implemented the Delft Virtual Window system to overcome visual handicaps.

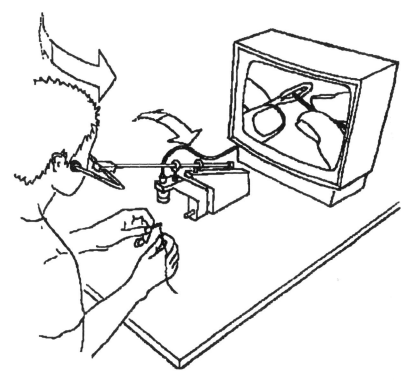

Figure 6.6. *Working principle for a video magnifier facilitating spatial tasks for visually handicapped people.*

---

[11]Needless to say, although we do not elaborate on them, those designs are subjected to extended performance measures and assessment techniques. We are conducting studies that consider the specific mission requirements of both industrial design engineering projects described later. This is being done in cooperation with industry. At the moment there are essentially no accepted standards for asserting the accuracy, repeatability, linearity, and so on, of teleoperating systems in general.

6.5.1. Helping the Visually Impaired

Leifer's (1983) review of the use of telerobotics for the disabled illustrated that most applications are designed for motor disabilities. One (famous) exception is the development of a teleoperated electromechanical guide dog for the blind (Tachi, 1981).

Poorly sighted people can be helped by an optical eye correction by means of spectacles or eye surgery. If this is not possible, they often can read with the aid of a so called video magnifier. This is a system consisting of a camera that registers a text and displays it extremely enlarged and eventually with enhanced contrast. This allows the poorly sighted to read and write. Yet, other tasks, such as sewing, tasks requiring depth perception, and eye–hand coordination, remain unassisted. However, we developed a hybrid of the video magnifier and the Delft Virtual Window system as depicted in Figure 6.6. Head movements steer the camera. The images displayed on the monitor are enlarged and with enhanced contrast. Exploratory research with a working prototype proved that this new device allows for reliable depth perception and manipulation as compared to a real-life setup as well as a comparable setup with a static camera, although it is based on monovision. Despite its limited range of action, it is seen as a striking and relevant perceptual aid by the visually impaired.

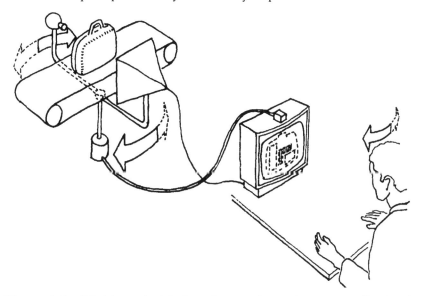

Figure 6.7. *Working principle for a luggage inspection system allowing for full depth perception.*

6.5.2   Luggage Inspection

An especially exciting challenge is to develop a luggage inspection teleoperator. For security reasons, the inspection of personal and freight luggage becomes more and more important. Part of this control consists of inspecting luggage with an x-ray camera. Luggage conveyed through such security devices can only be inspected from one viewpoint, that of the camera. Nevertheless, the safety attendant often would like to look in and around every hole and corner to solve perceptual ambiguities. A pistol, for instance, can be easily confounded with a lot of other objects when only looked at from one point of view. If the x-ray camera is steered by the head movements of the attendant, perceptual ambiguities disappear and depth perception is provided for. The working principle of this luggage inspection device, that can be used for freight as well as for personal luggage, is sketched in Figure 6.7. The security attendant can do the inspection task nearby or at a safe distance.

The advantages of this security system as compared to the existing security systems are greatly acclaimed by Amsterdam Airport's security service. Product development is under way.

6.6   Conclusions

What happened to the director of the transport company? I told him this story. As far as television was concerned, we could resolve his problems using the ecological theory of perception.

Research with and on teleoperators persuades us that the principles of the ecological theory of perception offer more than just a powerful engineering paradigm and also demonstrates that this applied research can help to clarify the nature of perception itself. For example, because telepresence is commonly claimed to be important for direct telemanipulation, a theory of presence/telepresence is sorely needed (Sheridan, 1989b). We think that the ecological theory of perception might be of help here, by explaining the difference between motion versus movement parallax as to their perceptual meaning, that is, as to their importance in allowing for adequate spatial behavior. We are talking about telepresence, yet, at the same time, we are, tentatively, trying to understand presence. How is it that we feel ourselves to be present, here and now?

Acknowledgments

Work mentioned in this chapter was supported by almost the entire staff of the laboratory of Form Theory at the Department of Industrial Design Engineering at Delft University of Technology. This is a multidisciplinary team with industrial design engineers, psychologists, and physicists working together to build an interdisciplinary approach to perception. I specifically want to mention and thank C. J. Overbeeke, P. J. Stappers, O. A. van Nierop, R. Wormgoor, A. M. Willemen, G. W. H. A. Mansveld, H. Subroto, A. C. M. Blankendaal, and J. P. Claessen.

Furthermore, I want to thank Bill Gaver who was, in 1992, a visiting lecturer in Delft, and with whom I discussed this chapter in full detail. I really enjoyed those discussions.

6.7   References

Andersen, G. J. (1990). Segregation of optic flow into object and self-motion components: Foundations for a general model. In R. Warren & A. H. Wertheim (Eds.), *Resources for ecological psychology* (pp. 127–141). Hillsdale, NJ: Lawrence Erlbaum Associates.

Bruce, V., & Green, P. (1990). *Visual perception: Physiology, psychology, and ecology.* London: Lawrence Erlbaum Associates.

Clocksin, W. F. (1980). Perception of surface slant and edge labels from optical flow: A computational approach. *Perception, 9,* 253–271.

Cutting J. E. (1986). *Perception with an eye for motion.* Cambridge, MA: MIT Press.

Davies, M. N. O., & Green, P. R. (1988). Head bobbing during walking, running and flying: Relative motion perception in the pigeon. *Journal of Experimental Biology, 138,* 71–91.

Ellard, C. G., Goodale, M. A., & Timney, B. (1984). Distance estimation in the Mongolian gerbil: The role of dynamic depth cues. *Behavioral Brain Research, 14,* 29–39.

Friedman, M. B. (1975). Visual control of head movements during avian locomotion. *Nature, 255,* 67–69.

Frost, B. J. (1978). The optokinetic basis of head-bobbing in the pigeon. *Journal of Experimental Biology, 74,* 187–195.

Gibson, J. J. (1962). Observations on active touch. *Psychological Review, 69,* 477–491.

Gibson, J. J. (1966). *The senses considered as perceptual systems.* Boston: Houghton Mifflin.

Gibson, J. J. (1979). *The ecological approach to visual perception.* Hillsdale, NJ: Lawrence Erlbaum Associates.

Gibson, E. J., Gibson, J. J., Smith, O. W., & Flock, H. (1959). Motion parallax as a determinant of perceived depth. *Journal of Experimental Psychology, 58,* 40–51.

Graham, M., & Rogers, B. (1982). Simultanuous and successive contrast effects of depth from motion-parallax and stereoscopic information. *Perception, 11,* 247–262.

Kobrinskii, A. (1960). The thought controls the machine: Development of a bioelectrical prosthesis. *Proceedings of the First International*

Federation on Automatic Control World Congress on Automatic Control. Moscow.

Koenderink, J. J. (1986). Optic flow. *Vision Research, 26* (1), 161–180.

Koenderink, J. J. (1990). Some theoretical aspects of optic flow. In R. Warren & A. H. Wertheim (Eds.), *Resources for ecological psychology* (pp. 53–68). Hillsdale, NJ: Lawrence Erlbaum Association.

Koenderink, J. J., & van Doorn, A. J. (1975). Invariant properties of the motion parallax field due to the movement of rigid bodies relative to an observer. *Optica Acta, 22* (9), 773–791.

Koenderink, J. J., & van Doorn, A. J. (1976). Local structure of movement parallax of the plane. *Journal of the Optical Society of America, 66,* 717–723.

Koenderink, J. J., & van Doorn, A. J. (1981). Exterospecific component of the motion parallax field. *Journal of the Optical Society of America, 71* (8), 953–957.

Lee, D. N. (1976). A theory of visual control of braking based on information about time to contact. *Perception, 5,* 437–459.

Lee, D. N. (1980). Visuo-motor coordination in space-time. In G. E. Stelmach & J. Requin (Eds.), *Tutorials in motor behavior* (pp. 281–293). Amsterdam: North-Holland.

Lee, D. N., & Reddish, P. E. (1981). Plummeting gannets: A paradigm of ecological optics. *Nature, 293,* 293–294.

Leifer , L. (1983) Interactive robotic manipulation for the disabled. *Proceedings of the 26th IEEE Computer Society International Conference COMPCON* (pp. 46–49). San Francisco.

Longuet-Higgins, H. C. (1984). The visual ambiguity of a moving plane. *Proceedings of the Royal Society of London, B 223,* 165–175.

Longuet-Higgins, H. C., & Pradzny, K. (1980). The interpretation of a moving retinal image. *Proceedings of the Royal Society of London, B 208,* 385–397.

Mace, W. M. (1977). James J. Gibson's strategy for perceiving: Ask not what what's inside your head, but what your head's inside of. In R. Shaw & J. Bransford (Eds.), *Perceiving, acting, and knowing* (pp. 43–65). Hillsdale, NJ: Lawrence Erlbaum Associates.

Marr, D. (1982). *Vision.* San Francisco: Freeman.

Norman, D. A. (1988). *The psychology of everyday things.* New York: Basic Books.

Overbeeke, C. J., & Stratmann, M. H. (1988). *Space through movement.* Unpublished doctoral thesis, Delft University of Technology, Delft, The Netherlands.

Overbeeke, C. J., Smets, G. J. F., & Stratmann, M. H. (1987). Depth on a flat screen II. *Perceptual and Motor Skills, 65,* 120.

Overbeeke, C. J., & Smets, G. J. F. (1990, June 7-8). Spatial versus temporal resolution in pattern recognition: Theoretical aspects and technological implications. *Abstracts of the First European Workshop on Ecological Psychology.* Marseille, France.

Owen, D. H. (1990). Lexicon of terms for the perception and control of self-motion and orientation. In R. Warren & A.H. Wertheim (Eds.), *Resources for ecological psychology* (pp. 33–50). Hillsdale, NJ: Lawrence Erlbaum Associates.

Pepper, R. L., Cole, R. E., & Spain, E. H. (1983). The influence of camera separation and head movements on perceptual performance under direct and TV-displayed conditions. *Proceedings of the Society for Information Display, 24/1,* 73–80.

Pradzny, K. (1980). Egomotion and relative depth map from optical flow. *Biological Cybernetics, 36,* 87–102.

Proshansky, H., Ittleson, W., & Rivlin, L. (1970). *Environmental psychology: Man and his physical setting.* New York: Holt, Rinehart and Winston.

Rogers, B., & Graham, M. (1979). Motion parallax as an independent cue for depth perception. *Perception, 6,* 125–134.

Sedgwick, H. A. (1980). The geometry of spatial layout in pictorial representation. In M. Hagen (Ed.), *The perception of pictures* (Vol 1, pp. 33–90). New York: Academic Press.

Sedgwick, H.A. (1986). Space perception. In K.H. Boff, L. Kaufman & J.P Thomas (Eds.), *Handbook of perception and human-performance* (Vol., pp. 21.1–21.57). New York: Wiley.

Sheridan, T. B. (1989a). Telerobotics. *Automatica, 25* (4), 487–507.

Sheridan, T. B. (1989b). Merging mind & machine. *Technology Review, 92* (7), 32–40.

Smets, G. J. F., & Overbeeke, C. J. (1992). Trading off spatial against temporal resolution: The importance of active exploration. *Proceedings of the Second International Conference on Visual Search* (pp. 387–397). Durham, NC.

Smets, G. J. F., & Overbeeke, C. J. (1991). *Trading off spatial against temporal resolution: Replication experiment* (Internal Rep.). Delft, The Netherlands: Delft University of Technology.

Smets, G. J. F., Overbeeke, C. J., & Stappers, P. J. (1991). *The ecology of human-machine systems III. The importance of actively controlled movement: Toward richer operationalizations of optic array & optic flow* (Internal Rep.). Delft, The Netherlands: Delft University of Technology.

Smets, G. J. F., Overbeeke, C. J., & Stratmann, M. H. (1987). Depth on a flat screen. *Perceptual and Motor Skills, 64,* 1023–1024.

Smets, G. J. F., Stratmann, M. H., & Overbeeke, C. J. (1988). Method of causing an observer to get a three-dimensional impression from a two-dimensional representation. *U.S. Patent 4,757,380.*

Smets, G. J. F., Stratmann, M. H., & Overbeeke, C. J. (1990) Method of causing an observer to get a three-dimensional impression from a two-dimensional representation. *European Patent 0189233.*

Stappers, P.J. (1992). *Scaling the visual consequences of active head movements: A study of active perceivers and spatial technology.* Unpublished doctoral thesis, Delft University of Technology, Delft, The Netherlands.

Tachi, S. (1981). Guide dog robot, feasibility experiments with MELDOG Mark 3. *Proceedings off the 11th International Symposium on Industrial Robots* (pp. 95–102).

Torrey, C. (1985). Visual flow and the direction of locomotion. *Science, 227,* 1064.

Ullman, S. (1979). *The interpretation of visual motion.* Cambridge, MA: MIT Press.

Vertut, J., & Coiffet, P. (1984). *Robot technology. Vol. 3a: Teleoperations and robotics, evolution and development.* Englewood Cliffs, NJ: Prentice-Hall.

Wallace, G.K. (1959). Visual scanning in the dessert locust *Schistocerca gregaria. Journal of Experimental Biology, 36,* 512–525.

Warren, W. H. (1984). Perceiving affordances: Visual guidance of stair climbing. *Journal of Experimental Psychology: Human Perception & Performance, 10,* 683–703.

Warren, R. (1990). Preliminary questions for the study of self-motion. In R. Warren & A. H. Wertheim (Eds.), *Resources for ecological psychology* (pp. 3–32). Hillsdale, NJ: Lawrence Erlbaum Associates.

Wertheim, A. H. (1990). Visual, vestibular, and oculomotor interactions in the perception of object motion during egomotion. In R. Warren & A. H. Wertheim (Eds.), *Resources for ecological psychology* (pp. 171–217). Hillsdale, NJ: Lawrence Erlbaum Associates.

Woodworth, R. S., & Schlosberg, H. (1962). *Experimental psychology.* New York: Holt, Rinehart and Winston.

Chapter 7

Representation of System Invariants by Optical Invariants in Configural Displays for Process Control

John Paulin Hansen

Cognitive Systems Group
Risø National Laboratory
Roskilde, Denmark

7.0   Introduction

The supervisory tasks in modern, high-technology systems are multidimensional, requiring coordinated manual and automatic regulation of interdependent parameters. If the operator is only supported with "one measurement–one indication" displays (Goodstein, 1981), coordination will depend on his or her ability to understand the multivariability of the system, the interdependencies of the parameters, and the interactions between manual and automatic controls. Because even minor malfunctions are potentially hazardous, the operator's understanding must develop without trial-and-error learning. This seems to be a very difficult learning task and, given the inevitable complexity of the system dynamics, the solution to the control problem is to structure the individual data as goal-relevant information. Whenever data visually combine as a group with the new quality relating directly to a meaningful task in the environment, it is more than a simple sum of data and is called *information* (cf. Pomerantz, 1986). The central design issue of this chapter is how to combine data on graphical displays, instead of forcing the operator into a mental integration. By which principles can the combination be done, and what kinds of combinations are particularly perceptually salient?

7.1   System Engineering and Ecological Psychology

In order to integrate the individual data in a meaningful way, the ecological setting of the control task and the essential information must be identified prior to any attempt at representation. As an example, the information needed to land an airplane has been specified in such an ingenious way that it has become possible to represent this information with a singular arrangement of landing lights for night flights. Landing lights potentially reduce information compared to the information in the optical array reflected from a runway in daylight, but it is sufficient functional information for the control task of landing. In general terms, the arrangement of landing lights enhances the relevant goals and constraints for the task by providing the pilot with real-time, spatial and dynamic error signals. This analogy synthesizes the central idea of ecological interface design: to make the invisible goal-relevant constraints in the interior structure of the work domain visible by direct specification, using a homeomorphic, real-time mapping from the domain onto the interface (Rasmussen & Vicente, 1989, 1990; Vicente & Rasmussen, 1990).

Gibson (1979) used "invariants" to refer to the properties of the optical array the perceptual system may attune to. Hereby, he meant permanent physical structures that normally remain the same across transformations. In the same sense, technical systems possess invariants given by their natural laws. An example is the thermodynamic conservation of energy in a power plant.

Normally, the attunement to task relevant invariants happens by encountering the constraints on successful behavior. But because high-technology systems do not allow for a free trial-and-error exploration of the system constraints (except, e.g., on computer simulations), the invariants may remain unexploited as they cannot be related to the specific constraints. A solution to this problem is to map the invariants of the technical process onto invariant optical structures of an interface that make the task constraints visible.

The identification of the constraints must be based on analyses by system engineers. In this way, the construction of ecological interfaces becomes a thoroughly interdisciplinary enterprise: Good display design ensures that the optical invariants of the representation, specified by perceptual psychologists, capture the actual constraints of its ecology, specified by system engineers.

As an example of ecology-related information within a technical domain, Rasmussen (1986) specified the various levels of control tasks in a power plant and the contents of an information representation necessary for carrying out the tasks (Rasmussen & Goodstein 1988; see Table 7.1. Information from all the levels of this means-ends analysis are needed in order to control complex thermodynamic systems, especially in rare or unanticipated situations (cf. Vicente, 1991). The information to be represented for the operator describes the functionality of the system as a nested set of constraints deriving from the physical laws that govern the process.

Means-Ends Levels of Description	Typical Control Tasks in Power Plants	The Contents of Information Representations
Goals, purposes, and constraints.	Monitor production and safety specifications of the customers.	Functional purposes and goals in terms of *energy, material flow, and distribution*
Abstract function; flow of mass, energy, and information	Control of the flow of energy through the plant from source to electrical grid; monitor major mass and energy balances for plant protection.	Mass, energy flows and balances in terms of *underlying generic functions*
General functions	Monitor and control individual functions such as coolant circulation, steam generation, power conversion from steam to electricity	Information on cooling, heat transfer, regulations in terms of *connected pieces of equipment*
Physical process of equipment and components	Adjust process parameters in order to align operational states of components and equipment to match requirements and limitations	Performance data on physical equipment in terms of *information on major components.*
Form, location, and configuration of equipment	Connect and disconnect components; change anatomy and configuration of equipment and installations to match requirements of physical processes and activities	Installation, maintenance information on components: *take a part diagrams, illustrations*

Table 7.1. *Levels of tasks and contents of a representation for power plant control.*

Although this framework has proven to be useful for specifying the information content and structure in several complex work domains (Rasmussen, 1988), this chapter addresses the general question of a suitable information form for revealing the constraints.

7.2   The Representation of System Invariants by Optical Invariants

The graphical design of Interfaces for complex systems is a great challenge to ecological psychology, because the theory of direct perception of the natural environment in which humans have developed their perceptual skills needs to be reframed in terms that can serve as guidelines for the construction of mediated information. Reframing the "ecological laws of surfaces" (Gibson 1979, pp. 23–24) in light of their applications to computer interfaces gives some indication of the potentials of surface invariants for the interface representation of abstract, functional system invariants. The fact that all natural solids have surfaces, and all surfaces have a layout, constitutes the fundamental possibility for surface invariants to be natural representatives for an abstract substance. The surface resistance to deformation and disintegration, depending on its viscosity and cohesion, gives a variety of representational possibilities by individual systematic movements, deformations, and disintegrations, provided that the changes follow the laws of natural, ambient optic arrays that specify, for example, perspective, dynamic occlusion (Gibson, 1979) and time-to-contact (Lee, 1980). As a well-known example, the desktop metaphor for Macintosh computers shows the abstract, computational function of closing a file as a gradual shrinking of its frame until it vanishes in the folder containing it.

The perception of the persisting identity and changes of things are fundamental to the perception of individual stimuli qualities (Gibson, 1979, pp. 246–248). What is needed to guide the design of displays for complex dynamic systems is an approach that starts out by giving the highest priority to the visualization of persisting domain constraints and the invariants in system changes. Here, the term *changed* is used as defined by Gibson: "change means to become different but not to be converted into something else,...the saying emphasizes the fact that whatever is invariant is more evident with change than it would be without change" (p. 73).

As one of the earliest examples of a display that utilizes graphical integration in order to support the perception of system invariants, Coekin (1970) suggested the display of eight analogue signals as an octagon (i.e., an eight-pointed star). The idea was to utilize the human ability for form perception and recognition found among operators interpreting sonar displays. The scales in the octagon were arranged so that normal values of the parameters fell on the circumference of a "mean-circle" and abnormalities showed up as deviations from that form. This presentation form had several applications within technical supervisory control (Goodstein, 1981; Woods & Roth, 1988).

In a display for medical monitoring, Goldwyn, Farrell, Friedman, Miller, and Siegel (1973) suggested a similar 11-dimensional object. They commented that "the ultimate goal is to reduce a large body of complex physiologic data to an information base that is relatively small and simple so that abnormal patterns may be exposed in a manner that can be directly interpreted" (p. 230).

Such a "direct interpretation" is possible because the polygon provides a visual integration of indications on the physiological parameters that will facilitate the perception of "the state of health." This is achieved by mapping persistence and changes onto preservation and disturbance of the optic array on the monitor. Thereby, it allows for a direct perception of disturbances as symmetry breaking and focuses further interpretation on the deviating parameter(s). In addition, it may support the discovery of possible interrelations among the changing parameters as particular shapes associated with particular types of traumas. The invisible, invariant relations in process control need a similar representational form that will allow the operator to apprehend the abstract system relations by the highly developed intelligence of human object and space perception. This point might seem obvious but, based on all the numerical and textual performance data found in conventional control interfaces (see, e.g., Goodstein, 1981), it is not.

Some designers of new process displays do recognize the importance of graphical information on system functions and behavior. For instance, the designers of the "Steamer" interface (Hollan, Hutchins, & Weitzman, 1984) wrote:

> Dynamic systems are particularly difficult to explain in language. However, relationships that are difficult to describe unambiguously in language are often easily depicted graphically. Putting a layer of interface computation between a user and a quantitative model provides a qualitative view of the underlying model. Such a qualitative graphical interface can operate as a continuous explanation of the behavior of the system being modeled by allowing a user to more directly apprehend the relationships that are typically described by experts. (p. 20).

7.3   Configural Interfaces

Interfaces that use dynamic surface geometry as their main representation of event data can be defined as *configural*. They consist of shapes that are changed according to incoming data.

Ecological interface design gives the highest priority to the visualization of preservation and disturbance in order to capture the distinctive, functional process invariants by mapping the goal related persistence and system changes over to the distinctive identities and changes of the form geometry. This overall intention makes the ecological interface different from mimic interfaces and metaphorical interfaces, although ecological interfaces may contain both mimic and metaphorical elements as well.

A *mimic* display is an interface in which the geometry of the representation has a direct resemblance to physical features of the work domain. It is always an abstraction of the work space, as it is never as rich in detail as the actual work space. The scaled diagrams of a process control plant, for example, piping and instrumentation diagrams, will generally not preserve the exact relations of the spatial layout of the components, and the components are often displayed by icons that only display the distinctive silhouette of a turbine or storage tank for example. Over the course of time, some of the objects on mimic displays have become purely conventional symbols that bear almost no similarity to the actual component form, but refer to the physical process represented by the component (cf. Table 7.1). As an example, a valve is often depicted as two symmetrical triangles with a silhouette of a tunable regulator at the axis. Hereby, the representation moves one level up in the abstraction hierarchy from the form/location to the physical process involved. In contrast to a configural display that represents the constraints related to system functions and changes its surface form according to the functional flow state, the valve symbol always looks as if it is permanently stopping the flow, even though the regulator in the middle symbolizes the variability of the obstruction (which is, in fact, the general function of a valve). If data on the present flow are provided, they are often in a purely numerical form.

By mapping functional invariants onto geometrical properties, the configural display begets an idiom that may look most like an abstract, cubist picture to the naive observer (see Figure 7.1 for an example). This association highlights important features that configural displays have in common with abstract paintings. Hester (1977) argued that recognition and identification in modern art are concerned with very general schemata of shapes, forms, planes, and lines in the picture space. The painting thereby activates a general knowledge of categories of things instead of activating a genuine identification and categorization related to the knowledge of particulars and types. In the same sense, it is the purpose of a configural interface to activate and generate general knowledge on the abstract invariants in the thermodynamic laws. The operator then gets an aid to discover the system behavior by his or her pattern recognition capabilities that would have been difficult to apply on a fragmented mimic representation of the system state at the level of physical form. This is not to say that mimic representations should be avoided; indeed, they are most important for several tasks (see Table 7.1) and should therefore be represented as well, but nested within the configural form.

The use of particular metaphors in displays will activate knowledge of particulars and types foreign to the work domain. The purpose of, for example, the desktop metaphor is to increase the initial familiarity with actions, procedures, and concepts in office information systems by letting them appear as well-known objects and actions. The objects of a metaphorical definition are what Lakoff and Johnson (1980) called "natural kinds of experience" (p. 117). Hereby, they meant experiences that are a product of our bodies, our interaction with the physical environment, or our interaction with other people.

A specific affordance (Gibson, 1979) of an object is always specified by the object's effectivity in an action context. A hammer normally affords amplifying the blow, but it can also be used as a hook, if one wants to reach out for a box of nails. When a particular object is used as a metaphor for computerized "tools," it does not retain the functional complexity and multiple affordances it can have in natural experiences. A metaphorical representation of a delete function by a rubber icon might be used to erase a line, but—much to the user's surprise—not to delete a file name.

So, ironically, the force of the pictorial metaphor—that it imports particular understandings from natural experiences—is also its weakness: These understandings smuggle in the multiplicity of effective actions the object makes possible in natural experiences. In complex, technical domains, in which the possible system states cannot be foreseen, novel reasoning using imprecise metaphorical symbols can be catastrophic. Therefore, it does not seem like a good idea to reframe control tasks into, for example, computer game metaphors, as proposed by Carroll and Thomas (1980).

Configural displays attempt to map the functional structure from the target domain onto the geometry of abstract forms. In a sense, their metaphorical domain is simply geometry. Lakoff and Johnson (1980) recognized the overall principle of transforming abstract relations into figures with properties such as spatial orientation, entity, substance, and containment as an important part of the prelinguistic natural intelligence that governs the formation of metaphorical transformations in language:

> *Spatial orientations like up-down, front back, on-off, center-periphery, and near-far provide an extraordinarily rich basis for understanding concepts in orientational terms. . . .Our experience of physical objects and substances provides a further basis for understanding—one that goes beyond mere orientation. Understanding our experiences in terms of objects and substances allows us to pick out parts of our experience and treat them as discreet entities or substances of a uniform kind. Once we can identify our experiences as entities or substances, we can refer to them, categorize them, group them, and quantify them—and, by this means, reason about them. (p. 25)*

Although particular, pictorial metaphors associate to a foreign domain object, the abstract space and object metaphors in configural displays are supposed to be almost neutral in their generality. It is assumed that their lack of specific connotations makes a direct mapping to the work domain events possible, as they do not demand considerations about the interfering objects in the mapping process itself. Metaphorical objects have one-to-many mappings from their representation to natural experiences, but only one (or a few) correct mapping(s) back to the target domain. So the interpretation of inherently imprecise particular metaphors may force the operator to consider their intended meaning in order to map them back from the foreign domain to the plant situation. Even if the operator gains a high familiarity with the imprecise nature of particular metaphors, he or she may still have to suppress some of the obvious, but "false" associations. In short, the less the load from foreign meaning, the higher is the likelihood of a direct mapping. A configural representation that does not look at all familiar is most likely to be associated with nothing but the real-world events that make it change.

It is important to recognize that, although not intended, abstract configural interfaces may activate intuitive idiosyncratic interpretations because they rely on human perception skills that have developed in the natural environment. Their relation to natural experiences may first become clear when a configural interface representation violates the laws of the four dimensional Newtonian space–time world (cf. Winograd & Flores [1986] notion of "breakdowns" and Alexander's [1964] notion of "misfits"). For instance, a display that ends up representing mass increase by a shrinking form or a particular temperature increase by a form moving downward will most likely be counterintuitive. Unfortunately, the precise nature of these common, governing intuitions is not known in detail. Lakoff and Johnson (1980) pointed at some important features of natural intelligence, and some research has been concerned with the naive physical interpretations of the abstract functionality of (simple) mechanical systems (diSessa, 1983; Gentner & Gentner, 1983; Kleer & Brown, 1983; McCloskey, 1983). But there is no general "basic cognitive semantics" available that can be used directly as a design guideline for the selection of appropriate geometrical representations. Moreover, chemical reactions, for example, will not always follow the laws of material kinematics, but may follow nonlinear and higher order laws. So, at some point, a dynamic representation may have to make exceptions from the Newtonian laws anyway. In fact, as these exceptions presumably are difficult to grasp intuitively, they should be clearly emphasized in the interface, making it educational as well as perceptually salient.

A partial solution to these problems, suggested in the next section, is to choose a highly restricted set of simple geometrical principles as a graphical syntax that creates a coherent space–time world following its own underlying geometrical relations. By recurring use of a few principles, it is intended to create a visual language that only uses graphical features that (in principle) can be tested for their individual perceptual salience and connotations. Even so, it is not believed that the visual complexity of large scale system representations and the specific connotations of operators in an actual work domain can be foreseen. This makes the interface development intimately linked to the cognitive task analysis and points at the inevitable need for user feedback in an iterating system development process (see, e.g., Gould & Lewis, 1985).

7.4   The Nesting of Information

The structuring principles for display surface invariants must facilitate the pickup of information at the individual levels of control (see Table 7.1), without violating the possibility for decomposition and aggregation of information between the levels. According to Gibson (1979), direct perception of the natural environment possesses the quality of utilizing a lot of nested information that forms a kind of hierarchy without categorical bounds. This is exactly the feature that should be achieved when multilevel information is displayed simultaneously in ecological interfaces (cf. Flach, 1988, 1990; Gaver, 1991).

In the following section, some general, graphical heuristics for the nesting of information by display geometries are suggested. The principles are inspired from well-known illustration principles (see Cleveland, 1985; Tufte, 1983, 1990) and modified to capture the specific values within process control (cf. Table 7.1):

1. Goals as figural goodness: *Goodness* is a very vague term from Gestalt psychology (Zusne, 1970), but there might not be a better word for the aesthetic quality it refers to. Here it implies that when the system is in a goal state, the display representation should show as much integrity, equilibria, linearity, alignment, symmetry, and/or balance as possible within the given format. On the other hand, deviations from the goal state should be clearly visible as, for example, nonlinearity or symmetry breaking. Examples include landing lights and the Coekin star.

2. Constraints as containers: The constraints identified in the task analysis map onto permanent forms of the background. The constraints may range from overall goals (e.g., production goals) to the individual component capacity, for example, a storage tank symbol with an area analogous to its volume capacity.

3. Dynamics as figural changes: On top of this background, the dynamics of the process are mapped to geometrical figures that change according to incoming data. Through the operator's control actions, the dynamic figures are to be kept within the borders of the constraints, for example, mass flow represented as a changing area in a storage tank.

4. Functional relations as connections: Contact in the form of lines, intersections, or coincidenesses are used to indicate a functional connection, for example, piping diagrams indicating possible mass and energy transportation.

5. Pictorial symbols as component representations: The actual physical appearance and configuration of the equipment may be represented by a detailed, mimic form, for example, valve symbols.

6. Alphanumerical signs as additional support: The nesting of text and numbers within the configural format can increase the data resolution and provide means for communication and symbol-based reasoning in novel situations, for example, numerical

information on a pump performance or maintenance information on components.

7. Time as perspective: The gradient of density (Gibson, 1950) can be used to represent temporal changes by mapping time onto depth, for example, the time tunnels (Hansen, 1989) described in section 7.3 of this chapter.

All real world design has its tradeoffs, according to Alexander (1964). So will interface design based on the seven principles just suggested. It is not the individual suggestions (except, maybe, 7) that are really new, but the ambitious aim of: Integrating (1 and 3) and putting them on top of (2), and letting (4) decide the placement of (2) without spoiling the integration of (1 and 3) and making a natural relation/transition between (2 and 5), and show (3) by (7) with (6) added when needed.

This ideal might be impossible to achieve for a given set of invariants and constraints from a control domain. But in at least one complex domain it has been partly possible, namely Lindsay and Staffon's (1988) interface for a nuclear power plant (see Vicente & Rasmussen, 1990, for a detailed description). The next section illustrates the author's attempt to apply these principles.

7.5   An Example of a Configural Display

Figure 7.1 is a sketch of a possible representation of the temperature constraints in a conventional power plant domain that will reveal the invariants in the conservation of energy. It is inspired by an interface developed by Lindsay and Staffon (1988) in which the incoming temperature data, following principle 3, forms a Rankine thermodynamic cycle (the closure figure with thick lines), originally suggested by Beltracchi (1987) as the structuring principle for a Breeder nuclear reactor display. The Rankine cycle shows the thermodynamic relationship between entropy (x axis) and temperature (y axis). It is composed of four general-process functions: (a) heat addition to the water, which results in a phase change from liquid to steam; (b) a reversible expansion of the steam, which does work upon a turbine; (c) a reversible heat-rejection process of condensation, which results in a phase change back to water: and (d) a compression of the water by means of pump work (Beltracchi, 1987).

The thermodynamic cycle is contained within the related constraints (principle 2) of the subsystems that generates it and within a graphical representation of the basic thermodynamic relationship between temperature and entropy for water (the sugar top-like curve). At the left side of the curve, the thick line represents water, within the curve, a two-phase mixture of water and steam, and on the right side, dry steam. Engineering analysis of the thermodynamic laws have clarified that it is critical that burning and pressure are regulated so that water boils within the curve. If this constraint is violated, the boiler may be filled with steam or filled with water. Both cases will make its control unsafe. Saturated steam and saturated water in the boiler are indicative of normal operation, whereas the presence of superheated steam is indicative of abnormal operation (Beltracchi, 1987). Even though operators do not use entropy as a critical variable—actually, the entropy dimension gets dropped in the display implementation (cf. Beltracchi, (1987)—the advantage of the temperature-entropy diagram for display is that constant pressure lines, which is a critical operational goal indicating a normal boiling process, becomes horizontal in the two-phase region in this presentation format. Horizontal lines are an example of figural "goodness" (principle 1) that maps onto a goal state, and the use of the entropy figure is an example of how the constraints identified in analysis of the abstract process functions are made available to the people controlling it (principle 2).

Another advantage of this format is that the area of the Rankine cycle represents the net energy of the system. The area from temperature point 21 to 303 and down to the baseline in the condenser shows the heat energy of liquid water, the area within the curve and down to the baseline shows the energy added by boiling, and the area under the section from the right side of the curve to the top point (576° C) shows the energy added in the superheater. The distance from the top point to the condenser point (46° C) represents the work being carried out by the turbines (cf. Knak, 1982).

The various physical subsystems involved in the energy production is nested as boxes around the Rankine cycle. Besides being a reference to the individual subsystems (preheater, boiler, superheater, turbines, and condenser), the size of the static boxes indicates the upper and lower temperature constraints of the subsystems. Moving down to the level of physical form (cf. Table 7.1) could then be done by choosing one of the boxes for various detailed pictorial component information (blueprints, CAD drawings, video pictures, etc.) following principle 5.

In present control rooms for conventional power plants, the most direct and fastest indication of the burning process itself is provided by the luminescence of the flames. This crucial information is nested inside the Rankine cycle as bar graphs below the boiler temperature indication and on top of the coal mill representation in order to resemble the elemental relationship between fuel feeding, burning, and energy generation. If the flame signals from the burners at one coal mill become weak, as shown at coal mill 3, the most likely cause, namely, a lack of coal or an unbalance in the air–coal masses, can be checked at the coal mill box. Inside this box, the actual coal load is shown as a gray surface, and if the invariant relation between coal and air is not maintained, the area will either fail to meet the borders, indicating that too much air is blown in (as is the case on coal mill 3 in Figure 7.1), or it will exceed the border of the coal mill box, indicating a lack of air for the amount of coal fed (principles 2 and 3). The sketch is only meant to illustrate some possibilities for the nesting of constraints and invariance representation in configural interfaces and is not recommended for any implementation in its present, rudimentary form.

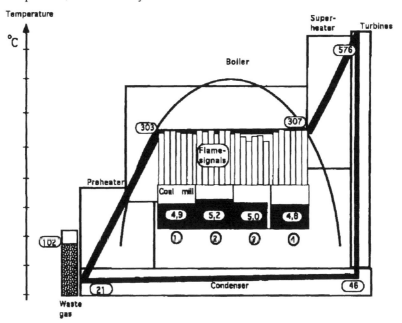

Figure 7.1. *A sketch of a configural display for control of conventional powerplants.*

7.6   General Discussion of the Nesting Principles

This section provides further theoretical and empirical arguments as supporting evidence for the nesting principles suggested in section 7.4, and discusses some practical aspects of their application.

7.6.1   Configural Displays of Abstract Data Relations

The suggested principles of figurality and display of functional relations as spatial contact between interacting elements seem to have a primacy in the ontogenesis of perception. Reviewing investigations in infant perception, Spelke (1990) summarized the general findings: "Infants divide perceptual arrays into units that move as connected wholes, that move separately from one another, that tend to maintain their size and shape over motion, and that tend to act upon each other only on contact" (p. 29).

Applying these principles as visual means by which system states are experienced should insure that they have a fundamental relation to human cognition and can be perceived with well established perceptual skills.

Some of the suggested graphical means such as size, closure, and symmetry were recognized by early Gestalt theory and used as explanations of, for example, the figure–ground phenomena (Zusne, 1970). As they mainly were related to perception of static two-dimensional figures, the importance of pattern changes during events have not been discussed in greater detail within this tradition. In general, the Gestalt tradition has identified some important principles for perceptual integration and illusions, but tended to ignore the close link between form perception and invariants in change. One of the main principles of ecological interface design is to preserve and utilize this link by means of direct visual feedback on all actions (Rasmussen & Vicente, 1989).

According to Pomerantz's (1986) widespread definition, it is some specifiable emergent feature, dependent on the identity and arrangement of the parts (but not identifiable with any single part), that "pops out" and configures parts into wholes. Closure and symmetry are examples of emergent features. In addition to this phenomenological definition of emergent features, he suggested that figures with emergent features are perceived faster in divided attention tasks, whereas all parts have to be taken into account and perceived slower in selective attention tasks, when responses are contingent on some individual parts, as compared to displays without emergent features.

Experiments showed configural displays to be more rapidly perceived than separable displays in data integration tasks. For instance,

Carswell and Wickens (1984) found that a triangular display was perceived approximately 800 msec faster than three separate bar graphs. Recently, Wickens and Andre (1990) compared an area object display with three bar graphs and obtained a reaction time (RT) superiority of the same magnitude.

Sanderson, Flach, Buttigieg, and Casey (1989) demonstrated that it is not the objectness per se that causes faster RT for these kinds of displays, but their exploration of emergent features. They found that three bar graphs, traditionally viewed as a typical example of a separated display, with ascending or descending linearity as an emergent feature in the target situations, was perceived faster in a data integration task than a triangular display. This experiment is particularly interesting because the ascending or descending linearity was not actually shown, but mentally imposed by the subjects themselves as subjective contours. It means that emergent features may arise from invariant higher order cues that map onto a required response, and that the discovery of these cues becomes an important factor for the development of efficient visual scanning strategies.

There seems to be no a priori, formal way to "speed up" the perception of complex interfaces, as their emergent features are partly beyond discretionary control, determined as they are by the representation, the task objective, and the observer. But the fact that emergent features on a given interface phenomenologically "pop out" and thus can be verbalized by the users (cf. Buttigieg, 1989) makes it possible to evaluate if they are at all available and if they match the system goals.

7.6.2   The Importance of Digital Information

One could claim that alphanumerical signs and symbols always are a compensation for the lack of direct specification of the affordances in the visual form. This point has been made within architecture (cf. Warren, 1994), and it applies to process interfaces as well. But alphanumerical signs do have some important qualities as means for rule-based behavior (Rasmussen, 1986); they can be communicated verbally to (remote) collaborating human agents, and they can hold a higher individual data resolution than pictorial representations (Hansen, in press). So, textual and digital signs are not to be totally removed from ecological interfaces, but used as additional information when the task analysis points to the importance of their use.

Several experiments have pointed at the lack of precision in the judgments based on graphs. Brunswik's (1956) classical investigation of the differences between perception and thinking involved problem solving on the basis of two different representations of the same task, each inducing one of the two processes. The graphical representation led to a large amount of responses close to the correct answers with a low variability, whereas the analytical representation gave more precise correct answers, but with a much higher standard deviation as a result of several extreme misjudgments.

There is evidence of a definite limitation on the resolution of configural displays. For instance, Veniar (1948) found that horizontal or vertical distortion of a square shape will not be perceived when the distortion is less than 1.4% of the original length. If the control task demands information precision below this threshold and if an up scaling of the configural displays will spoil the area dispositions of the overall interface, adding digital information seems to be a natural way to overcome this shortcoming.

The exclusive use of configural information types might not be satisfying from a safety point of view. In unanticipated situations (i.e., rare events), the high precision of digital information might eventually turn out to be crucial for control actions based on knowledge-based reasoning (Rasmussen, 1986). For instance, when a small leak has been correctly diagnosed, the operator might want to calculate its exact amount in order to compensate precisely for the loss.

Most real-life control systems have a mixture of mimic, graphical (i.e., curves or bar graphs), and digital information. So, human factors experiments in touch with reality should not threaten the discussion on graphical versus analytical (e.g., numerical) interfaces as an "either / or issue," but consider ways to improve their integration. As it is the numerical data that drive the graphical display, they are computationally accessed anyway and may be displayed in addition to the graphic at a very low cost in terms of occupied space. It is commonly assumed that when distributed in the spatial, dynamical state space of the display (see Figure 7.1), the digits association to the actual physics will be strengthened. In contrast, it will be much more difficult to add graphical figures to a fundamentally numerical display layout, due to the constraints imposed by the alphanumerical tradition of serial or table representation formats.

The previous discussion can be summarized as a threshold scaling problem: How much difference will "make a difference" in the actual domain and should therefore be made perceivable? Is it important to hold a very high degree of precision obtained by analytical representations, or will close estimations induced by graphical representations be sufficient? This kind of question emphasizes the close coupling between graphical interface design and task analysis pointed out in section 7.1.1 of this chapter.

7.6.3   Graphical Remembering: Making the Past Present in Order to See the Future

Although most of the suggested principles can be seen in other types of interfaces, the idea of displaying temporal changes by the use of perspective (principle 7) is believed to be novel (Hansen, 1989). A representation of the fluctuation of plant data over time is needed for the supervision of trends, gradients of changes, process-response time, and other time-dependent events (Boël & Daniellou, 1984; Leplat & Rocher, 1984). Within a technical domain, most of the higher order, compound invariants in system behavior are revealed in a temporal structure, for example, heat transfer across subsystems or mass and energy conversions, and they are controlled by timed sequences of operational actions. Many process invariants will exhibit cyclical patterns or recurring events. Discovering their temporal structure will increase the feeling of familiarity with a complex plant and the likelihood of an early detection of deviants from the normal system behavior. Attunement to this kind of higher order invariant behavior is an important part of the ongoing development of the operator's control competency and should not suffer from a fragmented representation.

An interface addressing this problem will have to integrate, in one single representation format: (a) the *objective clock time*, which is the semantic-free, discrete, and totally ordered invariant used for the construction of the control system itself (including alarms) and the unchangeable constraint of all (macro) physics; (b) the *evolution of the variables* of the process, for instance, temperature, and (c) a *configural form* with a symbolic reference to the functions being represented. As individual design goals, the objective clock time can be represented by a watch. The evolution can be traced by decaying shades of former indicator positions shown as an afterglow, and the system functions can be represented by two-dimensional figures of theoretical physical explanations found in textbooks and instruction manuals.

To integrate these representations, it is suggested to start out with a two-dimensional figure of the invariants, for example, the Rankine cycle, as it captures the constraints of physical laws that cannot be changed. Then show the changes of parameters as changing forms within these constraints. Information to trace the development in the form of afterglows does not show the objective clock time through which the development takes place, because no monotonous, one-directional process movement can be assumed. Neither do displays of evolutions by fast (video-) playback of the changes, as this breaks up the natural temporal structure by which the invariants are normally seen in the changing interface geometry and therefore demands some time indication on the display (e.g., a clock) in order to reconstruct the temporal dimension mentally.

By giving time a whole dimension by itself, it can be specified directly. As a matter of fact, 75% of all public statistical information do reserve one dimension for time, according to Tufte (1983): "With one dimension marching along to the regular rhythm of seconds, minutes, hours, days, weeks, months, years, centuries, or millennia, the natural ordering of the time scale gives this design a strength and efficiency of interpretation found in no other graphic arrangement" (p. 28).

But often the two dimensions will already be occupied by other physical unities; the Rankine cycle, for instance, needs them for temperature and entropy. This leaves no other alternative than applying the third dimension. Hereby, the basic form can be preserved, the evolutions of the form can be traced by comparisons between former or proceeding figures, and the invariant structure of perspective can show the invariant, objective time.

The use of depth to represent time I have termed *time tunnels* (Hansen, 1989). This is a modification of the Lindsay and Staffon (1988) interface that showed the information of four incoming data as a quadrangle in front (see Figure 7.2), and that pushed nine older sets one step "backward" at each updating, using perspective to map time over to depths. This gives the imprerssion of a tunnel, for example, like looking backward from the last car of a roller coaster as the props pass by. As the process moves further in time, the figure shrinks into the distance and new measurements appear in the foreground. Symmetry or asymmetry in the pattern over its depth reveals invariants or changes in the relationship among the four variables.

The intention is to facilitate the perception of higher order invariants in the process state (e.g., "being in a steady state" or a "unstable state") and the process history (e.g., "increasing") by the use of the natural perceptual skills for perceiving the invariants in perspective and changes in patterns. Note that constant time intervals between each update are a prerequisite for the direct specification of the time span displayed.

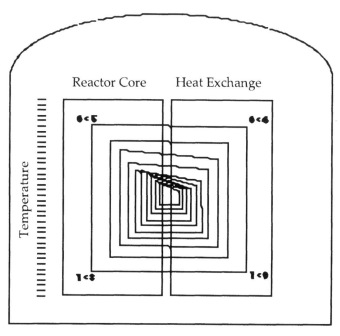

Figure 7.2. *A time tunnel depicting the increase of temperatures in the intermediate heat exchange of the primary coolant cycle from Lindsay and Staffon's (1988) display format. Four temperature measurements from the water going out of the reactor core (upper left corner), into the intermediate heat exchange (upper right corner), out of the exchange (lower right corner), and back to the core (lower left corner) naturally expands a quadrangle. Older measurements are shown inside this box, shrinking according to the gradient of density in natural perspective. It integrates information on the evolution of parameters over time, the objective clock time, and a symbolic representation of the function. This information is supposed to be accessible at "one glance" due to its exploitation of a natural invariant in the optical structures.*

A recent experiment (Hansen, in press) compared the perception of trends (i.e., simultaneous decreases on eight parameters) on the time tunnel display with other separated and integrated formats. Three sets of data with different rates of change and bandwidth were used. The utilization of the third dimension in the time tunnels was found to increase the number of correct responses in conditions with noisy data, in which changes had to be extracted from nonspecific, dysfunctional information. The experiment also revealed a general superiority of integral, configural formats compared to separable formats and a lower reaction time when numerical information was nested inside the graphical information (see the previous section).

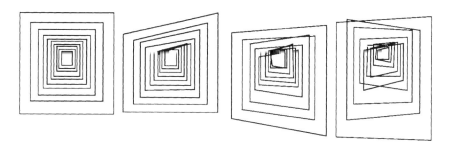

Figure 7.3. *Four time tunnels showing a possible development from a stable plant condition (left) to an unstable condition (right).*

The time tunnel display violates the invariant laws of ecological optics in three important ways: First of all, there is no dynamic occlusion of what would have been hidden frames, as there are no surfaces between the frames. Secondly, the lack of surfaces excludes the information in texture density, and the frame lines have the same thickness down through the tunnel sketch. Third, the significance of changing perspective with (head-) locomotion is ignored, as the arrested perspective in the time tunnel display only specifies what Gibson (1979) called an *artificial perspective*. An artificial perspective, as it is found on pictures, requires that the representation be viewed with one eye at a unique station point. The limited numbers of frames used, namely, 10, may also cause some perceptual problems. Gibson, Purdy, and Lawrence (1955) presented subjects an optical tunnel composed by alternating white and black plastic sheets with an increasing density of the contrast from the periphery to the center. They found that two-thirds of the objects saw a tunnel when 19 contrasts were displayed, but only one-third did so if the contrast was reduced to 9. So, it might well be that a more direct perception of the time tunnels can be achieved if the numbers of frames is increased. Even with these violations, most subjects actually expressed a phenomenological percept of "moving through a tunnel" when they were presented the time tunnel display. Similarly, Johansson (1975) found a perceptual tendency toward abstract projective invariances so strong that a highly complex and "unnatural" motion in three dimensions was preferred to a simple two-dimensional track trace by two moving spots. He provided experimental evidence that the visual system automatically prefers invariants of figure size, obtained by inferring motion in three-dimensional space. This preference was supposed to be in effect when perceiving the time tunnels as well, and to be strong enough to overrule the violations.

The violations and shortcuts were caused by the computational and display limitations of a standard personal computer. In the near future most of these limitations will disappear. Hansen (1992) discussed the representational potentials, with regard to control displays, of the new visualization techniques emerging within the so called virtual realities.

7.7   Summary and Conclusion

This chapter argued that interfaces for complex dynamical systems should seek to represent the system constraints identified by system engineers in an integrated configural display form, which allows for a direct perception of the invariant system behavior from display geometries changing according to incoming process data. Seven principles for the nesting of information in a configural fashion were suggested. The application of the principles was illustrated with a display sketch from a conventional power plant, and further arguments for the principles were given in the general discussion.

The overall goal of ecological interface design is not to eliminate the need for interpretation, but to save the highly limited cognitive processes from being unnecessarily occupied with the kind of data that the computer could just as well have integrated graphically. This integration must be guided by a scientific understanding of the basis for human perception of the natural environment in order to insure that the integration is compatible with man's highly developed perceptual skills. The ecological approach seems particularly fruitful for the further development of complex interfaces, as it provides a theoretical understanding of the invariants in the structures of the ambient optical arrays. This understanding makes it possible to utilize the invariants in display representations of mediated information. One example suggested in this chapter is the use of the gradient of density to map time over to depth. The invariant specifying time-to-contact (Lee, 1980) could be another candidate for a natural mediator of complex system information. Hopefully, this chapter showed that ecological psychology now has a chance to demonstrate the value of its insights in practical applications within process control.

Acknowledgments

The author thanks Jens Rasmussen and Len Goodstein, Risø National Laboratory, Denmark; Steen Folke Larsen, Aarhus University, Denmark; John M. Flach and Kevin Bennett, Wright State University, Ohio; and Kim Vicente, Georgia Institute of Technology; for stimulating discussions and useful comments to this chapter. The Time Tunnel software was developed by Carl. V. Skou, Riso National Laboratory, Denmark. Parts of this chapter have been presented at the first European chapter meeting on Ecological Psychology, Marsailles, June 1990, and at the Sixth International Conference on Event Perception and Action, Amsterdam, August 1991.

7.8   References

Alexander, C. (1964). *Notes on the synthesis of form.* Cambridge, MA: Harvard University Press

Beltracchi, L. (1987). A direct manipulation interface for water-based rankine cycle heat engines. *IEEE Transactions on Sytems, Man, and Cybernetics, 17,* 478–487.

Boël, M., & Daniellou, F. (1984). Elements of process control operator's reasoning: Activity planning and system and response-times. In *Ergonomics problems in process operations* (Institution of Chemical Engineers Symposium Series No. 90, pp. 1–8). Oxford: Pergamon Press.

Brunswik, E. (1956). *Perception and the representative design of psychological experiments.* Berkeley: University of California Press.

Buttigieg, M. A. (1989). *Emergent features in visual display design for two types of failure detection tasks.* (Tech. Report EPRL-89-04). Urbana: Department of Mechanical and Industrial Engineering, University of Illinois, Urbana-Champaign.

Carroll, J. M., & Thomas, J. C. (1980). *Metaphor and the cognitive representation of computer systems* (Tech. Rep. RC 8302). IBM Watson Research Center.

Carswell, C. M., & Wickens, C. D. (1984). Stimulus integrality in display of system input-output relationships: A failure detection study. *Proceedings of the Human Factor Society, 28,* 534–537.

Cockin, J. A., (1970). An osilloscope polar coordinate display for multidimensional data. *The Radio and Electronic Engineer, 40,* 97–101.

Cleveland, W. S. (1985). *The elements of graphing data.* Monterey, CA:

Wadsworth.

diSessa, A. A. (1983). Phenomenology and the evolution of intuition. In D. Gentner & A. L. Stevens (Eds.), *Mental models* (pp. 15–34). Hillsdale, NJ: Lawrence Erlbaum Associates.

Flach, J.M. (1988). Direct manipulation, direct engagement, and direct perception: What's directing what? *Proceedings of the Human Factors Society, 32,* 1355–1358.

Flach, J.M. (1990). The ecology of human-machine systems I: Introduction. *Ecological Psychology, 2,* 191–205.

Gaver, W. (1991). Technology affordances. In S. P. Robertson, G. M. Olson, & J. S. Olson (Eds.), *Reaching through technology, CHI '91 Conference Proceedings* (pp. 79–84). SIGCHI.

Gentner, D., & Gentner, D. R. (1983). Flowing waters or teeming crowds: Mental models of electricity. In D. Gentner & A. L. Stevens (Eds.), *Mental models* (pp. 99–129). Hillsdale, NJ: Lawrence Erlbaum Associates.

Gibson, J. J. (1950). *The perception of the visual world.* Boston: Houghton Mifflin Company.

Gibson, J. J. (1979). *The ecological approach to visual perception.* Boston: Houghton Mifflin.

Gibson, J. J., Purdy, J., & Lawrence, L. (1955). A method of controlling stimulation for the study of space perception: The optical tunnel. *Journal of Experimental Psychology, 50,* 1–14.

Goldwyn, R. M., Farrell, E. J., Friedman, H. P., Miller, M., & Siegel, J. H. (1973, May). Identifying and understanding patterns and processes in human shock and trauma. *IBM Journal of Research and Development,* pp. 230–240.

Goodstein, L. P. (1981). Discriminative display support for process operations. In: J. Rasmussen & W. B. Rouse (Eds.), *Human detection and diagnosis of system failures* (pp. 433–449). New York: Plenum.

Gould, J. D. & Lewis, C. (1985). Designing for usability: Key principles and what designers think. *Communication of the ACM, 28(3),* 300–311.

Hansen, J. P. (1989, June). Perspectives on analog interfaces. *Proceedings of the 8th European Annual Conference on Human Decision Making and Manual Control* (pp. 227–247). Institute of Automatic Control Systems, Technical University of Denmark. Copenhagen.

Hansen, J. P. (1992). *Information nesting in configural interfaces for process control* (Risø-R-616(EN)). Roskilde, Denmark: Risø National Laboratory.

Hansen, J. P. (in press). *Spatial and temporal analog interfaces: An experimental investigation into the importance of integration and temporal information*. Paper submitted in partial fulfillment of the requirements for the degree of Doctor of Philosophy in Psychology, Risø National Laboratory, Roskilde, Denmark.

Hester, M. (1977). Visual attention and sensibility. In R. Shaw & J. Bransford (Eds.), *Perceiving, acting, and knowing*. Hillsdale, NJ: Lawrence Erlbaum Associates.

Hollan, J. D., Hutchins, E. L., & Weitzman, L. (1984). Steamer: An interactive inspectable simulation-based training system. *AI Magazine, 4*, 15–27.

Johansson, G. (1975). Visual motion perception. *Scientific American, 232* (6), 76–88.

Kleer, J., & Brown, J. S. (1983). Assumptions and ambiguities in mechanistic mental models. In D. Gentner & A. L. Stevens (Eds.), *Mental models* (pp. 155–190). Hillsdale, NJ: Lawrence Erlbaum Associates.

Knak, C. (1982). *Stationære kraftanlæg* [in Danish]. Copenhagen, Denmark: G.E.C. Gads forlag.

Lakoff, G., & Johnson, M. (1980). *Methaphors we live by*. Chicago: University of Chicago Press.

Lee, D. N. (1980). The optic flow field: The foundation of vision. *Philosophical Transactions of the Royal Society of London, Series B: Biological Sciences, 290*, 169–179.

Leplat, J., & Rocher, M. (1984). Ergonomics of the control of simultaneous process: Case study in biochemical industry. In *Ergonomics problems in process operations* (Institution of Chemical Engineers Symposium Series No. 90, pp. 9–19). Oxford: Pergamon Press.

Lindsay, R. W., & Staffon, J. D. (1988, Nov.). *A model based display system for the experimental breeder reactor-II*. Paper presented at the joint meeting of the American Nuclear Society and the European Nuclear Society, Washington, DC.

McCloskey, M. (1983). Intuitive physics. *Scientific American, 248*, 114–122.

Pomerantz, J. R. (1986). Visual form perception: An overview. In H. C. Nusbaum & E. C. Schwab (Eds.), *Pattern recognition by humans and machines. Vol. 2. Visual perception* (pp. 1–30). Orlando, FL: Academic Press.

Rasmussen, J. (1986). *Information processing and human-machine interaction: An approach to cognitive engineering*. New York:

North-Holland.

Rasmussen, J. (1988). A cognitive engineering approach to the modeling of decision making and its organization in process control, emergency management, CAD/CAM, office systems, library systems. In W.B. Rouse (Ed.), *Advances in man-machine system research* (Vol. 4, pp. 165–243). Greenwich, CT: JAI Press.

Rasmussen, J., & Goodstein, L.P. (1988). Information technology and work. In: M. Helander (Ed.), *Handbook of human-computer interaction* (pp. 175–201). Amsterdam: Elsevier.

Rasmussen, J., & Vicente, K. (1989). Coping with human errors through system design: Implications for ecological interface design. *International Journal of Man-Machine Studies, 31*, 517–534.

Rasmussen, J., & Vicente, K. (1990). Ecological interfaces: A technological imperative in high tech systems? *International Journal of Human Computer Interaction, 2(2)*, 93–111.

Sanderson, P. M., Flach, J. M., Buttigieg, M. A., & Casey, E. J. (1989). Object displays do not always support better integrated task performance. *Human Factors, 31*, 183–198.

Spelke, E. S. (1990). Principles of object perception. *Cognitive Science, 14*, 29–56.

Tufte, E. R. (1983). *The visual display of quantitative information.* Cheshire, CT: Graphics Press.

Tufte, E. R. (1990). *Envisioning information.* Cheshire, CT: Graphics Press.

Veniar, F. (1948). Difference thresholds for shape distortion of geometrical squares. *Journal of Psychology, 26*, 461–476.

Vicente, K., & Rasmussen, J. (1990). The ecology of human-machine systems II: Mediating "direct perception" in complex work domains. *Ecological Psychology, 2(3)*, 207–249.

Vicente, K. (1991). *Supporting knowledge-based behavior through ecological interface design.* Unpublished doctoral thesis, Department of Mechanical and Industrial Engineering, University of Illinois, Urbana-Champaign.

Warren, W. (1994). Environmental design as the design of affordances. In J. M. Flach, P. A. Hancock, J. K. Caird, & K. J. Vicente (Eds.), *An ecological approach to human-machine systems I: A global perspective.* Hillsdale, NJ: Lawrence Erlbaum Associates.

Wickens, C. D., & Andre, A. D. (1990). Proximity compatibility and information display: Effects of color, space, and objectness on information integration. *Human Factors, 32*, 61–67.

Winograd, T., & Flores, F. (1986). *Understanding computers and cognition: A new foundation for design.* Norwood, NJ: Ablex.

Woods, D. D., & Roth, E. M. (1988). Cognitive systems engineering. In: M. Helander (Ed.), *Handbook of human-computer interactions* (pp. 3–43). Amsterdam: Elsevier.

Zusne, L. (1970). *Visual perception of form*. New York: Academic Press.

Chapter 8

Human Exploration and Perception in Off-Road Navigation

Leslie A. Whitaker

University of Dayton

V. Grayson CuQlock-Knopp

Army Research Laboratory

8.0   Introduction

Navigation is a task requisite for human survival in many activities. Use of environmental cues for navigational guidance is at least as old as the nomadic tribes of primitive man. However, two-dimensional representations of those cues (e.g., maps and charts) are relatively recent resources for navigational guidance. From the astrolabe and hand-drawn maps, which allowed navigators to travel beyond the sight of land during the Age of Exploration, to global-positioning satellites, which are the most recent tool in the navigator's arsenal, the use of abstract representations have supplemented the navigator's environmental cues.

Global-positioning satellites and digitized map databases provide a wealth of data for the navigator (Kiernan, 1991; Pollack, 1991; Yoder, 1991). Digitized maps (stored in object-oriented databases) can provide virtually any information to the navigator (King, 1990). The problem has now become one of using these data to provide information as it is needed by the navigator. Without some means of selecting only useful information from such a comprehensive map database, the navigator will be overwhelmed with irrelevant data, such as roof composition material; tree species, age, and growth rates; and property ownership data. The designer's challenge is to develop tools that provide information on an as-needed basis for the task at hand. The result can be a powerful information source to aid navigators under a broad range of adverse circumstances, including decreased visibility, minimal navigational skill or experience, and altered environmental cues. Our hypothesis is that the navigator will benefit from a tool that selects only relevant information from this map database.

We are presently involved in a 3-year effort to develop a model that specifies the *relevant* information needed by off-road navigators. Relevance is not an issue easily resolved in the abstract, nor can universal rules be described for extracting relevant information from the world (Dennett, 1987; Dreyfus, 1972; Fodor, 1990). However, humans are extremely skilled at extracting such information in domains with which they have had sufficient experience (Klein, 1989, 1993). We have taken advantage of this ability during our development of a navigational information model. In this chapter, we describe our research program for developing such a model and its results to date. This research considers the interaction of the human-task environment to be its appropriate focus. Our research methodologies include naturalistic observation, retrospective interviews, and laboratory studies in a multiyear effort to discover some of the processes by which humans navigate in off-road environments. The final product of this effort is a prototype of a navigational aid that will serve as a proof-of-concept for the navigational model.

This model development employed a variety of methodologies and explored several questions relevant to off-road navigation. In order to describe this development process, this chapter is divided into five sections: The first describes the nature of off-road navigation; the second, literature findings about individual differences in navigational skills and a study we completed to extend this literature to off-road navigation; the third explains the various research methodologies we employed; the fourth presents the resulting information-and skills taxonomy that we developed from this research; and the final section describes the prototype of a navigational aid that is being developed as a proof-of-concept for this model.

8.1   Off-Road Navigation

Understanding the process of navigation can be aided by examining a variety of subskills related to the activity. There is extensive literature dealing with these subskills: air navigation (Aretz & Wickens, 1990; Wickens, 1984, 1991), map navigation (Murakoshi, 1990), information processing (Chase & Chi, 1980; Seiler, 1985), physical locomotion through man-made environments (Evans & Pezdek, 1980; Thorndyke, 1980), individual differences (Just & Carpenter, 1985; Shepard & Cooper, 1982), plus some work on off-road navigation (Kaplan, 1976; Thompson et al., 1990).

The term *off-road navigation* means movement through natural environments. It refers to movement across countryside and through wooded areas instead of in a man-made or urban environment. Movement may occur on foot, in an off-road vehicle, or as "remote movement" through teleoperator control of a drone vehicle. Trails may or may not be present. For the most part, paved vehicle roads are absent. For the present discussion, we limit the topic to ground navigation instead of air or water navigation, which have unique information requirements of their own (see, e.g., Wickens, 1991).

Various activities require off-road navigation for work or pleasure. Military scouts complete reconnaissance missions; rangers patrol the backcountry of National Parks; hikers and backpackers explore wilderness terrain; mountaineers climb remote peaks; orienteers engage in meets for timed competition; cross-country runners compete in track-and-field events; children explore the woodland trails of state parks. All these people must use off-road navigational skills to some degree.

Much has been published describing the ways in which people navigate within the built environments of our modern communities. Wickens (1984) reviewed the psychological literature on space perception, maps, and navigation. Thorndyke (1980; Thorndyke & Hayes-Roth, 1982) described the process of learning to navigate in a new area. Many authors explored the issue of cognitive maps. Tolman (1948) coined the term *cognitive maps* in an effort to explain evidence that environmental information is remembered and used in the context of the person's (or rat's) total concept of the environmental area. Others explored the development and use of cognitive maps and cognitive models (Aretz & Wickens, 1990; Cadwallader, 1979; Couclelis, Golledge, Gale, & Tobler, 1987; Evans & Pezdek, 1980; Evans, Skorpanich, Garling, Bryant, & Bresalin, 1984; Foley & Cohen, 1984; Goldin & Thorndyke, 1982; Harvey, 1980; Kozlowski & Bryant, 1977; Leiser, 1987; McNamara, 1986; Rossano & Warren; 1989a, 1989b; Sholl, 1987; Wickens, 1990). This is not an exhaustive list of the publications that have explored the issues of cognitive models and their influence on navigation. Further work has been done by cartographers or developers of maps and map aids in attempts to clarify the ways in which navigators use maps and map aids as navigational tools (e.g., Blades & Spencer; 1987; Boys, 1986; Hunt, 1984; Monmonier, 1991; Salichtchev, 1978; Streeter, Vitello, & Wonsiewicz, 1985).

Considerable attention has been given to the ways by which humans navigate in built environments, but little has been published that describes the cognitive models of people navigating on the ground in natural settings (however, see Murakoshi [1990] and Seiler [1985] for some initial research on this topic). Thompson et al. (1990) published a theoretical paper discussing a specific off-road navigational problem known as "the drop-off problem." There is a need to know more about both navigation in natural settings and about the navigational aids that might be helpful in supporting those tasks.

There are a priori reasons to be cautious about attempting to generalize research results from built environments to natural ones. The two environments (built vs. natural settings) differ in several important dimensions. A few of these dimensions can be used to illustrate this point:

1. Built environments make extensive use of straight lines and square corners. Fletcher (1972) noted that square corners are so rare in natural settings that he was able to use that feature as a distinguishing clue for recognizing the archaeological remnants of Anasazi settlements in remote areas of the Grand Canyon.

2. Built environments in developed nations make extensive use of signing and linguistic cues. Streets are signed; maps use names of features, and these names will actually be found in the environment. One of the authors traveled in Moscow recently by using a street map and hand-lettered labels bearing the names of subway lines and stops, despite the fact that she knew very little of the Cyrillic alphabet or the Russian language. She traveled by matching the linguistic (in this case, a set of letters defining a word) information from her map aids to the same linguistic information displayed in the real world. This type of navigation is not possible in a natural setting because the label, for example, Mt. Shasta, on the map is not displayed in neon letters across the actual mountain.

3. Built environments allow movement in a motorized vehicle, whereas natural settings generally preclude the use of any means of locomotion except human or animal power. The exception to this generalization is the use of all-terrain vehicles

or the modern military "jeep" called a HUMMER. Even using such off-road vehicles, distance is covered much more slowly in natural settings than in built environments.

Initial exploration of the literature showed that navigation is a complex issue involving the full range of human information processing—perception, cognition, learning, and motor responding—coupled with the specifics of the environmental and task demands. These interact to provide the navigator with a continuing stream of information. Selecting from and acting on this information allows skilled navigators to accomplish their tasks.

Personal experience with the task of off-road navigation has confirmed that this task is often data rich from a variety of sensory input sources. The task becomes one of selecting the relevant cues and combining them into meaningful patterns. These patterns are then matched to the information provided by topographic map and compass in order to accomplish various navigational tasks.

8.2   Individual Differences in Navigational Skills

Individual differences contribute to differences in navigational skill. Some factors that produce these differences are prior training, specialized knowledge of a particular location, or more inherent spatial ability parameters. Previous work examining navigation in built environments (e.g., Streeter & Vitello, & Wonsiewicz, 1985) reported significant individual differences in map-reading ability. Those people who report themselves to be good navigators tend to use maps and differentially value landmarks along their routes. In contrast, people who report themselves to be poor navigators do not use maps and value all landmarks equally, regardless of their navigational value.

Sholl (1987) found that self-report of a "good sense of direction" is correlated with ability to manipulate spatial information mentally. People with a poorer sense of direction took longer to point in the direction of unseen targets. This did not seem to be related to poorer ability to handle spatial information, but instead to difficulty imagining themselves in different orientations.

Hardwick, Woolridge, and Rinalducci (1983) found a relationship between the manner in which individuals organized maps of a familiar environment and their selection of landmarks in an unfamiliar environment. They examined the landmarks mentioned in a survey of people with varying levels of experience in a particular environment.

The authors reported that people with less extensive spatial knowledge of a particular environment selected highly salient but spatially ambiguous landmarks. People with more extensive spatial knowledge of that environment tended to select landmarks located at changes in heading (i.e., intersections). This finding is consistent with Streeter and Vitello's finding that people with better navigational ability made better use of critical landmarks.

Thorndyke studied individual differences in navigational skills within built environments. He proposed that one's use of landmark cues changes in a systematic fashion as one becomes more familiar with an environment (Thorndyke, 1980). Furthermore, he found a correlation between spatial ability and subjects' performance on a map-learning task (Thorndyke & Stasz, 1980). Finally, Thorndyke and Hayes-Roth (1982) reported the effects of map learning versus route experience. Their results showed that some aspects of spatial knowledge could be acquired through exposure to maps of the environment; however, others required the direct experience of navigating in that environment.

Our own research examined the extent to which spatial ability is related to a person's ability to complete one type of spatial task; that of rotating the viewed world to match a remembered orientation of that world.

One navigation task that often requires rotating the viewed world occurs when the navigator is exploring or scouting new territory. He or she may find that this exploration involves crossing and then recrossing the same areas. The ability to recognize previously crossed terrain is fundamental to navigation in off-road environments and requires the ability to compare what the navigator is presently seeing with a remembered image of that terrain *that* often had been viewed from a different perspective. Terrain recognition under these circumstances requires that the navigator be able to imagine how terrain would look when viewed from various orientations.

This matching task can be described as mentally rotating the scene to match the orientation of the mental image. This is the type of mental rotation usually associated with Cooper and Shepard's experiments (Cooper & Shepard, 1973; Shepard & Cooper, 1982). In those studies and for our terrain-recognition problem, individuals must be able to identify an image when seen from various angles. We predicted that people who test high in spatial ability will be better able to recognize terrain sites from different orientations than people who test low in spatial ability.

We photographed several woodland scenes from each of eight

compass directions. We tested military subjects who had completed the Army Navigation course on the following task: A pair of photographs was shown to the subjects. The two photographs were taken from different compass positions. The subject's job was to determine whether the photographs were of the same woodland area or of two different areas.

All subjects were tested on a battery of standardized tests of spatial and analytical ability (Cognitive Laterality Battery; Gordon, 1987) in order to obtain a measure of these abilities for the subsequent analyses. Our interest in individual differences was motivated by the problem of personnel selection for teleoperator or degraded visual display tasks (such as night reconnaissance). Either of these tasks requires the operator to use visual information that has somewhat reduced levels of resolution. Therefore, we tested our subjects at several levels of resolution. We predicted that the high spatial ability subjects would show a greater advantage over the low spatial ability subjects when both groups were trying to resolve degraded, fuzzy images with few identifiable details than when high-resolution images full of easily labeled details were used. Hence, we predicted that lower resolution images would show greater differences between the two spatial ability groups. Such images had lost much of their details, and recognition must be accomplished primarily by using contours. This task was similar to Shepard's matching task using line drawings and a task employing computer-generated three-dimensional images (Barfield, Sandford, & Foley, 1988).

To test the interaction of spatial ability and the visual quality of the displayed scenes, the photograph pairs were shown at three levels of visual resolution to vary the amount of feature detail that would be available. Our prediction was that high and low spatial ability groups might perform equally well when the stimulus resolution was high; however, low spatial ability subjects would give poorer performance than high spatial ability subjects when the visual resolution of the photographs was degraded.

The terrain rotation task proved to be too difficult for some of the 108 Army Noncommissioned Officers who were tested. Therefore, we dropped all subjects who could not do better than chance on the high resolution photographs. Subjects who were dropped and those who remained in the study did not differ significantly on either spatial or analytical scores. Seventy-seven subjects remained in the analysis.

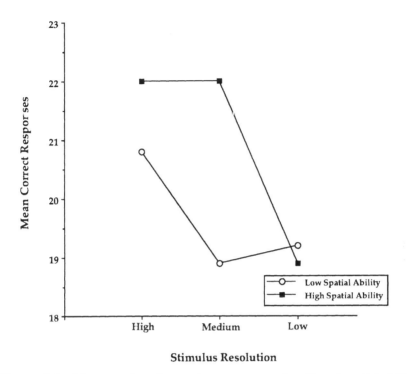

Figure 8.1. *Mean number of correct responses as a function of spatial ability and stimulus resolution.*

The results of this analysis were exactly as predicted. The following findings were statistically significant. Spatial ability (Spatial Composite Score) was correlated with task performance (Terrain Matching), whereas analytical ability (Analytical Composite Score) was not. All subjects did better on the higher resolution photograph pairs. High spatial ability subjects made fewer terrain matching errors than did low spatial ability subjects. Finally, there was a significant interaction of spatial ability and photograph resolution: Subjects low in spatial ability made significantly more errors than did the subjects high in spatial ability (see Figure 8.1). This interaction was the critical effect predicted from the hypothesis that the Spatial Ability Composite score would mark subjects who could perform the recognition task better when the display conditions decreased visual resolution.

8.3   Research Methodologies

We examined navigation from several perspectives, because the goal of this research program is to build a model of off-road navigation that considers the interaction of the navigator-environment task. Ours may be viewed as an ecological approach, or it may be seen as the committee of blind men trying to describe an elephant. By using multiple approaches to the question, we attempted to see as much of the elephant as possible. We employed various types of naturalistic observations and laboratory studies, as well as semi-structured interviews.

Naturalistic observation (Kantowitz, Roediger, & Elmes, 1991) provides an excellent way to study a new research area. By using unobtrusive observation, unobtrusive measurement, and case studies, it is possible to delimit the critical variables that influence behavior. This methodology is often seen as diametrically opposed to laboratory studies, but we have seen the two extremes as complementary to one another. For example, in initial discussions with teleoperators about their task of guiding drone vehicles in off-road environments, we learned of very tight restrictions on bandwidth transmission (Whitaker & CuQlock-Knopp, 1990). Television transmission of visual information used by teleoperators is severely restricted by expense, security, and distance. Therefore, the problem of a degraded visual image is a very real one. Hence, we were prompted to conduct the laboratory study of individual differences in spatial ability and its effect on the terrain-matching skill necessary during navigational exploration or reconnaissance.

One of the authors is a cross-country hiker, experienced in the use of topographic maps and compass. She has a library of maps, sketches, written logs, and slides that allow the correlation of visual cues with topographic terrain features. Consulting these resources increased the efficiency of developing and testing hypotheses about the importance of one type of visual cue versus another. Therefore, this research project allowed her to combine an avocational interest with a professional research project. The accumulation of decades of trail miles and high country off-road navigation provided insights into the issues surrounding this topic.

Laboratory research can also be a useful tool for the study of navigation. The study of individual differences described earlier began with naturalistic observations, a literature review, and informal interviews with teleoperator experts. It culminated in an experiment. That experiment confirmed the hypothesis that spatial ability does differentially affect terrain matching under degraded viewing conditions. We are planning to conduct additional laboratory research to study the effectiveness of the navigational aid NAVAID that we are prototyping. Although this aid may not be developed as a field tool; it can be evaluated as a prototype in the laboratory.

In contrast, no laboratory research is planned to study orienteers' skills in artificial or controlled environments. Instead, we used interviews as a knowledge elicitation technique with a sample from this population of navigators. Although we argue that we do know some factors that are important to off-road navigation (see section on taxonomy later), we do not believe that we know enough to simulate the navigation task in a laboratory experiment.

Interviews with navigational experts were a very valuable source of information. We conducted 16 in-depth interviews with a selected group of navigators, either civilian orienteers or military scouts. Orienteers are sportsmen and sportswomen who navigate set courses in woods or fields. Their success is measured by their speed and accuracy in locating fixed positions marked by a "control." The remaining navigators were military scouts who perform a similar navigation task when on a reconnaissance mission. Our goal was to determine what navigation skills are necessary for successful completion of this task, what strategies are employed by the navigators, and which visual cues are observed from map, compass, and environment to execute those strategies successfully.

Many well-developed skills (especially those involving nonverbal material such as visual pattern recognition) are difficult for an expert to describe. In the present interviews, we used a technique known as the Critical Incident Technique, which has been employed successfully to elicit such information (e.g., Crandall, 1989). The Critical Incident Technique is a semistructured interview that begins with a request such as: "Think of a time when your navigation skills were particularly challenged and describe how you found your way."

Each navigator then described a specific navigation incident in which his or her skill had been needed to navigate successfully. Orienteering is a sport that attracts people of both genders and all ages to pit their navigational skills against each other in different events. The critical incidents they described ranged from a trek across Polar Bear habitat in northern Canada to a timed course through wooded parkland in the midwestern United States. Subsequent probes focused on the visual cues that were used, the hypotheses about anticipated landmarks, mental maps, and so on., that the person remembered having used to solve his or her navigational problem. This method was particularly fruitful in exposing cues, strategies, and skills employed by these navigators. A taxonomy that lists this information is summarized in the next section.

8.4   Information and Skills Taxonomy

A listing of the visual cues, problem-solving strategies, and navigational skills was developed by coding the verbal protocols obtained from the 16 navigational interviews described earlier. The authors described further details of the verbal protocol analysis and the taxonomy development in previous publications (Whitaker & CuQlock-Knopp, 1991a, 1991c).

Visual Cues.   First, visual cues could be divided into four categories. By order of frequency of mention, these are *Man-made features*, *Contours*, *Water*, and *Vegetation*. Visual cues that could not be placed into one of these main categories constituted 8% of the cues.

1. *Man-made features* were valued for their uniqueness and their distinctiveness. Kaplan (1976) also mentioned this finding when working in off-road environments. Although such environments may contain some man-made features (e.g., trails, fences, ditches, an occasional building, horse corral, foundation, lean-to), these are not the norm. Therefore, finding such man-made features provides excellent location information when correlated with these same features marked on a map. This diagnostic quality of man-made features in off-road environments probably explains their high frequency of mention (33%).

2. *Terrain contours* are found on all topographic maps and exist in all but the flattest of terrains. Therefore, these experienced navigators found contours to be helpful visual cues when they were able to correlate this evidence with mapped information. Contours constituted 26% of the visual cues mentioned. There is no guarantee that less experienced travelers would be able to use physical land topography (contours) with a topographic map to aid their navigation. Topographic maps are notoriously difficult to read (Selwyn, 1987), However, in light of the importance of contours to experienced navigators, one of the goals of the present program is to understand how to make terrain contour a more useable tool for the less experienced

traveler.

3. *Water features* serve several functions for the navigator, they may be a barrier to travel, provide a distinctive feature, serve as a handrail (restricting travel to a correct path), or serve as a catching feature (preventing the navigator from going too far beyond his or her goal location). This collection of uses made water features useful visual cues for the navigators. Water features were 17% of the visual cues mentioned.

4. *Vegetation features* were rated as the least reliable cues by many navigators. They cited out-of-date maps, seasonal changes, and unreliable coding of vegetation as the primary reasons that vegetation cues were less useful to them. Vegetation was mentioned as 16% of the cues. A number of these mentions included more of a description of the obstructive nature of the vegetation or its unreliability as a mapped feature than its value as a visual clue.

Problem-Solving Strategies. Second, problem-solving strategies employed by the navigators to use these cues were coded into four types. By order of frequency, these were *prediction, recovery,* use of *catching features,* and *aiming off.*

1. *Prediction* can be used under two general circumstances: to check the accuracy of route finding or to test a location hypothesis. Good orienteers explained that they anticipate various route features as they follow a route. They use terms such as *predict, anticipate,* and *should be* to describe this process. By predicting various landmarks along their route, they are quickly able to determine if they have strayed from their intended route. When a navigator questions whether he or she is in the correct location, prediction is used to determine which landmarks should be visible from the hypothesized location.

2. *Recovery:* If the navigator does lose his or her way, some means of recovering the route (or location) must be used. These navigators reported that they seldom blindly continue through the countryside, hoping to see some feature that will orient them. Some of our orienteers physically retraced their steps to their last known position. Alternatively, they might mentally retrace their actions to develop an hypothesis of where they currently were. Finally, they might engage in landmark prediction to determine what position on the map would afford a view of the features presently visible: "I knew I was lost; it was

time to start from scratch—which is basically, get the map out, line it up, turn around, look and see where am I. Where can I identify a landmark?"

3. *Catching features* serve to guard the navigators from going too far beyond their goal location. One navigator described how he used a water feature (a stream) as a catching feature in the following way: "It's a big stream. If I keep going north, I am going to hit it" (see Figure 8.2).

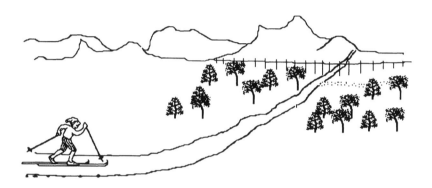

Figure 8.2. *Illustration of the problem-solving strategy of using a catching feature. The cross country skier wants to turn off the main trail at a small trail to the right. His map shows that this small trail occurs immediately before a powerline. Therefore, he can ski quickly until he approaches the powerline (his catching feature) where he must slow down to look for the small trail's intersection.*

4. *Aiming off* is the strategy of deliberately planning a route that will not bring the navigator to the exact goal location. It is usually used in combination with a linear feature, such as a river or a road, that will mark the navigator's general location near the goal. For example, "I would aim off slightly to the right of north. Last thing that I want to do is miss the stream crossing [a ford]." If this orienteer had planned a route directly north and he had not seen the ford when he reached the stream, he would not have known whether to run east or west to actually reach it. (see Figure 8.3 for an illustration of this aiming off strategy.)

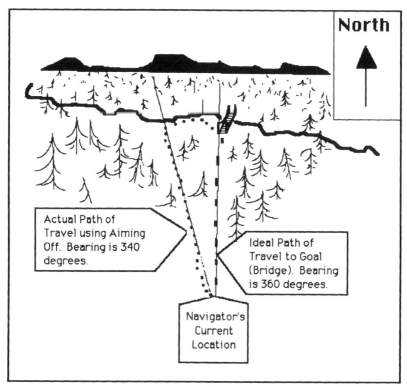

Figure 8.3. *Illustration of the problem-solving strategy of Aiming Off. The navigator wants to reach a bridge located due north of her current location. She aims off slightly to the west (by 20 degrees) to guarantee that she will know that she will reach the river with the bridge (goal) on her right.*

Navigational Skills. Third, navigational skills (or subtasks) were classified into four categories. By order of frequency of mention, these were *route choice, location, orientation,* and *route finding.*

1. *Route choice* is the process of selecting the route alternative that will maximize the navigator's goals. If these goals include maximum speed, then a short route will probably be selected. However, considerations of elevation change may alter this choice. One orienteer said that he queried his body to see if he felt strong enough to manage a route requiring a lot of up and down at high speed. If not, then he would opt for the longer, flatter route. Another consideration is security, military scouts may wish to avoid the possibility of visual exposure encountered by taking a direct ridge route. Safety is an

additional consideration; mountain trails above tree line are exposed routes and should be avoided during afternoon thunderstorms.

2. *Location* was the sine qua non of navigation for these interviewees. If they could not be sure of their location on the map, then they usually stopped to reconnoiter. The most experienced navigators reported that they began to feel uneasy if they had not secured their location after traveling for some time interval. This time interval was a function of the type of terrain and the difficulty of recovering from being lost; shorter times being specified in more complex terrain, and longer times being tolerated in flatter, more open terrain or when on a well-established trail.

3. *Orientation* describes the ability to know the compass directions from a current location. This is easily done by using a compass. The difficulty comes in using this information to accurately locate an unseen landmark from one's current position. The literature reports individual differences in subjects' ability to do this task (e.g., Sholl, 1987). This is consistent with our findings from the interview protocols. Some subjects felt very secure in their ability to know the location of an unseen landmark; others were not comfortable until they had gained a direct view of the landmark.

4. *Route finding* describes the actual process of following the chosen route. This task is difficult in off-road environments where urban humans are less skilled at discriminating subtle differences among natural landmarks (Kaplan, 1976). All orienteers reported that it was easy to convince themselves that they were following the correct route despite ample visual evidence to the contrary. The more experienced navigators reported that they used prediction strategies (see earlier) to guarantee that they were, in fact, following the route they intended to be following.

These three areas (cues, strategies, and skills) constitute the outline of the off-road navigator's task as we currently see it. The next step is to develop a navigational aid that uses this model as the basis for its aiding features.

8.5   Navigational Aid Prototype (NAVAID)

Our goal in this research program is to develop a model of off-road navigation that specifies the relevant information navigators need to accomplish their tasks. To the extent that we are successful in developing this navigation information model, we should be able to demonstrate the usefulness of that model. We plan to do this by creating a screen prototype of an adaptive decision aid for off-road navigation called NAVAID. The research we conducted to date focused on this goal

The development of the NAVAID prototype is important as a proof-of-concept for our navigation information model. For our navigation application, the relevance of information will be determined pragmatically, which of the information (provided by NAVAID) results in better navigation. Candidate information was obtained from experienced navigators and incorporated into a series of screen prototypes (developed in Hypercard). This prototype was then further refined in additional meetings with these experts. Future research plans include the testing of laboratory navigation problems using nonexperts.

In systems development, user interface screens (commonly known as storyboards) are a very valuable technique for communicating with subject matter experts, in this case, navigators (Whitaker & CuQlock-Knopp, 1991b). When we ask whether a particular feature would be helpful to them, it is much easier to show them the mockup of this feature on the computer screen than to describe it to them in words. We used Hypercard 1.2.5 to develop such screens. Both contour maps and horizon profiles were used to display the navigational terrain. Functions were developed to aid Route Choice, Location, and Route Finding skills. Man-made features, water, and contours were used as visual cues. Rules of thumb obtained from the interviews were used as explanations of route choices. These storyboards were shown to a subset of the interviewed orienteers. They were shown on a MacIntosh IIci with VGA monitor. The changes suggested by the orienteers were made as notes on a paper form of these screens. The changes were then incorporated in the storyboards and rechecked with the orienteers. By this iterative process, we developed the collection of user interfaces for the first prototype called NAVAID.

In conclusion, we are in the process of developing and testing a cognitive model of off-road navigation. This model will include the navigation tasks commonly encountered in this environment, the problem-solving strategies employed to accomplish those tasks, and the important visual cues navigators use as information features. We are in the process of instantiating this model as a prototype navigational aid that will serve as a proof-of-concept for this navigational model. This project focused on the human-task environment as the correct unit of analysis and required a broad range of research tools. It was challenging and continues to be enlightening with each iteration in the process.

Acknowledgments

This work was conducted under contract DAAA15-89-C-0505 from the Army Human Engineering Laboratory, Aberdeen Proving Ground, to Klein Associates Inc., Fairborn, OH.

8.6   References

Aretz, A. J., & Wickens, C. D. (1990, March). *Cognitive requirements for aircraft navigation* (ARL-90-3, NASA-90-2). Urbana: Aviation Research Laboratory, Institute of Aviation: University of Illinois at Urbana-Champaign.

Barfield, W., Sandford, J., & Foley, J. (1988). The mental rotation and perceived realism of computer-generated three-dimensional images. *International Journal of Man-Machine Studies, 29,* 669–684.

Blades, M., & Spencer, C. (1987). How do people use maps to navigate through the world? *Cartographica, 24,* 64–75.

Boys, R. M. (1986, May). Terrain-based information: A reason to integrate. *IEEE NAECON 86, Proceedings of the National Aerospace Electronics Conference* (pp. 888–893). Dayton, OH.

Cadwallader, M. (1979). Problems in cognitive distance: Implications for cognitive mapping. *Environment and Behavior, 11,* 559–576.

Chase, W., & Chi, M. T. H. (1980). Cognitive skill: Implications for spatial skill in large-scale environments. In J. Harvey (Ed.), *Cognition, social behavior, and the environment* (pp. 111–136). Hillsdale, N.J.: Lawrence Erlbaum Associates.

Cooper, L. A., & Shepard, R. N. (1973). Chronometric studies of the rotation of mental images. In W. G. Chase (Ed.), *Visual information processing* (pp. 75–175). New York: Academic Press.

Couclelis, H., Golledge, R. G., Gale, N., & Tobler, W. (1987) Exploring the anchor-point hypothesis of spatial cognition. *Journal of Environmental Psychology, 7,* 99–122.

Crandall, B. W. (1989). A comparative study of think aloud and critical decision knowledge elicitation methods. *SIGART Newsletter,* [Knowledge Acquisition Special Issue], *108,* 144–146.

Dennett, D. (1987). Cognitive wheels: The frame problem of artificial intelligence. In Z. W. Pylyshyn (Ed.), *The robot's dilemma: The frame problem in artificial intelligence* (pp. 41–64). Norwood, NJ: Ablex.

Dreyfus, H. L. (1972). *What computers can't do: A critique of artificial reason.* New York: Harper & Row.

Evans, G. W., & Pezdek, K. (1980). Cognitive mapping: Knowledge of real-world distance and location information. *Journal of Experimental Psychology: Human Learning and Memory, 6,* 13–24.

Evans, G. W., Skorpanich, M. A., Garling, T., Bryant, K. J., & Bresolin, B. (1984). The effects of pathway configuration, landmarks, and stress on environmental cognition. *Journal of Environmental Psychology , 4,* 323–335.

Fletcher, C. (1972). *The man who walked through time.* New York: Vintage Books.

Fodor, J. A. (1990). Modules, frames, fridgeons, sleeping dogs, and the music of the spheres. In J. Garfield (Ed.), *Foundations of cognitive science: The essential readings* (pp. 235–246). New York: Paragon House.

Foley, J. E., & Cohen, A. J. (1984). Working mental representations of the environment. *Environment and Behavior, 16,* 713–729.

Goldin, S. E., and Thorndyke, P. W. (1982) Simulating navigation for spatial knowledge acquisition. *Human Factors, 24,* 457–471.

Gordon, H. W. (1987). *The cognitive laterality battery.* Pittsburgh, PA: University of Pittsburgh School of Medicine, Western Psychiatric Institute and Clinic.

Hardwick, D. A., Woolridge, S. C. & Rinalducci, E. J. (1983). Selection of landmarks as a correlate of cognitive map organization. *Psychological Reports, 53,* 807–813.

Harvey, J. (Ed.). (1980). *Cognition, social behavior, and the environment.* Hillsdale, NJ: Lawrence Erlbaum Associates.

Hunt, M. E. (1984). Environmental learning without being there. *Environment and Behavior, 16,* 307–334.

Just, M. A., & Carpenter, P. A. (1985). Cognitive coordinate systems: Accounts of mental rotation and individual differences in spatial ability. *Psychological Review, 92,* 137–172.

Kantowitz, B. H., Roediger, H. L., & Elmes, D. G. (1991). *Experimental psychology* (4th ed.). New York: West Publishing.

Kaplan, R. (1976). Way-finding in the natural environment. In G.T. Moore & R.G. Golledge (Ed.), *Environmental knowing: Theories, research, and methods* (pp. 46–57). Stroudsburg, PA: Dowden, Hutchinson & Ross.

Kiernan, V. (1991). Guidance from above in the Gulf War. *Science, 251,* 1012–1014.

King, R.B. (1990). Exploitation of geographic information system (GIS) technology for decision aiding. *Proceedings of the 7th Annual Workshop on Command and Control Decision Aiding* (pp. 3.1–3.10). Dayton, OH: Air Force Institute of Technology, Wright-Patterson Air Force Base.

Klein, G. (1989). Recognition-primed decisions. In W. Rouse (Ed.), *Advances in man-machine systems research* (Vol. 5, pp. 47–92. Greenwich, CT: JAI Press.

Klein, G. (1993). A RPD model of rapid decision making. In G. Klein, J. Orasanu, R. Calderwood, & C. Zsambok (Ed.), *Decision making in action: Models and methods* (pp. 138–147). Norwood, NJ: Ablex.

Kozlowski, L. T., & Bryant, K. J. (1977). Sense of direction, spatial orientation, and cognitive maps. *Journal of Experimental Psychology: Human Perception and Performance, 3,* 590–598.

Leiser, D. (1987). The changing relations of representation and cognitive structure during the development of a cognitive map. *New Ideas in Psychology, 5,* 95–110.

McNamara, T. P. (1986). Mental representations of spatial relations. *Cognitive Psychology, 18,* 87–121.

Monmonier, M. (1991) *How to lie with maps.* Chicago: University of Chicago Press.

Murakoshi, S. (1990). Map reading beyond information given. *Scientific Journal of Orienteering, 6,* 10–25.

Pollack, A. (1991, February 6). War spurs navigation by satellite. *New York Times,* (p. C1 +).

Rossano, M. J., & Warren, D. H. (1989a). The importance of alignment in blind subject's use of tactual maps. *Perception, 18,* 805–816.

Rossano, M. J., & Warren, D. H. (1989b) Misaligned maps lead to predictable errors. *Perception, 18,* 215–229.

Salichtchev, K.A. (1978). Cartographic communication/Its place in the theory of science. *The Canadian Cartographer,* pp. 93–99.

Seiler, R. (1985). The psychological structure of information-seeking and decision-making in route-choice situations in orienteering: An empirical study with Swiss elite orienteers. *Scientific Journal of Orienteering, 1,* 24–34.

Selwyn, V. (1987). *Plan your route: The new approach to map reading.* London: David and Charles.

Shepard, R. N., & Cooper, L. A. (1982). *Mental images and their transformations.* Cambridge, MA: The MIT Press.

Sholl, M. J. (1987). Cognitive maps as orienting schemata. *Journal of Experimental Psychology: Learning, Memory, and Cognition, 13,* 615–628.

Streeter, L. A., Vitello, D., & Wonsiewicz, S. A. (1985). How to tell people where to go: Comparing navigational aids. *International Journal Man-Machine Studies, 22,* 549–562.

Thompson, W. B., Pick, H. L., Bennett, B. H., Heinrichs, M. R., Savitt, S. L., & Smith, K. (1990). Map-based localization: The "Drop-off" problem. *Proceedings of Image Understanding Workshop* (pp. 706–719). Sponsored by Defence Advanced Research Projects Agency, Information Science and Technology Office. Pittsburgh, PA.: Morgan Kaufmann Publishers, San Mateo CA

Thorndyke, P. W. (1980). Spatial cognition and reasoning. In J. Harvey (Ed.), *Cognition, social behavior, and the environment* (pp. 137–149). Hillsdale, NJ: Lawrence Erlbaum, Associates.

Thorndyke, P. W., & Hayes-Roth, B. (1982). Differences in spatial knowledge acquired from maps and navigation. *Cognitive Psychology, 14,* 560–589.

Thorndyke, P., & Stasz, C. (1980). Individual differences in procedures for knowledge acquisition from maps. *Cognitive Psychology, 12,* 137–175.

Tolman, E. C. (1948). Cognitive maps in rats and men. *Psychological Review, 55,* 189–208.

Whitaker, L. A., & CuQlock-Knopp, V. G. (1990). Adaptive decision aiding for off-road navigation. *Proceedings of the 7th Annual Workshop on Command and Control Decision Aiding* (pp. 3j: 1–11). Dayton, OH: Air Force Institute of Technology, Wright-Patterson Air Force Base.

Whitaker, L. A., and CuQlock-Knopp, V. G. (1991a). Command and control navigation: A decision support system. *Proceedings of the National Aerospace and Electronics Conference 91,* (pp. 1024–1030). Dayton, OH.

Whitaker, L. A., & CuQlock-Knopp, V. G. (1991b). Prototyping a navigational aid to supplement map and compass. *Proceedings of the 8th Annual Decision Aiding Conference.* Washington, DC: Joint Directors of Laboratory Technology: Department of Defense.

Whitaker, L. A., & CuQlock-Knopp, V. G. (1991c). Use of visual cues by orienteers: An analysis of interview data. *Proceedings of the Human Factors Society 35th Annual Meeting* (pp. 1566–1569). Santa Monica, CA: Human Factors Society.

Wickens, C. D. (1984). *Engineering psychology and human performance.*

Columbus, OH: Merrill.

Wickens, C. D. (1990). Navigational ergonomics. In E. J. Lovesey (Ed.), *Contemporary ergonomics*, (pp. 16–29). London: Taylor and Francis Ltd.

Wickens, C. D.(1992) *Engineering psychology and human performance*. (2nd Edition). Harper Collins:

Yoder, S. K. (1991) Tiny Trimble locates profits in Gulf War. New York Times, April 24, 1991.

Chapter 9

Topographic Map Reading

H. L. Pick, M. R. Heinrichs, D. R. Montello, K. Smith, and C. N. Sullivan

University of Minnesota

W. B. Thompson

University of Utah

9.0   Introduction: Nature and Context of Problem

Topographic maps are the relatively familiar form of geographic maps in which elevation is represented by contour lines connecting locations of equal elevation. These can be, and are, used as navigational aids for helping to solve quite sophisticated forms of way-finding problems. The present chapter is concerned with how such maps are used for solving localization problems. However, before addressing specifically how topographic maps are used, it is worthwhile to consider such technological aids in the general context of navigation.

Navigation can be considered to be the process by which an organism or machine finds its way through the environment. This obviously involves noting one's current position, planning a route to a desired location, and negotiating that route. At one extreme, navigation tasks involve small-scale spaces in which the entire space can be apprehended immediately and one's current position as well as the desired location can be simultaneously perceived. This is the case, for example, in locomoting around a room or an open field. Even in such instances a certain amount of route planning may be involved. One may have to decide whether to go over or under a barrier, through an aperture (Warren & Whang, 1987), over or under a table, and so on) Such decisions are often made so effortlessly that we tend to forget that they can be problematic. However, their potential difficulty is apparent when the behavior of immature organisms is examined, such as infants deciding how to negotiate a slope (Adolph Gibson, & Eppler, 1990) or deciding what size aperture to go through (Palmer, 1987), or when one observes visually impaired persons using a long cane to detect free paths (Farmer, 1980).

At the other extreme is navigation around large-scale spaces in which all parts of the space are not simultaneously perceivable. The current position or viewpoint of the navigator is presumably perceptually available, but the destination is not, nor are greater or lesser portions of possible routes. In this case the location of the destination needs to be specified in relation to the current position and possible routes determined. For many the problems that arise in everyday life in such situations are all too familiar.

In fact, such difficulties are so common that many cultures have developed aids for facilitating this kind of navigation. Some navigational aids such as optical or electronic beacons make destinations and landmarks more salient or involve sensors for detecting one's own position in the world. However, maps are one of the most useful and common navigational aids in our culture, providing a symbolic representation of spatial information about the environment. Cartographers have elaborated the science of map making for many purposes, but two kinds of maps are particularly useful for navigation. A route map is useful when the environment is structured to provide specific and constrained paths from place to place. In such environments the routes are so important and obvious that route maps often omit much other spatial information. In contrast, the topographic map is useful when the environment is not artificially structured with routes (e.g., roads) connecting locations. The spatial topography of the environment, that is, the elevation above sea level of all points of the environmental surface, is represented by contour lines. Thus, within the limits of map resolution (and map errors), the topographic map presents a two-dimensional depiction of a three-dimensional spatial layout of our environment.

Such topographic maps are widely used for both professional and recreational purposes. Geologists and agronomists among others use topographic maps for navigation in their work, and orienteers and hikers use maps in way-finding recreationally. An important navigational use of topographic maps is to help with problems of localization, that is, determining correspondences between particular locations in the environment and locations on the map. Commonly, the environmental location of interest is the current position or viewpoint of a navigator (i.e., "where am I?" problems). Such localization problems can be characterized in terms of how much a priori information is available about likely current positions. At one end of such an information continuum, *drop-off* problems involve substantial initial uncertainty in current position. (The name comes from the extreme case in which a navigator is literally "dropped off" into a totally unfamiliar environment. This can occur practically in an airplane crash or more generally in losing one's way under certain conditions.) Toward the other end of the continuum are updating problems in which the task is to maintain a sense of the current position with respect to a map as that position changes with locomotion.

Two aspects in the use of topographic maps for solving localization problems make it a particularly interesting cognitive–perceptual problem. First of all, as with other types of maps, there are two perceptual tasks: perception of the environmental scene and perception of the environment via the map. Perception of the scene—the terrain—is prototypical of perception of the natural environment. Perception via the map is what Gibson (1979/1986) termed *mediated* perception—perception of encoded information. Thus, as with pictures, blueprints, text, and so on, one obtains information about something else through an immediate medium of a different kind. In the case of a topographic map, the mode of representation of the environment is particularly interesting in that the X–Y or two-dimensional layout of the map is formally similar to the two-dimensional layout of the environment. In contrast, the elevation information on the map is a symbolic or encoded representation of elevation in the environment. Such a mixture of mode of representation might cause difficulties in map reading. A second intriguing aspect of topographic map reading is the radical difference in perspective between the view of the environment which is typically more or less parallel to the ground from eye level and the view of the map representation of the environment which is typically a bird's eye view with the map more or less perpendicular to the line of sight.

9.1   Background Research

Both psychologists and geographers (especially cartographers) have been active in investigation of map reading. The largest amount of this research is on the use of road and street maps or thematic and political maps. However, when topographic map reading has been investigated, the research has been mostly concerned with problems involving detection of correspondences between information about elevation represented by contour lines and information portrayed in a less encoded way. For example, subjects were asked, using a multiple-choice format, to select the one of several line shapes that best represented the elevation profile of a direction line on a map segment crossing a series of contour lines (Griffin & Lock, 1979). The line shapes varied in degree of curvature, whether they were concave or convex, and in direction of slope (left-to-right or right-to-left). Results indicated that lines of uniform slope were most easily identified, followed by concave lines, and then convex ones.

In the context of the issues of the encoded nature of elevation information and the change of perspective from map to environment, researchers have been concerned with the kinds of information processing that are involved in map reading. Taking a relatively direct approach to this question, Chang, Antes, and Lenzen (1985) analyzed the patterns of eye movements used by map readers examining a series of topographic maps in order to identify the high and low areas. Illustrative of their results was the finding that inexperienced map readers had a more uniform distribution of eye fixations across the map areas than experienced map readers whose fixations tended to be concentrated around the areas of high and low elevation.

A related question has been how navigators are able to determine the correspondence between maps and more iconic representations of the world. Another information processing paradigm, that of reaction time, was employed by Eley (1988) to determine the effects of differences in alignment between map and landform view on map reading. Subjects were asked to indicate whether a particular view of a landform matched what would be seen from a given station point on a map looking in a given direction. Typical mental rotation results were obtained. The greater the required viewing angle of the map deviated from the subject's own orientation to the landform view, the longer the reaction time. A second experiment in this study was directed at the question of difference in perspective of map and environment. The elevation angle of the subject above the landform view was varied, and it was found that a viewpoint 30° above horizontal was more effective than higher or lower elevation angles in terms of speed of processing.

Other research approached information processing less directly. In systematic psychometric research, Sholl and Egeth (1982) examined individual differences in information processing of map readers. They related performance on a number of map-reading tasks to several more general standard psychometric measures. The map tasks such as landform identification, slope identification, spot elevation, and terrain visualization were factor analyzed yielding two major factors, one being described as spatial visualization and the other as an altitude estimation factor. Surprisingly, standard tests of spatial ability were not highly related to the spatial visualization map-reading factor, whereas verbal-analytic measures were. (The standardized measure of mathematical ability was related to the altitude estimation map-reading factor; yet, finding the altitude of points on a topographic map or finding the highest and lowest elevations on a map would not seem to involve very sophisticated mathematics. The authors suggest that the relationship between mathematical ability and altitude estimation is probably due to arithmetic skills.) In general, the results seem to suggest that our standardized tests do not reflect very well the skills used in a practical task such as topographic map reading.

Overall, the research on topographic map reading is limited in extent, but it is interesting and tantalizing. The results suggest a rather sophisticated skill, but we do not yet understand, either through analysis of individual differences or through analysis of task process, the exact nature of this skill. One reason is simply that there has been relatively little research. Another is that the tasks used are artificial in two respects. They are artificial in the materials used. The samples of the maps themselves are real but are often only tiny segments of real maps. When the experimental tasks involve relating maps to the environment, they typically do not involve the real environment, but employ relatively impoverished sketches. These may, on the one hand, emphasize features that would not be as clear with natural terrain or, on the other hand, omit the incredible richness of natural terrain. The tasks are often artificial in the problems posed. Subjects may be asked only to find high and low spots, to judge the qualitative nature of a landform, and so on, and usually not even to solve a localization problem. (One source for research describing navigation in natural terrain is the *Scientific Journal of Orienteering*, edited by R. Seiler.[1])

The purpose of the remaining part of this chapter is to summarize a program of research on topographic map reading in which an attempt

---

[1]The *Scientific Journal of Orienteering* can be obtained from the International Orienteering Federation, Secretary General, P.O. Box 76, S-191 21 Sollentuna, Sweden.

was made to use more realistic materials and a more complex and realistic map reading problem than that of much of the previous research. The materials used involve relatively large segments of topographic maps, real environments or photographs of environmental scenes, and localization problems. The prototypic map-reading task examined in this research is the drop-off localization problem, as defined earlier. It was posed to experienced map readers in a real outdoor map reading situation. The data for this prototypic situation consist of the verbal protocols of these map readers as they think aloud working through the problem (as well as the correctness of their solutions). On the basis of these field protocols, additional studies were carried out, some in the laboratory under more controlled conditions, but still using relatively large map segments and photographic scenes. The goal of the research was to identify strategies that experienced map readers use to solve localization problems and the kinds of features in the terrain and on the maps that are important in their solutions.

9.2   Experimental Research

9.2.1   Field Experiment and Protocol Analysis

The approach involved collection and analysis of protocols of experienced map readers attempting to solve drop-off map reading problems in the field. The map readers were recruited from geology and geography departments, orienteering clubs, and other outdoor and wilderness organizations. The range of experience varied from professionals who use topographic maps daily on the job to experienced recreational users. Among the participants was one who hiked the length of the Brooks range in Alaska for recreation and another who consistently places in the top five in national orienteering competitions.

Each participant was driven blindfolded approximately 30 miles and led by foot to a station point on a hill in the generally rolling terrain of central Minnesota. (The majority of protocols were collected from two different sites.) The blindfold was removed, and the participant was given a portion of a topographic map of the area and asked to think aloud while trying to determine the position on the map that corresponded to the location in the terrain. The verbal reports of the participants were recorded, and they were videotaped as well, so as to be able to see what they were looking and pointing at as they engaged in solving this localization problem. The segment of the map they were

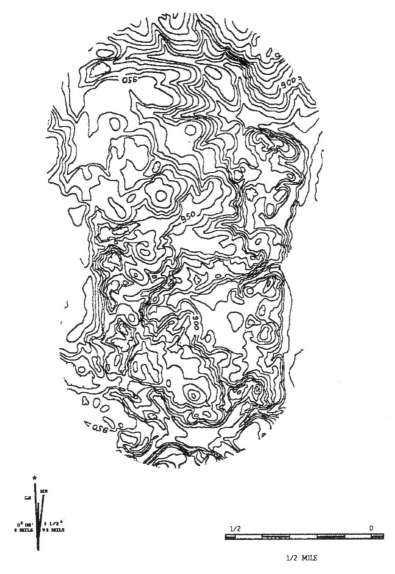

Figure 9.1. *Example of topographic map used for drop-off problem at Site A, O'Brien State Park, Minnesota.*

given was an irregular portion of a U. S. Geological Survey map from which all nontopographic information had been deleted. Thus, it contained no information about roads, railroads, fence lines, or distribution of vegetation, and so on. The relatively impoverished map was used to see whether and how the map readers could solve the localization problem with only topographic information. The map did have a distance scale marked on it and elevation values on some of the contour lines. An example of the topographic map for one site is shown in Figure 9.1.

How well did the map readers do? The problem proved to be quite difficult. Initially, the problem was posed to a group of 17 map readers with instructions not to move from the original station except for turning around and moving slightly to see around occluding bushes or brush. Of these, only one person arrived at a correct solution. A second group of 12 map readers was given the task with instructions that permitted free movement while performing the task. Six of these obtained a correct solution. A chi-square test indicates that this difference in performance between the two groups is statistically significant. However, examination of the actual behavior of the participants in the two groups suggests it was not actually the difference in freedom to move that made the difference. Some of the people in the freedom-to-move condition who solved the problem correctly, in fact, only moved after they had arrived at a correct solution; they moved to confirm the solution that they had already arrived at. The analysis of the verbal protocols indicated that those who arrived at a correct solution employed a particular set of strategies that were at least partially missing in those that failed. We now turn to the protocol analysis.

The goal of analysis of the verbal protocols was to identify the kinds of information that map readers used in attacking this localization problem and to characterize the components of the problem-solving process itself, insofar as the method of protocol analysis permitted. To this end the verbal protocols were carefully examined for each subject, and a coding scheme developed that placed each statement into an "information" trace, on the one hand, and a "process" trace, on the other.

Figure 9.2 is an example of an information trace. The horizontal axis indicates the chronological order in terms of the sequence of statements of the map reader. (Each statement is a coherent utterance with a single focus of attention.) The vertical axis specifies the focus of attention, that is, the source of information, whether it is the map or terrain, and three categories of information—features, relations, and attributes. *Features* are the individual topographic objects that are typically identified with a familiar count noun, such as hill, valley, pond, and so on. Each map

Figure 9.2. *Portion of the "information" trace of a map reader solving the drop-off problem at O'Brien State Park.*

reader's lexicon tended to be small and consistent, although there was diversity between map readers in the specific terms used. The composite lexicon of all the map readers was compressed to form the taxonomy of topographic terms shown in Figure 9.2. *Attributes* are properties that modify individual features. These tend to be bipolar and qualitative and are used to differentiate among similar features, for example, narrow or wide, steep or shallow. Relations are connectives that conjoin two or more features into a single structural unity termed a *configuration*.

Figure 9.3 is an example of a process trace with the horizontal axis again indicating the same chronological order of statements as in the information trace. Hence, the two traces can be coordinated. To make this coordination easier, the vertical axis again contains a focus of attention category. In addition, the vertical axis includes processing categories of metastatement, reconnaissance, matching, other claims, viewpoint hypotheses, and conclusions.

A description of a few of these process categories provides a sense of what they are. *Reconnaissance* involves identification of features, attributes, and relations for subsequent processing. It occurs typically at the beginning of problem solving with broad scanning of terrain or map, but also is found later on in the problem solving process after hypotheses are made and/or tested. An example of reconnaissance focused on the terrain:

> So, umm..., standing on a slope here, it's sloping down on pretty much all the way, like 180 degrees sloping down that direction, so. And it looks like there might be a hill behind us, although it's hard to say if it goes down on the other side. But it looks like a pretty high spot in the terrain area, so it's probably one of the higher areas on the map, especially and higher over there. That's about it. (OA 3-5)[2]

Typical of map reconnaissance would be:

> "Ahh, looking at the maps, the map, it uh, doesn't show trees so as far as the wooded area and.... It doesn't show like the farms out in that direction, ah. I think the biggest thing is for me to use hopefully would be this long valley if it's a stream of a river system. Umm, looking at

---

[2]Coding in parentheses refers to subject and place in protocol from which the quote was taken.

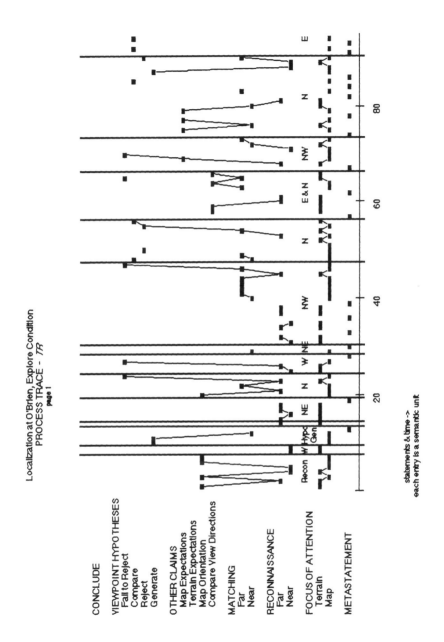

Figure 9.3. *Portion of the "process" trace for the same map reader as in Figure 9.2.*

*the map there, there appears to be a couple of things that could be a
stream valley, umm. This marks a depression with the slash marks.
(DJ 5-9)*

Feature matching, the major activity during the localization task,
involves matching features in the terrain to features in the map or vice
versa. Feature matching does not require the existence of a specific
hypothesis about viewing location. Such matching can establish general
correspondences between the environmental scene and map, facilitating
subsequent generation of specific viewpoint hypotheses. After
hypotheses are formed, feature matching plays a key role in their
evaluation.

Feature matching is based on a common identification and similar
characterization of topographic structures in map and scene.
Identification is done in terms of a set of labels and properties, often
specific to a particular geographic landform. In the geological area of
the present study, the most common features attended to were hills and
valleys. Attempting to find correspondences for the mere presence or
absence of a hill or valley was not particularly diagnostic of location.
Accordingly, map users more commonly attended to the attributes of
these features and the relations between them as noted earlier. Consider
the following examples:

*Then down there there's a big valley so I guess that could be this valley
going down here, and if that's the case, the high area we're seeing,
might be this ridge extending out here, and umm. (OA 15-16)*

*This area right here, ah, gently sloping while fairly flat on top, so
maybe look for some kind of plateau on the map, and, that drops off
relatively to my left to the water and to the front. There's a couple of
areas on the map that look gently rolling like this area here or over in
here, umm, both of them to have a water area off to the left. (DJ 22-26)*

*Hmmmm, I don't know. There should be a hill on the other side of
that, on this wet land right in there. There is a hill I see over there, a
grassy hill. Trees behind it. Could be, could be this hill here. It's kind
of steep slope, indicated by the closeness of these topo lines right here.
(RB 32-34—-hypothesis evaluation)*

Viewpoint hypotheses include the generation, comparison, and
evaluation of hypotheses about the viewpoint. An hypothesis posits a

distinct map location and direction as corresponding to the viewing position. The hypothesis is initially triggered by possible map-scene correspondences between a small number of features or configurations. Evaluation proceeds by examining other scene and map features or configurations using expectations derived from the hypothesis. Often a brief reconnaissance of a local region in the map and/or scene is required to identify additional features and configurations useful in the evaluation process. The strategies involved here have much in common with those used in other diagnostic tasks (e.g., Johnson, Moen, & Thompson, 1988). The following illustrates the generation and rejection of hypotheses:

> And that other open area that we just barely see, between the, that seems that could be this area here. Umm. But yeah, it looks pretty good. The other side, if that is the case, that we're actually down here now, that, if we were down here, that should be like umm, a ridge going out. I guess there is sort of like a ridge right down there. A little ridge. I don't see it bending to the right though. There's definitely a valley going down there, but yeah. Oh I see. Maybe that valley is this valley. In that case, this makes us up, more... no that doesn't look right either, cause then it should be pretty flat, and it looks sloping more going down there. Hmm, maybe I'm in a completely different location on the map. Hmm, it goes... (OA 32-40)

> So what I'm looking at, is that we're kind of on a hump that kind of comes out quite a ways. Maybe I should, seems like we're coming around in direction up a hill like that. So maybe an oblong, more oblong shape hill. Probably something like this down here [referring to map]. Umm, there's more of a ravine on that, but this, this hill here doesn't look like it's big enough. This is the big, seems like the biggest hill in the area, and so to me, that isn't a big enough contour or a big enough hill on this map to signify that's where I'm at. (JP 50-57)

To *conclude* the problem solving, hypothesis evaluation leads to the tentative rejection or confirmation of hypotheses that have been generated. A final step in the localization process produces the best estimate of actual location and viewing direction. Depending on the search strategy used, this may be based on a comparison of the likelihood of competing hypotheses or may simply be the identification of a single hypothesis that survived a sequential generate-and-test procedure. The subject may be satisfied (success) or unsatisfied (failure) with the final statement. An example of each follows:

> *Let's say these are about 10, 20, this is 30 feet above this line so these actually should be the same height. OK, so I probably wouldn't notice it too much. I think, ah, we're here on this ridge. Umm, let's just look at this one more time here. This is, umm, OK, I think we're here. (RB 96-99)*

> *This one doesn't match because it was too steep. What about this one? Maybe this one. I have to...Let's see. But then, umm, there should be a very sharp or steep valley here, but I don't see that at all, so it's probably not that place. And it's probably not on this....Unless it's down here, umm. Because it's very steep below that, and it's certainly not generally sloping here. So I think the best guess is that we're about here. That's my best guess. It doesn't match completely though. (OA 73-79)*

These excerpts are illustrative of the kinds of protocol statements received and their scoring. On the basis of the analysis we provide the following description of the kinds of information and strategies used in the problem solving of the localization task.

9.2.2   Information and Strategies Identified

The information that was attended to by map readers solving the localization problem was defined basically in terms of the features mentioned in the protocols. In general, these would be different for different geographic areas. In the rolling hills of central Minnesota, the features mentioned most frequently included hills, valleys, flat areas or plateaus, and ridges. However, features such as those seen from a particular viewpoint were not usually distinctive enough to uniquely specify a particular feature on the map. To help reduce this ambiguity, map readers took into account two kinds of constraints on the featural information they attended to. First, the features were qualified by denoting distinctive properties such as relative size, elevation, slope, and so on. These were the attributes coded in the information trace earlier and tended to be specified in bipolar qualitative terms. Features were described as large or small, narrow or wide, steep or shallow. Comparison among features was quite common. One feature was described as larger, broader, or steeper than another. It is noteworthy that metric descriptions in terms of units of distance or degree of slope were rarely seen, in spite of the fact that metric information was readily available on the map with distance scale and contour lines at standard intervals.

The second constraint on map readers' attention to features was their tendency to focus on assemblies of features, or configurations. Configurations were specified in terms of the features of which they were composed and the relationship among those features. Those relationships again tended to be qualitative and topological rather than metric (e.g., behind, in front of, next to, etc., although sometimes actual elevation was used when looking at the map). Attention to configurations reduced the number of individual items that needed to be considered, and in addition, configurations were more likely to be distinctive than individual features. There were fewer matches to "a hill with a dip and a ridge," than there were to individual hills, valleys, and ridges. The complexity of the configurations was usually relatively small, typically involving two to four individual features.

On the basis of the protocols, especially the information encoded in the processing trace, several strategies used by map readers were identified. These were particularly useful in establishing the correspondences between features in the terrain and on the map necessary for solving the localization problem. First, initial reconnaissance tended to be concentrated on the terrain rather than on the map. Generally, maps include an area much larger than that visible from any particular viewpoint in the terrain. Consequently, the majority of map features will not be relevant to any viewpoint determination, whereas most of the distinctive visual features of terrain will have correspondences on the map. Second, as noted earlier, features were organized into configurations resulting in a reduction of the number of items necessary to attend to and a decrease in ambiguity. Configurations are particularly useful when assembled along the line of sight. Such configurations have a viewpoint-independent property in that they have a linear representation wherever they appear on the map reflecting their linear alignment in the terrain. They also have a viewpoint-dependent property in that once found on the map they constrain the map location of the viewpoint to a line. If several such configurations around a viewpoint are found, they markedly constrain the map location of the viewpoint. Third, considerable attention was devoted to local terrain features around the viewpoint. Although this seems obvious, most approaches to landmark-based robot navigation do not pay special attention to local features of the immediate environment. If a feature or configuration near the viewpoint corresponds to particular features of the map, then the position of the viewpoint on the map is highly constrained. Furthermore, detailed determination of local features is often easier or more accurate than that of more distant features. Fourth, multiple hypotheses were generated and evaluated. A procedural goal is to select quickly a viewpoint hypothesis that can be evaluated against the current view of the terrain. Because terrain features are highly ambiguous, it is difficult to identify landmarks with certainty. Any single viewpoint hypothesis based on a small number of features has a high probability of being incorrect. In complex terrain, it appears necessary to develop a number of different plausible hypotheses for subsequent verification. All successful map readers employed the above four strategies. One or more of the strategies were omitted by those who did not arrive at a correct solution.

Two other strategies also appeared to be quite useful although not uniformly used. A fifth strategy was to compare hypotheses using a disconfirmation procedure. Validation of an hypothesis involves comparing the terrain view with expectations generated from the map based on hypothesized viewpoints. It is most important to note expectations that are not met. If one clear mismatch is found, then the associated hypothesis should be rejected. Because terrain features often look more or less the same, validation based on finding expected features is far less effective than rejecting hypotheses when expected features are not found. Map readers arriving at incorrect solutions often "explained away" incorrect evidence. (The recommendation to look for disconfirming rather than confirming evidence to test alternative hypotheses is a widely accepted component of normative models of inference and decision making [e.g., Platt, 1964; Wason & Johnson-Laird, 1972]. Such a recommendation attempts to counter the pervasive "confirmation bias" found in numerous empirical studies of decision making [Mitroff, 1974; Mynatt, Doherty, & Tweney, 1978; Wason, 1960]. The present work extends the generality of this finding to topographic map reading: Many unsuccessful subjects did not try to eliminate hypotheses. Instead, they tried to "prove" their hypotheses by seeking consistent information. These subjects often concluded by "explaining away" inconsistent evidence as they accepted an incorrect hypothesis. It appears, then, that decision making with the assistance of a topographic map is prone to the same bias and needs to be countered with the same prescription as decision making generally.) Finally, *changing one's viewpoint was important*. Movement to bring obscured features into view or to generate parallax sufficient to gain distance information has clear advantages and is obvious. However, often overlooked is the role of movement in verifying hypotheses specifically about the viewpoint. Viewpoint hypotheses are used to generate expectations about nearby features that can then be confirmed or disconfirmed by local movement.

9.2.3   Laboratory Simulation of the Localization Problem

The field study enabled the collection of protocols of experienced map readers solving a real drop-off localization problem. However, it was limited both in type of terrain examined and in the control of information available to the map readers during problem solving. A laboratory simulation of the problem provided a first attempt to deal with these issues. The localization problem was posed to map readers through photographs of terrain of different types, whereas the maps of the corresponding areas were masked in varying amounts to manipulate information available. For a subgroup of problem solvers, verbal protocols were collected while performing this task in order to determine just how the specific information affected the solutions.

The field problem, even with map readers able to scan all 360° of the terrain, was quite difficult, as previously mentioned. In order that the laboratory problem could even be possible with only photographs of the terrain, a forced-choice task was developed. Map readers were given map-terrain correspondence problems of two types. In one, the "map" task, they had to select the one of three direction lines (arrows) on the map that specified a view corresponding to a photograph of a scene projected on a large screen. In the other, the "scene" task, they had to select the one of three photographs that corresponded to the view that would be seen from a viewpoint marked on a map looking in the direction of an arrow emanating from that viewpoint. The geography sampled across these problems included more rugged hilly and mountainous terrain in Arizona and New Mexico in addition to the gently rolling hill areas of Minnesota.

For different groups of map readers, circular topographic maps were masked in differing amounts. One group had full or unmasked maps. For a second group, the "inner 1/3" masked condition, an area defined by one-third of the radius of the map directly surrounding the central station point was occluded. For a third group, the "outer 1/3" masked condition, the more distal third of the map's radius was masked leaving a central area corresponding to two-thirds of the radius unmasked. Finally, in the "outer 2/3" masked condition, the area corresponding to the distal two-thirds of the radius was masked leaving only a small central area directly surrounding the station point unmasked. As a consequence, the "inner 1/3" and "outer 1/3" conditions were equivalent in terms of the proportion of the radius masked, whereas the "outer 2/3" masked condition had the smallest amount of visible area (Figure 9.4).

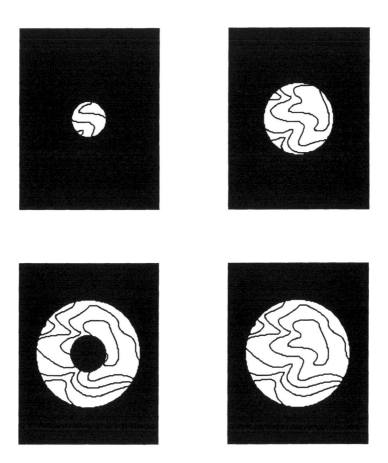

Figure 9.4. *Examples of the different levels of the masked condition used in the laboratory simulation task.*

This masked condition manipulation permits directly addressing the question of the amount of map information needed to solve the task, as well as whether particular areas were favored over others. If the amount of available map area is the only variable affecting performance, the full map condition should produce the best performance, the performance under the "inner 1/3" masked condition would be the next best, followed by the "outer 1/3" masked condition. Finally, worst performance should occur in the "outer 2/3" masked condition because it has the most map area masked. Deviations from these predictions would point to the areas richest in information, and examination of those areas would suggest the specific features most important for problem solution.

Table 9.1 presents response accuracy on the forced-choice tasks for each location tested as a function of masked condition. On the average, accuracy significantly exceeded chance performance in all masked conditions, although not always at each location. Average performance in the full map and the outer 1/3 masked conditions was equivalent and significantly better than performance in the 1/3 inner and 2/3 outer masked conditions [$t(14) = 2.80$, $p<.01$)], which was also equivalent. This pattern of response accuracy suggests that masking areas of the maps impeded the solution of the correspondence problems, only when the areas were close to subjects' locations on the map or when large areas of the maps were masked.

Masking did not uniformly disrupt performance at each location in this manner, however. As Table 9.1 shows, a variety of patterns of results occurred at different locations. For example, accuracy on the O'Brien A map task, like the pattern of average results, was high in the full map and outer 1/3 conditions, but it was low in the inner 1/3 and outer 2/3 conditions. At the least, this suggests that the outer 1/3 radius area did not contain necessary information to solve the task. Accuracy on the New Mexico map task, however, was very poor in all but the inner 1/3 masked condition, suggesting that map information within the 1/3 radius area was possibly misleading to subjects. On the Afton map and O'Brien B scene tasks, accuracy was high in all conditions except the inner 1/3 masked condition, indicating the importance of information within the 1/3 radius area for success on these tasks. At still other locations, accuracy was either uniformly high across conditions (O'Brien A scene task) or uniformly mediocre (Afton scene task). These results suggest that the entire area of a map representing part of the visible landscape is not typically necessary for the solution of correspondence problems. They also suggest that any particular place on the map (such as near the station point) is not consistently necessary to solve the problems.

Table 9.1

Percent Correct on Forced-Choice Tasks for Each Location by Map Masked Condition

Testing Location and Task Type	Masked Condition[1]				Signif. of contrasts among masked conditions
	Full Map	Inner 1/3	Outer 1/3	Outer 2/3	
O'Brien A Map Task	81	35	73	40	p<0.05
O'Brien B Map Task	56	60	33	40	
Arizona Map Task	81	47	47	20*	p<0.01
New Mexico Map Task	12	50*	13	7	p<0.01
Afton Map Task	69	18*	80	67	p<0.01
O'Brien A Scene Task	75	82*	67	73	
O'Brien B Scene Task	75	12	73	60	p<0.01
Arizona Scene Task	62	47	73	47	
New Mexico Scene Task	31	65*	47	47	
Afton Scene Task	50	41	53	53	
Mean	59	45	56	45	

(Notes. Underlined percentages are significantly greater than chance, 33%, at p < 0.05. Doubly underlined percentage is significantly less than chance at p < 0.05. [1]N = 16 in the full map condition, N = 17 in the Inner 1/3 masked condition, and N = 15 in the Outer 1/3 and 2/3 masked conditions. * Masked conditions used in the protocol study.

The variation in arbitrary masking by area done here is the crudest kind of manipulation to ascertain the important information. The eventual goal is to predict specifically the features and configurations that are critical for map readers. The results for individual scenes were examined to begin to get at this question. It is not simply the case that any distinguishing feature will be used if it is the only one available. Consider the O'Brien A map task, for example. The results for this problem fit quite closely the overall pattern of mean results with the performance on the full map condition and on the outer 1/3 condition quite good and performance on the inner 1/3 masked and outer 2/3 masked quite poor (essentially at chance level). In this problem, a distant, large valley on the left could have served as a distinguishing feature in choosing the correct line. This was visible in the inner 1/3 masked condition but not in the outer 1/3 masked condition, but apparently it was not used because performance was so poor. A similar result was obtained in the Afton map task. In the Afton scene, the foreground contained a very distinctive pair of hills that were visible in all the conditions except the inner 1/3 masked. Although the midground was not very informative, there was one very distinctive wide valley far away on the right of the picture that was clearly visible on the map in all the conditions. Although the performance in the two outer masked conditions was as good or better than the full map condition (confirming the importance of the two hills in the foreground), the performance in the inner 1/3 condition was even below chance, suggesting that this far away valley was once again not used.

Assuming that subjects do have a bias toward reliance on foreground features in solving the map tasks, this tendency may lead them into trouble in some situations. The New Mexico map problem is a case in point. Subjects performed best when the inner 1/3 was masked, but when the outer 2/3 was masked, that is, when only the inner 1/3 was visible, the subjects performed at a level significantly below chance. Subjects' comments at the time of testing suggest that they misjudged the foreground slope, perceiving it to be flat or inclining down, even though it actually was slightly rising.

The laboratory simulation task is most valuable for the hints it provides as to the specific information affecting the problem solution. Verification of these hints can be obtained from the verbal protocols collected from the additional map readers given a subset of the original simulation problems.

9.2.4   Protocol Analysis and Laboratory Simulation of Localization

Three map tasks and two scene tasks were selected from the original set of laboratory simulation problems. (The particular tasks and their masked condition are marked with an asterisk in Table 9.1.) Ten additional map readers solved the problems twice, first in a masked condition and then in the unmasked full map condition. As in the field study, subjects were asked to think aloud while solving the problem, this time in the laboratory. The overall results indicate 44% successful solutions in the masked map condition and 64% successful solutions in the corresponding full map condition (see Table 9.1). The performance on the full map condition is significantly above the chance level of 33%, whereas performance under the masked condition does not differ significantly from chance. However, these values approximate the average performance values from original simulation tasks considering the full map, inner 1/3 masked, and outer 2/3 masked conditions from which this subset of problems was taken. Thus, the overall results from these protocol map readers represent a reasonable replication of the original simulation task.

The protocols were scored in the same manner as in the earlier field study. The main scoring category that did not apply in the same way with these laboratory protocols was hypothesis generation. With the laboratory task the hypotheses were in a sense given to the subjects by means of the three choices of map direction or scene. Their task was testing the choices.

We now consider for detailed analysis two of the problems used in this laboratory simulation: the New Mexico map task and the Afton map task. As mentioned earlier and evident from Table 9.1, the New Mexico map task is one in which performance is paradoxically better under the inner 1/3 masked condition than on the full map condition. The results of the present experiment replicate the earlier findings. Six of the 10 new participants chose the correct one of the three map arrows under the masked condition, but only 1 of the 10 under the full map condition. A different pattern was found with the Afton map task: Originally, task performance under the full map condition was better than under the inner 1/3 masked condition. Again the results replicate these additional map readers: 9 of 10 gave the correct answer under the full map condition, while all 10 gave the incorrect answer under the masked condition.

The protocols help account for these patterns. In the inner 1/3 masked condition of the New Mexico map task, the mask covers most of the terrain presented in the slide. This pushed the participants toward a disconfirmation strategy with which they were generally successful. One incorrect direction arrow had a prominent hill in the background that the participants surmised was in the background of the slide. This permitted rejection of that arrow. Then they were able to guess between the other two. For example:

1. Umm. The slide is ah, the slide is a fairly flat area, I can't ah.... The map doesn't look particularly flat. Ah, O.K. I guess I could look for, I guess I could look for things in the distance and see if.... It probably isn't (arrow) 2 because if it was in the direction of 2 there's some sort of hill in that direction. And since I'm not seeing a hill in the distance, it probably isn't there, although the trees could be obscuring it. Umm... I guess ah, let's see now. It's hard to say. I'm just going to eliminate 1 for the same reason I guess. Well... 7400 ft. fairly close there, whereas, in that direction (arrow 3) there's also a 7400 ft. point but it's a little further off, so that would be more likely obscured on 3. So 1 is the best guess I can make. (Correct, JS New Mexico, Map-masked condition)

When they got to the full map condition, they chose the arrow that had the gentlest slope close to the station point. The terrain in the slide appeared to be almost flat, although in fact it was rising, thus accounting for their erroneous responses. Here, as in the field study, errors were caused by incorrectly assessing the terrain of the station point:

2. This one looks so flat it's hard to tell anything. I guess if anything those trees are maybe a little bit higher, ah.... It's really hard to tell ah, I guess maybe I can try and eliminate things, umm.... O.K. if I was looking in the direction of (arrow) 1, I would expect to be looking up, right in front of me. Well I don't know how steep a slope that is, but I guess it's a couple of contour lines. Umm, let's see, smaller lines are 20 ft. intervals so that would be up about 40 ft. in the space of 100 yds. It doesn't look like it's going up that much. I'm inclined to think that (arrow) 1 would be going up a little bit more than this one is. Ah, (arrow) number 2 there's generally sort of a... from the left to the right, it's going down. This seems so flat. Doesn't even seem like there's a slight downhill. Number 3 is I guess the flattest looking one. Ah...Hmm... I guess since number 3 looks the flattest looking and this looks so flat, I'm going to guess number 3. Cause there just doesn't seem to be... Number 2, it's too steep a hill going up. I'm sorry, number 1 (rejects arrow 1), and number 2 I'd expect to see a little more of a left to right, left to right downhill, some sort of angle. It seems so flat that I'll say number 3. That's a hard one. (Incorrect JS New Mexico Map full map condition)

The Afton Map task was one in which performance under the inner 1/3 masked condition was markedly deficient in comparison with the full map condition as evident in Table 9.1. In fact, the pattern of results for all the masked conditions would suggest that the crucial information for distinguishing among the arrows on the map was close to the station point. From examination of the map and scene, a nearby prominent hill would appear to be the primary critical distinguishing information. Two strategies were identified from the protocols: one in which more attention was paid to the map, and the other in which more attention was spent on the scene. When the center masked map was the focus of attention, the salient feature was a large river valley, and an attempt was made to see how this fit into the scene. Then subjects choose between two plausible direction arrows:

> 3. I am starting by looking at the map and am trying to determine the general shape of the terrain. I believe that this area here represents some high land and this is a river running in a quite deep gorge as indicated by the very close contour lines. This here represents I believe a valley... probably a stream valley which comes up between this high land some other high land on the other side. I am a little surprised looking at the picture because I expected that the land form, for instance, that this describes would appear steeper land than I, appears on the picture. If I was looking in this direction (arrow 3) I think I would be looking downhill and across this...I assume this is a river but maybe I...no I think it must be... and then on to some banks on the other side. If I was looking in this direction (arrow 2), I am looking constantly downhill. And though I don't what is out here, it doesn't appear to be what I am looking there. This (arrow 1) shows that it is slightly downhill and then over perhaps a high point there . Which I think is probably that. I choose direction 1. (Incorrect JD Afton Map task—-masked condition)

When the scene was the focus of attention in this masked map condition, subjects seized on the salient feature—the hill—and ignored the distance scale and incorrectly selected the direction arrow that showed a hill:

> 4. I guess I'm looking up a hill. From the map I was gonna guess that I was gonna be looking down on pretty much everything. There's looks like two trails going though. I don't have those on the map. I don't, looking for a river but I don't see that in there which should make me eliminate choice 3. Looking to see if there is another hill on here... It appears that number, ah, choice number 1 goes across a low area and then back up a hill to a higher point. And that would be the

*direction I would choose of the three geographic areas. Number 1.
(Incorrect TH Afton Map task—-masked condition)*

In the full map condition of the Afton Map task, all subjects focused on the hill feature and an attribute, orientation of the hill, or the distance of the hill from the station point. This readily yielded the correct answer:

*5. O.K. This is much easier. O.K. now for number 3 to be correct I
want to see a big re-entrant, two big valleys right ahead of me and
leading into a lake. I don't see that at all. So, 3 doesn't make any
sense at all. Now (arrow) 1. I would see a slight downhill and then a
smaller hill in front of me before it drops off into a big valley. There is
no indication of a big hill from what I am seeing on the slide to indicate
that. So that doesn't make sense. Number 2 does have...ummm... this
hill here, this big knoll could easily be that big hill on the map on the
slide. And it also looks like you could see some of the things that we're
seeing in the background—- the place where the road goes and comes
in a lower spot and goes around the hill. That could definitely be
around here. And you probably can't see anything off here because it
is just too far. So now I would say that it is number 2. (Correct PD
Afton Map task—-full map condition)*

The protocols help explain the particular patterns of results obtained for the different problems. In addition, they also illustrate a number of features that frequently occur in the problem solving of subjects in the field as well as in the laboratory simulation. One aspect was a tendency to focus on particular salient features. This occurred with the large river valley in response 3 above and with the hill in response 4 and 5. Even when the focus was on a salient feature, a second aspect of the problem solving involved attempting to find more reliable configurations or combinations of features as what happened with the attributes of the river valley in response 3 and the observation of the low area and hill going to a higher point in response 4. As mentioned before, a common source of error was incorrect registration of the area very close to the viewpoint which occurred in response 2. It was also the case that metric information was often ignored, which led to error as in response 4. However, often ordinal information about the relative heights of features or magnitude of distances was sufficient to decide between hypotheses. Finally, in testing hypotheses, especially in the laboratory simulation, subjects realized that detection of one clear difference

between a hypothesized position and what was visible in the terrain was
sufficient to rule out a hypothesis. This was exemplified by the
disconfirmation strategy in response 1. However, acceptance of an
hypothesis usually requires more converging evidence (Smith,
Heinrichs, & Pick, 1991), which is one of the results of attending to
configurations.

9.3   Conclusion

This chapter described how experienced topographic map readers
approach a particularly difficult form of localization task: the drop-off
problem. It is possible to describe the performance of these map readers
in terms of the information they attend to and the process strategies that
they use. The specific features attended to depend on the topography of
the particular problems. However, the commonly found ambiguity of
features is generally resolved by noting qualitative characteristics or
attributes of individual features and attending to relations among
multiple features.

It is noteworthy that feature attributes of the topography are
generally described with qualitative and ordinal terms instead of metric
values. Metric estimations of distance from the viewer to features and
distance among features as well as metric judgments of the steepness of
hills and valleys is potentially very useful. The maps contain
considerable metric information that could increase the precision of map
terrain matches. Some relevant psychophysical research, for example,
by DeSilva (1985) and Haber (1985), in outdoor environments, suggested
that sensitivity to distance across terrain is quite high. Haber, in
particular, found that subjects were very accurate in magnitude
estimations of distances between markers distributed around an open
field, although there was some underestimation of distances along the
line of sight (radial distance) relative to distances across the line of sight
(lateral distance). However, in almost all psychophysical studies of
distance in outdoor terrain, the terrain has been quite homogeneous and
flat. In contrast, data have been collected in connection with the present
map-reading research involving psychophysical judgments of distance
between locations across nonhomogeneous rolling terrain similar to that
of the current field study. These judgments turned out to be rather
inaccurate and unreliable. There have been almost no psychophysical
studies of the steepness of slope in outdoor terrain. Such data collected
in connection with the present study have also proved inaccurate and
unreliable. Thus, it would appear to be the case that, at least without

special training, people would not ordinarily be sensitive to the metric characteristics of the topographic features they are observing.

Information processes such as reconnaissance, feature matching, and so on. were identified from the protocols of the map readers solving the drop-off problem. Analysis of the use of these processes resulted in descriptions of several process strategies employed by the successful map readers. They include general reconnaissance focusing on the terrain, use of the relations among features to assemble configurations, attention to local terrain features around the viewpoint, and generation of multiple hypotheses about the location of the viewpoint. Testing hypotheses using a disconfirmation strategy is also commonly observed, as is changing one's position in order to gain additional information about the nature of one's own viewpoint.

Of particular interest is the absence in the protocols of any mention of a holistic global visualization strategy. In informal interviews some topographic map readers reported a process of global visualization of the terrain when looking at topographic map. There is practically no evidence for this in any of the protocols collected here. Rather, the processes reported were attention to features and feature or configuration matching. It is possible that the localization task biases the map readers in this direction, and that a task of studying a map for a more general purpose such as getting a feel for the land would lead to the global visualization.

Also of interest is the fact that the protocols provide practically no evidence for any quantitative geometric reasoning, such as triangulation processes, or even qualitative geometric reasoning, such as deciding which side of a line between a pair of features one is on. A conjunction of several such "landmark-pair-boundary" decisions, if appropriately chosen, can tightly constrain one's own position as shown by Levitt, Lawton, Chelberg, Koitzsch, and Dye (1988). It is, of course, possible or even probable that a crude form of triangulation is being carried out semi-automatically and does not appear on the protocols, or that qualitative decisions are being made about landmark-pair-boundaries. Levitt, Lawton, Chelberg, and Nelson (1987) and Sutherland (1992) showed that quantitative triangulation processes, even with only a single pair of landmarks, logically constrains one's position to a circle going through the two landmarks. Thus, if two landmarks are identified in the terrain, map readers should be able to locate themselves on a map at a point that subtends the same angle with the map landmarks as the visual angle subtending the terrain landmarks. There is a partial circle of such points whose circumference goes through the landmarks. This very tight constraint depends on accurate appreciation of the visual angles subtended by landmarks. Preliminary observations suggest that untrained observers have only the crudest sensitivity to the size of visual angles subtended by landmarks.

The material presented in this chapter may be useful in training both the qualitative and quantitative aspects of topographic map reading. On the qualitative side, the types of processes and strategies identified here could be incorporated into training programs. On the quantitative side, procedures could also be developed to improve distance and slope estimation over irregular terrain. Similar, training methods for increasing accuracy of estimation of visual angle should not be difficult to devise and evaluate.

The drop-off localization problem is one of the more difficult uses of topographic maps. A more common use of topographic maps is for updating when one's initial position is more or less precisely known. It will be important to determine what aspects of the processes and strategies identified here will also be useful for the updating problem. The identification and matching of features will almost certainly maintain a central role. Along with terrain reconnaissance, it is very likely that accurate registration of one's own locomotion will be important. This is an aspect of navigation in what was referred to in the introduction as small-scale spaces. Presumably keeping track of one's own movement depends partially on the continuous optical flow stimulation that specifies how far, how fast, and in what direction one is going. This information must be remembered and periodically integrated with map information. Thus, the updating localization problem is a way of relating navigation in large- and small-scale spaces in the context of topographic map reading.

Acknowledgments

The research reported in this chapter was supported by Grant AFOSR-88-0187 from the Air Force Office of Scientific Research and by Grant IRI-8901888, a joint NSF-DARPA Image Understanding initiative, to the University of Minnesota.

The authors are indebted to Bonnie Bennett and Elizabeth Stuck for critical evaluation of an earlier draft of the present chapter and to Karl Rosengren for considerable preliminary observations.

9.4 References

Adolph, K. E., Gibson, E. J., & Eppler, M. A. (1990). *Perceiving affordances of slopes: The ups and downs of toddlers' locomotion* (Emory Cognition Project, Rep. #16). Atlanta: Psychology Department, Emory University.

Chang, K-T., Lenzen, T., & Antes, J. (1985). The effect of experience on reading topographic relief information: Analyses of performance and eye movements. *The Cartographic Journal, 22,* 88–94.

DaSilva, J. A. (1985). Scales for perceived egocentric distance in a large open field: Comparison of three psychophysical methods. *American Journal of Psychology, 98,* 119–144.

Eley, M. G. (1988). Determing the shapes of land surfaces from topographical maps. *Ergonomics, 31,* 355–376.

Farmer, L. W. (1980). Mobility devices. In R. L. Welsh & B. B. Blasch (Eds.) *Foundations of orientation and mobility* (pp. 357–412). New York: American Foundation for the Blind.

Gibson, J. J. (1986). *The ecological approach to visual perception.* Hillsdale, NJ: Lawrence Erlbaum Associates. (Original work published 1979)

Griffen, T. L. C., & Lock, B.F. (1979). The perceptual problem of contour interpretation. *The Cartographic Journal, 16*(2), 61–71.

Haber, R. (1985). Toward a theory of the perceived spatial layout of scenes. *Computer Vision, Graphics, and Image Processing, 31,* 282–321.

Johnson, P.E., Moen, J.B., & Thompson, W.B. (1988). Garden path errors in diagnostic reasoning. In L. Bloc & M.J. Coombs (Eds.), *Computer expert systems* (pp. 395–427). Hiedelberg, Germany: Springer-Verlag.

Levitt, T. S., Lawton, D. T., Chelberg, D. M., & Nelson, P. C. (1987). Qualitative navigation. *Proceedings of DARPA Image Understanding Workshop* (pp. 447–465). Los Altos, CA: Morgan Kaufmann.

Levitt, T., Lawton, D. T., Chelberg, D. M., Koitzsch, K. V., & Dye, J. W. (1988). Qualitative navigation II. *Proceedings of DARPA Image Understanding Workshop* (pp. 319–326). Los Altos, CA: Morgan Kautmann.

Mitroff, I. (1974). *The subjective side of science.* Amsterdam: Elsevier.

Mynatt, C. R., Doherty, M. E., & Tweney, R. D. (1978). Consequences of confirmation and disconfirmation in a simulated research environment. *Quarterly Journal of Experimental Psychology, 30,* 395–406.

Palmer, C. (1987, July). *Infant locomotion through apertures varying in width.* Paper presented at the International Conference on Event Perception and Action, Trieste, Italy.

Platt, J. R. (1964). Strong inference. *Science, 146,* 347–353.

Smith, K., Heinrichs, M. R., & Pick, H. L., Jr. (1991). Similarity judgment and expert localization. *Proceedings of the Thirteenth Annual Conference of the Cognitive Science Society* (pp. 706–719). Chicago. Hillsdale, NJ: Lawrence Erlbaum Associates.

Sutherland, K. T. (1992). Sensitivity of feature configuration in viewpoint determination. *Proceedings of DARPA Image Understanding Workshop* (pp. 315–320). San Mateo, CA: Morgan Kaufmann.

Warren, W., & Whang, S. (1987). Visual guidance of walking through apertures: Body-scaled information for affordances. *Journal of Experimental Psychology: Human Perception and Performance, 13,* 371–383.

Wason, P. C. (1960). On the failure to eliminate hypotheses in a conceptual task. *Quarterly Journal of Experimental Psychology, 12,* 129–140.

Wason, P. C., & Johnson-Laird, P. N. (1972). *Psychology of reasoning: Structure and content.* Cambridge, MA: Harvard University Press.

Chapter 10

On the Specification of the Information Available for the Perception and Description of the Natural Terrain

Robert R. Hoffman

Richard J. Pike

Adelphi University

U.S. Geological Survey

> *The first step . . . is to describe the visible terrestrial environment.*
> *—Neisser (1991, p.15)*

10.0   Introduction

As that "first step" to which Neisser refers, we attempt some interdisciplinary mediation by presenting research that links cognitive science with environmental science. Our work began with the problem of developing expert systems for a process called aerial photo interpretation or "terrain analysis," which we describe in more detail later. Needed for that project was a knowledge base incorporating the terminology, concepts, and principles involved in this domain of expertise. To "pick the brains" of the experts, Hoffman (1984, 1987) utilized methods from sociolinguistics (structured interviewing), methods from cognitive psychology (picture perception tasks), methods from ergonomics (task analysis), and methods from psycholinguistics (propositional analysis of verbal protocols). In this chapter we report on the results of the psycholinguistic research—an analysis of the terminology that is used, by experts and nonexperts alike—as they view and attempt to describe the natural terrain. The results point to the basic problem—the specification the information provided by both the direct visual exploration of terrain and by the stereoscopic aerial photos utilized in terrain analysis.

Why does the world look the way it does? This question is typically raised regarding theories of perception and the problem of explaining the appearance of the phenomenal world. In that context, the Gibsonian realist's answer is, "Because the world is the way it is." The elaboration of that answer, which involves much research within the paradigm of ecological psychology, has yielded a new corpus of laws collectively called "ecological optics" (Gibson, 1979; Johansson & Borjesson, 1991). What if we took the Gibsonian's answer seriously in the context not of analyzing the perceiver, but of analyzing the world? What do we perceive when we view the natural landscape, and why? In his 1950 book, Gibson illustrated perceptual principles using aerial and surface-perspective photographs. Of particular interest to Gibson was the analysis of surfaces and objects, their edges and textures. Gibson's photographs from terrestrial and near-terrestrial perspectives included fields of grass and tilled soil, roadways, and buildings; his aerial photos were of hilly terrain, airplane landing fields, military barracks, and beaches. Gibson again used aerial photos in subsequent works (1966, chap.10; 1979, chap. 1), and Eleanor Gibson (1969, chap.1) cited remote sensing (infrared photography and sonar) images to illustrate perceptual learning.

Some research was conducted on the perception of natural scenes, specifically, psychophysical judgments of the distances of objects in open fields (e.g., DaSilva, 1985; Haber, 1985). Research in environmental psychology investigated how people perceive the natural environment (e.g., Craik, 1972; Stringer, 1975), but much of the emphasis is sociological. How do people feel about pollution? Are some landscapes rated as more tranquil than others? Are some natural disasters rated as more semantically similar than others? Recent research in "environmental cognition" examined spatial reasoning, cues for path finding, short-term memory for routes, distortions in cognitive maps, and so on. Potentially, all such research relates closely to the general topic of terrain and map understanding, but that literature too seems to have skirted a key point: If it is true that "research on cognitive mapping is basically concerned with describing and understanding people's awareness of the physical phenomena surrounding them" (Golledge, 1987, p. 144), then why not begin with a detailed description of the physical phenomena? In concluding his survey of the field, Golledge seemed to recognize this need (echoing Neisser): "Perhaps the fundamental question that all researchers would raise and that has barely been touched is, how do variations in the environment itself influence the nature of cue selection and the storage of environmental information?" (p.164). Most of the extant research on environmental cognition is not immediately relevant to the question of terrain perception. Researchers in psychology have not posed the basic question: What's there in the natural terrain to be perceived (or judged) in the first place? It is this question, with special reference to topography, that we address here.

10.1 Terrain Analysis

In civil and military engineering, the process of terrain analysis (TA) consists of an elaborate analysis of stereoscopic aerial photographs—adjuncted by maps and other sources of information—to determine soils, rock types, and other pertinent features of terrain.[1] The process of aerial photo interpretation is not only very time-consuming, but by tradition must yield correct answers—What if that dam was sited on unsuitable bedrock? What if the analyst had been wrong about those missile launchers in Cuba? The need for both accuracy and completeness in aerial photo interpretation is just as explicit as in medical diagnosis. No wonder that it takes decades to become a TA expert, let alone one of its "legends." Expertise in aerial photo interpretation involves an enormous amount of detailed declarative knowledge of geological and geobiological processes. As an example, Table 10.1 presents just a tiny fraction of the pertinent knowledge of climate, landforms, rock types, drainage patterns, and other aspects of terrain (from Hoffman, 1984).

The declarative knowledge of the expert terrain analyst goes well beyond the examples in Table 10.1, to include literally thousands of propositions. To one unfamiliar with terrain analysis, the process would therefore seem to serve as an example of the kinds of elaborate information processing typical of many domains of expertise (for a review, see Hoffman, 1992). However, expert aerial photo interpreters have such vast experience that a process—call it "perceptual learning," "concept formation," "automatized pattern recognition," or the "declarative-to-procedural shift"—has resulted in their being able to directly perceive terrain facts that escape the novice. Consider, for example, the following case:

---

[1]Most photos used for TA are taken from airplanes at altitudes of about 10,000 feet, although specialized craft fly much higher. Preferred scales for the photos are 1:20,000, 1:30,000, and 1:40,000.

Table 10.1.

Example Entries from the Terrain Analysis Database.

Climate Context #1: Tropical
    Very rugged topography.
    Lush vegetation.
    Highly eroded in deforested areas.
    Little agriculture in less populated areas.
    Deep soils except on slopes
Rock Forms #3: Domes
    Raised rock defined by closed topography.
    Can be small.
    Can be circular,linear,or ellipsoid in shape.
    Can be compound.
    Can be clustered.
    Structural disturbances at the flanks imply
        tilted beds, faulted beds, hogback ridges,
        rugged mountains.
    Can have radiating fractures.
    Undissected implies young.
    Can be salt, gypsum, intrusive bedrock.
Rock Types #1: Flat Shale
    Gently rolling, irregular plain.
    Rounded contours.
    Symmetrical finger ridges.
    Branching rounded hills with saddle ridges.
    Usually lowlands.
    Uniform gradients imply uniform erosion.
    Tonal bands imply bedding.
    Compound slope gradients imply thick bedding.
    Flared base implies thick bedding
    Scalloped hill bases
    V- and U-shaped gullys.
    Landslides.
    Escarpments, very sharp ridges, steep slopes
        and pinnacles, V-shaped gullys and a
        medium to fine drainage imply sandy soils.
    Humid Climate:
        implies valleys, rounded hill.
        implies dendritic drainage.
        implies fine drainage net, ponds,
        meanders, especially if bedded.
        implies forested, dense on hill slopes
        implies row crops in rectangular
        arrays.
        implies intense agriculture.
    Arid Climate
        implies steep, rounded hills
        implies intermittent drainage.
        implies asymmetrical slopes.
        implies steep gullys.
        implies barren or shrub land.
        implies light or mottled soil tones.
Soil Type #2: Silt
    Light tones.
    Silky texture.
    High water-holding capacity.
    U-shaped gullys
    Highly erodible.
    Less permeable than sand.

Climate Context #3: Arid
    Rugged topography.
    Thin soils.
    Mottled soil tones.
    Rapid erosion.
    Surface mineral deposits
Rock Forms #2.1: Faults
    Long, wide.
    Possible differences in drainage,
        topography, texture, and tone of
        the two sides.
    Talus at base.
    Offset beds, streams.
    Possible folding
    Linear vegetation patterns in arid
        regions, possibly perpendicular to gullys.
    Implies displacement of rock.
Rock Types #3.6 Tilted, interbedded sedimentary
    rock with limestone predominating
    Very rugged topography.
    Rough, knobby texture.
    Massive hills with rounded summits
    Uniform gradients imply homogeneous rock.
    Blocky hills imply thick beds.
    Long, parallel ridges and valleys.
    Slope asymmetry, long gradients.
    Straight midsection of sides.
    Sawtooth ridges.
    Talus at bases.
    Narrow valleys.
    Dendritic, rectangular drainage net.
    V-shaped gullys.
    Humid Climate:
        implies rounded ridges.
        implies dark to medium soil tones.
        implies forested.
    Arid Climate:
        implies stairsteps, blocky hills.
        implies sharp ridges.
Soil Type #4: Gravel
    Igneous, metamorphic or sedimentary rock
    Usually contains some sand, which
        can be mixed or layered.
    V-shaped gullys.
Arid Climate:
    implies shrubland or grassland.
    implies desert varnish tones.
Aeolian Landforms #3: Barchan Dunes
    Distinct crescent shapes.
    Asymmetrical slopes.
    Broad ridges and horns.
    Horns and steeper side point downwind
    Usually clustered, with parallel axes
    Usually 30 to 100 meters in relief.
    In areas of little sand and flat topography
    Internal drainage, but possibly a few V-
        shaped gullys.

Based on what I see in these photos, this is a semi-arid climate. Over here you can see there's a gently rolling, irregular plain. The gullys down here are V- and U-shaped. Lots of lowlands . . . rounded contours. But over here you see some escarpments. The gradients are fairly uniform, so it's homogeneous rock. About midway down the slopes is a relatively thick tonal band, so there's a thick bed there, pretty thick, and at the base it's flared and a bit scalloped. This terrain is flat shale, with loess topsoil. So what else you wanna know?

The story, concocted from Hoffman's experience of "picking the brains" of TA experts, makes a number of points. First, the expert does indeed possess a staggering amount of knowledge (Exactly how does she know that it's a semi-arid climate?). Second, although the expert relies heavily on perception (Show me again, where does she see those scalloped hill slopes?), the expert's judgments seem to go beyond the information given (exactly how does she know that the shale is flat lying?). Third, the expert's reasoning is nonobvious to the novice (which statement is a premise and which is a conclusion?).

What appears to the outsider or the novice to be elaborate inference making is largely perceptual and is immediate if not "direct." The immediacy was made salient in one of the experiments conducted by Hoffman (1984). An expert was shown photos of a region of Panama and within moments asserted that any people going to that region would need to anticipate certain bacterial infections. Hoffman asked, "How do you know that?", to which the expert replied that he could "see" it. See bacteria floating on a pond in a photo taken from 40,000 feet? Well, the rock type was limestone, fractured in a certain pattern leading to stagnant ponds and drainage gullies. Given the tropical climate, the vegetation type would include leguminous plants. All this amounts to trouble for personnel.

This explanation, after the fact, appears to be an elaborate inference chain. But the expert's perceptual understanding of the terrain came first—and came quickly. The expert perceived *geobiological dynamics*— the complex of natural events and processes that creates the manifestations of terrain forms observed at any given time (even though they also have things to say about the literal photograph—see Gibson, 1971). Accordingly, terrain analysts and earth scientists distinguish generic forms of topography from landforms. Generic terrain forms (hill, mountain, etc.) originate by nonspecific, putative, or unknown dynamics or processes (uplift, sedimentation, erosion, etc.). Landforms are terrain forms that originate through specific, known dynamics or processes. Dunes (classified by characteristic shape), glacial features (such as eskers), and fluvial features (such as flood plains), for example, are landforms.

The task of automating all or part of the terrain analysis process must involve linking perception to language, just as the expert does. Forming the links will require: (a) an analysis of the language used to describe terrain, and (b) an analysis of the optical information about terrain that is available to the perceiver. In this chapter we focus on the psycholinguistic side of the equation, that is, the types and meanings of the terms used by experts and laypersons in the description of natural terrain. Our ultimate goal is to generate specific, computable definitions for terrain terminology. Our efforts along these lines so far have revealed not only the enormity of the computational problem of defining terrain forms, but have also suggested some remedial measures.

10.2   Machine Analysis of Terrain

The field of remote sensing draws on many specialties and applications to extract information from pictures taken above the Earth's surface. These include analysis of multispectral satellite images to address issues in environmental science, the study of visible-light aerial photos for civil engineering, and automated cartography and geographic information systems to manipulate spatially based data (Lillesand & Kiefer, 1979). The digital pictures ("images") that are the common currency of such work usually are analyzed, in part, by machine methods. These computer-generated pictures comprise square-grid matrices of (essentially) context-free atoms called picture elements or pixels. Based on the analysis of pixels, the digital techniques of raster image processing (RIP) can readily distinguish among some types of terrain: water surfaces versus vegetated land, urban, suburban, and rural areas. But despite the many tools—image rectification, spatial filtering, convolution, merging, region segmentation, edge enhancement, Gaussian maximum-likelihood classification, and so on—machine methods of image analysis still find it difficult to consistently distinguish such common features as roads. RIP has indeed met with much success in its efforts to "increase the visual distinctiveness between features in a scene in order to increase the amount of information that can be visually interpreted from the data" (Quattrochi & Pelletier, 1991, p. 10). However, to date, the digital information processing systems cannot outperform the human expert when it comes to perceiving.

Artificial intelligence has come to the rescue (Hoffman, 1984)—or has it? Imagine that a TA trainee is attempting to interpret a stereo aerial photo with the aid of a computer software tool, or rule-based "expert system," and a topographic map. The computer system might ask the interpreter a series of questions about the photo coverage and the map information, say, about the relief (difference in elevation) in a particular area. As a result of its analysis, the computer might inform the trainee that the coverage consists of "gently rolling hills and plains." But how does the computer system, or the human interpreter for that matter, know what a "rolling" plain is (or, by implication and perhaps even more to the point, what a "nonrolling" plain is)? The question has confounded us for decades (e.g., Wolfanger, 1941).

The earth's surface is so complex that its features have long resisted precise description and classification. Obviously, any computer system designed for terrain analysis must be able to take remotely sensed data and perform operations (via neural net algorithms?) that allow it to apply generic topographic terms and descriptors, in order to tell the human what to look for. Conversely, from the human's description of the visual impression of terrain forms (e.g., "It looks like rolling plains") and other information, the system would need to derive hypotheses about rocks and soils (via a rule-based system?). In other words, both human and machine must relate topographic information about relief and slopes to perceptual information about the appearance of landforms, linguistic terminology, and cognitive information about the ecology.[2]

10.3   The Basic Problem

The language that nonspecialists use to describe terrain involves concept terms such as *peak, valley,* and *plateau,* and it also uses descriptors such as *rounded, steep,* and *rolling.* As one might expect, formal terrain analysis relies heavily on some special jargon—dozens of less familiar terms (e.g., *cirque, lineament, aeolian, centripetal*). But the language used by expert terrain analysts, on the main, consists of common terms: dune, gorge, and plain. Such familiar descriptors as *rugged* are relied on rather heavily in the literature on terrain analysis and aerial photo

---

[2]For a recent discussion, from the ecological psychology perspective, of whether a computer can "know" anything at all or "possess" any meaning—unless it has the ability to perceive—see Neisser (1991).

interpretation (see, for instance, Fairbridge, 1968; Lobeck, 1939; Parker, 1984). However, reference works on terrain analysis, aerial photo interpretation, and earth science rarely offer explicit definitions for terrain terms. Of hundreds of topographic terms and descriptors, only a few, such as *hill*, have operational definitions that involve numbers (i.e., a high hill becomes a low mountain once it's relative relief exceeds 320 m or so). Not all operational definitions of terrain terms must be stated quantitatively (although such parameters are essential to RIP and to automated terrain analysis based on digital elevation data). However, we do join others (Pike & Rozema, 1975; Wolfanger, 1941; Wood & Snell, 1960) in urging that topographic terms be defined more usefully—right now, for example, about all we can say concerning a faulted plain is that "it is a plain that has faults."

With few exceptions, definition-like statements of the meaning of topographic terms are definitional only insofar as they describe perceptual appearance: "A kame is a short, irregular ridge, hill, or hillock of stratified glacial drift" and "An esker is an elongated, serpentine ridge" (Frost, 1953, pp. 192, 194). Generic descriptors such as *blocky*, *broad*, *rugged*, and the like, are relied on heavily in the definitions, but are themselves never defined. Thus, one is left to imagine what *ribbed topography*, *bench-like elevations*, *irregular star-shaped hills* or *canoe-shaped* mountains should look like. *Elongate*, *domical*, and other generic descriptors are evidently too general for inclusion in earth science glossaries, although similar terms—*rounded*, *angular*, and *equant*—appear because they apply to another subdiscipline of earth science (Bates & Jackson, 1987). An *exfoliated* terrain form has undergone "breaking or spalling off of thin concentric shells, scales, or lamallae from rock surfaces" (Way, 1978, p. 384). But how does this look in an aerial photo? What is "thin"? One millimeter, one centimeter, one meter?

In their guide to aerial photo interpretation, Rinker and Corl (1984) explicitly acknowledged the role of perception (the visual impression of shape) and experience. Although they too relied heavily on generic terms and descriptors (e.g., *closely spaced hills*, *teardrop-shaped hills*, *sharply rounded summits*, etc.), they attempted some verbal definitions. For example, "basins and valleys are depressions that are sufficiently large to provide a significant separation between adjacent higher elevations" (p. 17). A *fault* is a "perceptible displacement between the sides of a fracture along the fracture plane, ranging from small cracks to transverse continental lineations" (p. 33). *Plains* are defined as "any relatively flat surface of sufficient extent to be a mappable unit" (p. 63). Rinker and Corl even offered an occasional definition of a descriptor—

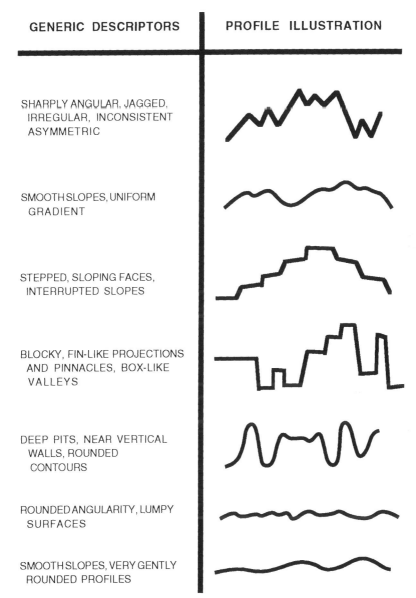

GENERIC DESCRIPTORS	PROFILE ILLUSTRATION
SHARPLY ANGULAR, JAGGED, IRREGULAR, INCONSISTENT ASYMMETRIC	
SMOOTH SLOPES, UNIFORM GRADIENT	
STEPPED, SLOPING FACES, INTERRUPTED SLOPES	
BLOCKY, FIN-LIKE PROJECTIONS AND PINNACLES, BOX-LIKE VALLEYS	
DEEP PITS, NEAR VERTICAL WALLS, ROUNDED CONTOURS	
ROUNDED ANGULARITY, LUMPY SURFACES	
SMOOTH SLOPES, VERY GENTLY ROUNDED PROFILES	

Figure 10.1. *Visual appearance of continuous topography as sampled by terrain profiles. Contrasting profiles adapted from Air Photo Analysis and Feature Extraction, p. 1-67. by J.N. Rinker & P.A. Corl, 1984, Ft. Belvior, VA: Engineer Topographic Laboratory, U.S. Army Corps of Engineers. Copyright 1984 by . Reprinted by permission.*

lumpiness is "small- (e.g., fine-) scale roundness". They also provided ostensive definitions—clear-case examples in stereophoto form and profile diagrams, illustrated here in Figure 1.

Although Rinker and Corl's attempt to describe terrain is a definite step forward, the basic problem remains. We can show photos of clear-case examples, draw some profiles, and come up with a few adjectives, but ambiguities in the taxonomy abound; for example, A ridge can be long, but so can a stream gully, and a canyon. How does *long* for a canyon differ from *long* for a gully? A mountain range can be *rugged*, but so can a *plain* (e.g., the surface of a lava flow). Can a plain be defined so that the word *rugged* can be applied to it without negating the meaning of *plain* as a form that is flat overall?

10.4   Method

Our first step in addressing the problem of topographic perception and description was to analyze frequencies of occurrence of terms and descriptors in the Terrain Analysis Database (TAD), a systematic compendium of 1,233 verbal propositions about the character of terrain (Hoffman, 1984, 1985). This material was derived by:

1. Extracting the informative or definitional statements from standard reference works on terrain analysis and aerial photo interpretation (Fairbridge, 1968; Frost, 1953; Lobeck, 1939; Mintzer & Messmore, 1984; Parker, 1984; Rinker & Corl, 1984; Way, 1978).

2. Refining and extending the statements in a series of structured interviews with a leading, if not legendary, expert in aerial photo interpretation and terrain analysis (for details, see Hoffman, 1984, 1987).

The TAD does not exhaust all propositional knowledge in the field of terrain analysis. However, its extent and (especially) the refinements stemming from the interviews assure that the compilation does represent much of the necessary information (e.g., it covers all the major rock types, landforms, soil types, etc.). Furthermore, we believed that a thousand propositions are sufficient for a start—no one had ever attempted anything quite like this.[3]

The TAD has several major categories: Climate Context, Rock Forms, Rock Types, Soil Types, Landforms, Drainage Patterns, Gully Shapes, Agricultural Patterns, and Cultural Features (Cultural Features are omitted because they were not the major focus of the research). The examples in Table 10.1, which were selected to represent both the familiar and the not so familiar, reveal that the TAD is essentially a long hierarchy, in places as many as four levels deep.

Our analysis began by assessing each TAD statement in terms of the predications it contains. A *predication* was defined, in the spirit of predicate calculus, as any binary statement consisting of a property asserted of some object. For example, one TAD statement asserted that a particular rock type results in hills that are "smooth, long, oval, rounded, cigar-or teardrop-shaped and low-lying." This statement contains seven predications: "The hills are smooth," "The hills are long," and so on. Counting adjectives is not equivalent to counting predications, however, because one adjective can serve more than one predication. For example, the TAD statement that a particular landform "has terraced hill slopes, canyon walls, and gorges" contains three predications in which it is asserted that the hill slopes, canyon walls, and gorges are all terraced.

Armed with this simple psycholinguistic method, we asked: How many predications occur in the statements? Or, how "dense" are they with meaning? We asked, How many adjectives are used in the statements? In other words, how much reliance is there on perceptual (i.e., relativistic) judgments in contrast with the mere assertion of the presence or absence of features? We answered these questions by counting frequencies. The process was fairly straightforward, but evaluating the hundreds of predications and terms in the TAD required two weeks of concentrated effort, including double-checking.

To order the information content of the TAD thus obtained, we applied a classification scheme. Initially, we relied on the taxonomy of "pattern elements" that is customary in terrain analysis (cf. Reeves, 1975), wherein features are classified into seven groups: Tones, Vegetation, Drainage Patterns, Gully Shapes, Relief (or "form"), Land Use, and Associated Implications (also called Special Features—e.g.,

---

[3]At the time of this research, most expert systems operated with inference engines of perhaps a few hundred rules and a few dozen core concepts, although a few systems were more complex.

"playas can include beach ridges"). Subsequently, we departed from this seven-fold classification in some important respects, mandated by the data themselves:

1. Vegetation and Land Use categories were collapsed because propositions on land use invariably referred to vegetation or agriculture.
2. The Drainage Pattern and Gully Shape categories were combined because both refer to drainage.
3. A category Soil Types was added because many propositions referred to soil types and not just soil tones (note: "tones" can also refer to rock).
4. We added a Formation Processes category because many TAD propositions cited geologic or geobiological dynamics (process). For example, a proposition under the heading of Arid Climate, "evaporation causes surface mineral deposits," refers not only to the presence of a feature, but to its mode of origin.
5. We also added a Location category because some propositions referred to the location of forms. For example, an entry under the heading of Outwash Plain asserted that they "occur at the borders of glacial forms." Under the Glaciers category, a TAD statement asserted that "moraines are at the glacier edge."
6. Because we chose to analyze the TAD for its meaning content, each Associated Implication was counted as such, but was also counted as an instance of any of the other categories to which its content referred. For example, the statement "arid climate implies blocky hills and ridges" was counted both as an Associated Implication and as a reference to Relief.

The eight-fold taxonomy that seemed a good fit for the TAD suggests that the seven categories customarily used in terrain analysis may indeed not be true to their own spirit. One goal of this study was to determine if terrain analysts practice what they preach: Are all terrain forms (e.g., plateaus) always defined by their tones, vegetation, drainage patterns, and so on? Reference works on TA emphasize that a thorough evaluation of photos should always include separate, systematic analysis of all seven categories (e.g., preparation of clear film overlays with contour lines indicating boundaries and regions). The training manuals may say, "Use these categories, use only these categories, and always use all of them," but the reference works and interpretation guides for carrying out the actual work may not use all of them all the time, and may rely on other, implicit, and possibly critical categories.

Especially intriguing to us was the need for the "Formation Processes" category, forced by what we recognized as a critical link between this category and the expert's direct perception of natural processes. Table 2 contains some example TAD propositions and indicates how they were scored on each of the eight categories we adopted.

Table 10.2

Scoring Scheme for the TAD Statements

PROPOSITION	ADJECTIVES	PREDICATIONS	CATEGORY
Usually within a rolling prairie.	1	1	L,R
Undulating, linear bands of hills.	2	2	R
Small, concentric knobs, ridges, and depressions.	2	6	R
Resulting from material pushed ahead of glacier. [†]	0	0	F
Many ponds, swamps, and hedgerows.	1	3	D, V
Deranged drainage.	1	1	D
Narrow, V-shaped gullys	2	2	D
Rounded, steep ridges, light tones, saucer gullys, and forests imply coarse soils.	5	5	A, R, T, D, V, S

(<u>Note</u>. The example entry "terminal moraine" is a type of glacial landform. The eight categories are R = relief, D = drainage, A = associated implication, T = tones, V = vegetation, S = soils, L = location, and F = formation processes.)

[†]Technically, the fourth statement has no predications because it contains no explicit adjectives. However, it predicates the landform with a statement of its mode of origin.

10.5 Results

We organized our frequency counts for the TAD into two lists. (Both compilations are far too long to reproduce here in their entirety.) Main entries in the first list were the descriptor adjectives (e.g., alluvial, barbed, centripetal, displaced, fringed, granular, hook-shaped, intermittent, jagged, knobby, level, many, narrow, oval, pitted, repeated, scattered, terraced, uplifted, vertical, winding, young, etc.). Each descriptor entry was accompanied by a list of all term topics (generic terms and landforms) to which the descriptor was applied (e.g., hills, plains, dunes, valleys, domes, deltas, etc.). Accompanying each descriptor-term pair was a frequency count. The counts ranged from 1 to 61, 61 being that for the phrase *light tone* (descriptor) *soils* (term). The listing contains 249 adjectival expressions. The most common descriptors (frequencies of 20 or more) were: broad, coarse, cultivated, curvilinear, dark, dendritic, equal, few, flat, forested, grassy, light, little, long, medium, parallel, rectangular, rolling, round, steep, uniform, and V-shaped.

Main entries for the second list were the nouns, both generic terrain terms and landforms (e.g., basin, delta, dune axis, escarpment, hill, orchard, rock, terrace, upland, water table, etc.). Accompanying each of the 99 noun expressions was a list of all the descriptors that were applied to them (e.g., mature, intensive, estuarine, nonuniform, small, stabilized, interconnected, etc.). Again, a frequency count accompanied each pair. The most common noun terms were: beds, deposits, drainage patterns, dunes, fields, gullies, hills, land, plains, ridges, rock, slopes, soils, tones, topography, and valleys.

Table 10.3.

Summary of adjective-descriptors and terms in the TAD

DESCRIPTORS		TERMS	
FREQUENCY IN THE TAD	NUMBER WITH THAT FREQUENCY	FREQUENCY IN THE TAD	NUMBER WITH THAT FREQUENCY
1	66	1	22
2	49	2	17
3	21	3	7
4	18	4	8
5	8	5	5
6-10	38	6	8
11-15	18	7	5
16-20	12	8	4
21-25	6	9	0
26-61	13	10	2
		11-15	6
	TOTAL 249	16-20	1
		21-25	2
		26-30	1
		31-60	2
		61-80	2
		81-100	2
		101-192	5
		TOTAL	99

Table 10.3 presents the resulting frequency distributions for both adjectival descriptors and (noun) terms, and Table 10.4 breaks down predications by our eight categories. Table 10.4 separates predications that contain adjective-descriptors from those that do not (i.e., predications that merely assert the presence or absence of a form or features). Entries in Table 10.4 consist of the frequency and the relative frequency of predications of each category.

Table 10.4.

Breakdown of Frequencies in TAD by Category

CATEGORY	PREDICATIONS FROM STATEMENTS WITH ADJECTIVES	PREDICATIONS FROM STATEMENTS WITHOUT ADJECTIVES
DRAINAGE	292 = 19.56%	39 = 18.31%
TONES	136 = 9.11%	2 = 0.94%
VEGETATION / LAND USE	181 = 12.12%	17 = 7.98%
RELIEF	401 = 26.86%	59 = 27.70%
SOILS	136 = 9.11%	8 = 3.76%
FORMATION PROCESSES	7 = 0.47%	14 = 6.57%
ASSOCIATED IMPLICATIONS	309 = 20.70%	58 = 27.23%
LOCATION	31 = 2.08%	16 = 7.51%
TOTAL	1493	213
	TOTAL	1706

Note. *Predications from statements without adjectives merely asserted the presence or absence of some particular form or feature. Predications from statements with adjectives asserted some feature or property.*

The 25 most common (10 or more occurrences) descriptor-term pairs identified in the TAD were: angular gullys, banded tones, barren land, dark tones, dendritic drainage, equal elevations, few gullys, fine drainage (density), flat topography, grey tones, interbedded rock, internal drainage, light tones, medium tones, mottled tones, parallel ridges, rectangular field pattern, rounded drumlins, saucer gullies, shrub land, steep slopes, tilted beds, uniform slopes, U-shaped gullies, and V-shaped gullies. The number of TAD predications (the grand total in Table 10.4 of 1,706) exceeded the number of statements (1,233), because a given statement could contain more than one predication and because adjective-containing predications in the Associated Implications category were counted twice—once as an Associated Implication and again in other categories depending on the specific reference.

The number of different adjective descriptors found in the TAD is 249, but the total number of adjectives only, summing frequencies for all 249, is 1,493 (column 1 in Table 10.4). Because the total number of statements in the TAD is 1,233, there are about 1.2 adjective descriptors per statement (1,493/1,233). On the average, 2.5 different descriptor adjectives were applied to each term (249/99), and each term was described by an average of 15.1 adjectives (1,493/99). Of all the TAD statements, 1,156 (93.8%) contain adjectives and 77 (6.2%) do not. (These figures account for the 213 predications without adjectives indicated in column 2 of Table 10.4.) If we compare the total number of adjectives to the total number of statements that contain adjectives (1,156), we find that each predication-containing statement has an average of 1.3 adjectives.

In sum: (a) each TAD statement tends to include at least one adjectival predication; (b) statements that do contain adjectival predications tend to have about one or two adjectives; (c) each adjective descriptor is typically used to describe more than one topic; and (d) each topic is described using two or three adjectives. An example of a statement containing a relatively large number of descriptors is found under the category of Schist Rock Type: "Long, deep, parallel, U- or V-shaped gullies with light, banded tones."

Most statements in the TAD rely on adjectival descriptors, verifying the hypothesis that generic descriptors are essential to terrain analysis. The adjective-free statements tend to be references to Relief (e.g., hills have relief less than 300 m), to Drainage (e.g., drainage patterns in a till plain are controlled by the underlying rock), and to Associated Implications that merely asserted the presence of a form or feature (e.g., ponds are present in outwash plains). Because it is unlikely, or apparently difficult, to refer to image tones without using adjectives, the number of adjective-free predications referring to Tones is very small. Conversely, it is relatively easy to refer to Formation Processes without using adjectives (see Table 10.4).

Some of the more frequent generic terms also appear as the more frequent adjective descriptors. Thus, the TAD includes statements about hills and statements about terrain that is hilly, statements about soil that is eroded and statements about erosion, statements about rock that is folded and statements about folds. Other terms that appear in both noun and adjectival form are: band(ed), bed(ed), cluster(ed), contour(ed), displacement (displaced), dome(d), elevation (elevated), fan(-shaped), fault(ed), flat (flats), forest(ed), fringe(ed), intersection (intersected), joint(ed), meander(ing), mountain(ous), peak(ed), slope(d), stairstep(ped), streak(ed), surface (surficial), and terrace(d).

A few terms carry more than one distinct meaning. The best example is *relief*, a general reference to the land surface—as in "The relief was uninteresting" (e.g., Bates & Jackson, 1987). Such alternatives as *terrain* or *topography* do not alter the intended meaning. *Relief* also indicates a specific measure, relative (or *local*) relief: the difference in elevation between the highest and lowest points in a particular area or on a particular terrain form. In the following discussion, we use the term only in this latter, specific sense. Similarly, *slope* has two distinct meanings that are expressed separately here. *Slope* refers to a particular attribute of terrain geometry (change in horizontal extent divided by change in elevation), and *side* indicates the (sloping) side or flank of a terrain form such as a hill or valley.

The eight-fold categorization of TAD statements revealed something quite interesting about the customary TA pattern element called Special Features. In our analysis, predications that would ordinarily be referred to as Special Features were counted as Associated Implications and were also counted in other categories, depending on the nature of their referents. Because the percentage of such predications (309 or 23%) is higher than in any other category save Relief, the category of Special Features may not afford much power in discriminating aspects of terrain: It serves as a catch-all for any type of Associated Implication. Rather, it is the specific meaning content of various "special features" that is important. Certainly, every landform or terrain type will have certain "special" features associated with it. However, these features are special not just because they are associated with certain forms, but because they refer to such attributes as relief, tones, and so on, and thereby refer to perceptible qualities.

The one category of TAD predications that we found unexpectedly low in frequency was that of Formation Processes. Examinations of the perceptual skills of expert terrain analysts (Hoffman, 1984) led us to believe that experts perceive formational process, and that it is their perception and conceptual understanding of the underlying geobiological dynamics that allows them to make what appear to novices to be rapid yet complex inferences. Accordingly, reference to formational processes was expected to be more dominant in the TAD. In retrospect, it is obvious why this was not the case. It is not the purpose of photo-interpretation keys and reference manuals to explain why terrain forms appear the way they do, but merely to describe their appearance. To use the terminology of artificial intelligence (AI), this other task would be a part of the "explanation" component of an expert system, and not a part of its working database. Clearly, much information will have to be included in the explanation component of any expert system devised for aerial photo interpretation, and the data on formational processes in the TAD represent just a beginning.

Analysis of the TAD yielded a large corpus of generic terms and descriptors, including a listing of which descriptors were applied to which terms, and how often (Tables 10.3, 10.4). Although the TAD statements themselves do not exhaust the propositional knowledge in terrain analysis, we believe that the term and descriptor listings derived here do exhaust the basic terminology used in the field. The next step was to use these listings as the starting point in an attempt to define the terms and descriptors, by identifying the information that links the topographic specification of terrain to the perceptual specification of terrain. The definitions will have to satisfy constraints. Each adjective can be applied to many terms, and each term can be described using a number of adjectives. For example, the adjective *flat* modifies 12 different forms (deltas, plains, ridges, etc.), and the term *hill* is elaborated by 37 adjectives (e.g., conical, long, rugged, etc.). Any proposed definition of a term must work when the term is modified by each of the applicable descriptor adjectives. Conversely, the definition of each descriptor must work when applied to a given term. Thus, the definition for *hill* must work even when the hills in question are blocky, bold, jointed, knobby, and so on. The definition of *flat* must work when it is applied to ridges as well as plains. Fashioning the needed definitions is a major challenge. For assistance in meeting that challenge, we turned to Gibson.

10.6   Gibsonian Theory and Terrain Perception

Based as it was on evolution and organism-environment coadaptations, mutualities, and reciprocities, Gibson's (1979) theory of perception began with a description of the natural environment and its size-dependent structure: "Canyons are nested within mountains, trees are nested within canyons. There are forms within forms both up and down the scale of size" (p. 9). Then Gibson went right to the heart of the matter:

[This] would constitute a hierarchy except that this hierarchy is not categorical but full of transitions and overlaps. Hence, for the terrestrial environment there is no special proper unit in terms of which it can be analyzed once and for all. The unit you choose for describing the environment depends on the level of the environment you choose to describe. (p. 9)

As is the case of size and texture, there is structure at all durations (Gibson, 1979, p. 11), both fast events (e.g., human alteration of the landscape, eruption of volcanic cinder cones) and slow events (e.g., uplift of the Tibetan plateau, faunal succession, death of coniferous forests). The causal covariation of transitional hierarchies of both time and space is the operational or analytic manifestation of what we call *geobiological dynamics*. The appearance of terrain, essentially static measured by the time scale of the human lifespan, in fact, results from complex persistences and changes that are governed by the causal laws of geology and biology.

Surfaces of the terrestrial environment—their orientation, microtexture, and colors—add structure to the ambient light (Gibson, 1979). The light thus manifests information that is specific to the terrain, allowing for the perception of such aspects as "blockiness" and "plateauness." Indeed, the information specifying the perceptible form pervades the terrain, just as it pervades the optic array. That information is specific to surfaces, not to individual points, pixels, or rays of light. "We see by means of optical structures, but we do not see the structures themselves" (Neisser, 1991, p. 20). Perception is of properties of surfaces and the events (fast or slow) in which they participate (e.g., occluding edges) and of the affordances of the terrain (e.g., is that hill too steep or too gravelly to be climbable?). The causal linkages are depicted in Figure 10.2.

Photo interpreters have long debated such questions as "What do we mean when we say that a hill is "blocky"? The usual answer is to throw up one's hands and say "It's blocky because it *looks* blocky!" Indeed, it does look blocky, and we claim that people can agree it looks blocky because sufficient information is available in the optic array (or captured in aerial photos) to specify the perceptual judgments. This reasoning suggests that the perception of such aspects of terrain as "blockiness" is not necessarily buried layers of inference deep in the cognitive apparatus. Indeed, perception of terrain is direct and requires no levels of inference or elaborative computation. The *labeling* process, specifying that a terrain form looks blocky, requires conceptual recognition and elaboration, and TA trainees may labor in their explicit inference making and conceptual analyses. However, the perceptual act itself is direct in that the information available in the optic array directly specifies the shape of the terrain. But if the information is available in the optic array, then it can be specified in topographic terms.

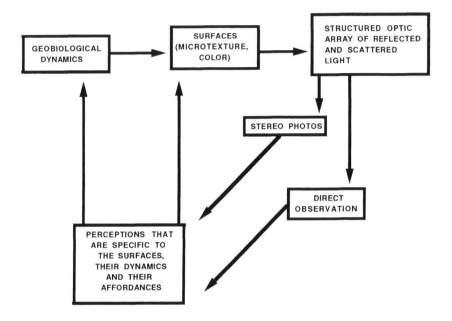

Figure 10.2. *Conceptual model of causal linkages in the perception of terrain, following Gibson's (1979) theory of the relations of surface properties, the structure in the ambient optic array, and the perception of surfaces.*

This foundation in Gibsonian theory may enable us to define generic terms and descriptors, and to do so one must start at the appropriate level of analysis, the "surface" and its "texture."

10.7   The Role of Scale in Terrain Perception

In discussing the uses of remote sensing in environmental science, Quattrochi and Pelletier (1991) enumerated various attributes for landscape ecology. Their perspective dovetails with that of Gibsonian psychology—a convergence that might be remarkable were it not simply forced on us by the very existence of geobiological dynamics.

Quattrochi and Pelletier strongly emphasized dynamics: Remote sensing is change detection, landscape is a process (see also Forman & Gordon, 1986). Indeed, remotely sensed data are often used to model the cause-and-effect linkages that define the structure, function, and change in landscape evolution, "to measure the dynamics of ecological variation and the processes that drive them" (Quattrochi & Pelletier, 1991, p. 15). Scalar relativity is a consequence of this, or is perhaps another way of saying the same thing. "The remotely sensed landscape is multidimensional, multitemporal, and multiscaled" (p. 8).

A surface is defined relative to what the observer is looking for and what the observer sees, and these are determined in large part by level of generalization, or scale. Research in landscape ecology spans many scales involving both grain size (resolution) and the distance of the observer from the terrain (which tend to covary). Scales range from microtopographic (i.e., pebbled beaches or pahoehoe lava surfaces) and that of local terrain (i.e., a square mile of glaciated topography with drumlin landforms), to the regional (10 square miles of dissected terrain bordered by a broad river floodplain), to that of the physiographic province (hundreds of square miles)[4], and beyond that to continental extent (Hoffman, 1990; Pike & Rozema, 1975; Quattrochi & Pelletier, 1991; Wall, Farr, Muller, Lewis, & Leberl, 1991; Weszka, Dyer, & Rosenfeld, 1976)

Choosing the "right" or characteristic space-time scale for a particular terrain surface is critical to any remote-sensing investigation (Davis et al., 1991). For most work, pixel sizes or image resolution can range from square meters up to square kilometers. The chosen resolution depends on the phenomena of interest. The spatial and temporal resolving power must capture both homogeneity and heterogeneity of the landscape with respect to its important features and the research objectives (e.g., the size of crowns of particular diseased tree species versus the extent and spectral signature of a salt flat for minerals prospecting).

There is no one "right" way to approach image analysis. The method chosen depends on the type and scale of the data, the intended

---

[4]A physiographic province is analogous to a landform in that its terrain forms arise from formational processes governing that area's geologic history. A landform is one specific terrain form, whereas a physiographic province, depending on its size, includes many landforms. An example would be the Great Basin and Range Province in the western United States.

application, and the computer resources available (Quattrochi & Pelletier, 1991). Many sensor systems provide aerial (integrated) information as well as point information. Computer graphic displays of remotely sensed information reveal that the spatiotemporal units represented or depicted by each pixel vary enormously, depending on the kind of spectral signatures being sought. For example, the LANDSAT image resolution of about 30 m permits integration over a number of electromagnetic frequencies. At this resolution RIP methods can distinguish soils from vegetation, healthy from stressed vegetation, soil and vegetation moisture content, and mineral and rock types as well as many cultural features. (For a detailed presentation of the spatial and temporal scales used in remote sensing, see Davis et al., 1991; Hoffman, 1990.)

At any given scale, a surface is defined in terms of its edges and its texture. A particularly apt example of the surface–texture relation or duality is a photograph in Gibson (1950, chap. 6, p. 83). From a near-terrestrial perspective, one sees a large array of wooden barrels packed side by side; their flat round tops serving as the texture elements, and their sides and the shadows serving as texture gaps. The analogy carries over directly into terrain perception. At local and regional scales, texture is the microtopography (e.g., pebbles are the texture elements on a beach; in a forested expanse, they are individual tree crowns), and the surfaces are terrain forms (e.g., a beach as a flat or low-relief area; a forest as an area of rolling hills of moderate relief). At the continental scale, terrain forms and landforms become the texture (e.g., a broad mountainous region, in which individual peaks are the texture elements), and the surfaces are large areas of contrasting types of terrain (e.g., the mountainous plateau bordering a dissected plain).

If one area can be discriminated from others, either because of relief contrasts, tone or color features, or texture differences, then that area constitutes a surface—at least for that observer. Granted, a particular observer may be wrong about a given surface relative to the knowledge of the expert. An observer might lump two areas together and call them one, and be incorrect. The expert might say, "Look here, at this subtle tonal boundary; there are two different soils—there are actually two different landforms here." But for each observer, a surface is whatever is seen as a surface.

It is important to keep in mind that even though scales are relative, they are not arbitrary. Because formational processes operate up and down the scales, even microtopographic texture, for example, can be used to classify images by automated means. Weszka et al. (1976)

applied RIP techniques (e.g., grey-level difference statistics, Fourier transforms, etc.) as texture measures to aerial photos and had some success in distinguishing among urban area, scrubland, swampland, orchard, forested land, and so on. Clearly, more such research into RIP for texture–terrain relations is called for on the part of earth scientists, just as there is a need for psychologists to conduct research on how people perceive real terrain and terrain textures.

10.8   The Computation of Terrain Forms

Some of the concepts and observations developed in our effort to systematically define generic topographic terms and descriptors can be implemented by linking perceptible attributes with topographic features. Invariably this endeavor leads to the description of topography in numerical terms. Although relatively new to psychologically oriented research (e.g., Savitt, Henderson, & Colvin, 1992), *morphometry*—the quantification of landscape morphology—has a long, rich history in the earth and applied terrain sciences (Carr & Van Lopik, 1962; Evans, 1972; Hammond, 1964; Mark, 1988; Morisawa, 1985; Neuenschwander, 1944; Zakrzewska, 1967). With the advent of computer technology, "it should be possible to simulate the visual perception of topographic form using numerical methods and digital elevations" (Pike, 1988, p. 492). We need not go into this field in detail here. To illustrate salient points raised by our analysis, we developed some simple descriptors of surface features using outlines (plan views and profiles) of terrain forms and statistics of elevation and slope angle.

10.9   Outlines as Samples of Terrain

One might begin with the observation that hills and other terrain forms are complex three-dimensional objects. Should not, then, it be necessary to use complex equations or even fractal modeling to build a complete model of the terrain and thereby define such attributes as blockiness? There are two sides to this question: (a) Will a partial description of terrain forms suffice rather than a complete mathematical specification? and (b) Suffice for what purposes? Certainly, terrain forms range from the geometrically simple (e.g., a drumlin as a longitudinally truncated prolate spheroid) to the complex (e.g., a highly dissected range where one mountain's slopes blend into another's). As we show, however, ecological optics and geobiological dynamics co-imply the following proposition: A complete geometric and topological description of terrain forms need not be generated to define generic descriptors.

The information available in the optic array "pervades" the landforms. Thus, merely sampling a terrain form should enable one to describe its surfaces in enough detail to determine whether such terms as *hill* and *plateau* apply, and whether descriptors such as *blocky* and *rugged* apply. Because terrain surfaces structure the ambient optic array, the information that allows for perception of such properties as blockiness is the same information that allows topographers to describe terrain adequately by sampling. In other words, to determine that a hill is blocky or rugged, one need not compute the total three-dimensional shape. Not surprisingly, a method long used in terrain analysis and cartography can effectively sample terrain form by "slicing" it up into topographic profiles (e.g., Rinker & Corl, 1984).

The attributes of topographic form (surface, area, etc.) needed for this exercise are defined operationally in Table 10.5. The angles-of-regard (Table 10.6) locate the observer relative to the photo coverage or terrain surfaces under observation. The *slice* determinants, defined in Table 10.7, show how both vertical and horizontal outlines of topographic features may be generated, given the angle-of-regard.

Three terms—*plan view, longitudinal profile,* and *transverse (cross) profile*—are fairly standard in terrain analysis and earth science. Together they define the "normal" (ground) perspective on terrain forms. However, they also depend on the observer's angle-of-regard. This dependence is illustrated graphically in Figure 10.3, a topographic map of North Menan Butte, a small extinct volcano in southern Idaho. Because all points on a contour line lie at the same elevation, each contour defines a plan view at that elevation (as defined in Table 10.7). The shape of each plan view is a descriptor of the terrain form sampled as a horizontal slice through the topography.

In order to identify longitudinal and transverse profiles, a terrestrial angle-of-regard is established by locating the observer (open circle) at a point on the terrain (arbitrary in Figure 10.3). Relative to this location, the line labeled "1" defines the vertical plane along which a longitudinal profile could be drawn, and line labeled "2," the plane of a transverse profile. In the "normal" perspective, the observer's location is shifted so that line 1 is longer than any other line drawn across the form. Line 2 is then drawn perpendicular to line 1.

The "normal" perspective is laden with relativities, the most important of which is the observer's location. Positioning the observer to generate the longest possible longitudinal profile presupposes that

Table 10.5.

*Some Attributes of Topographic Form and its Perception
Defined at an Ecological Scale*

ATTRIBUTE	DEFINITION
SURFACE	Interface of terrain and atmosphere. Divisible into subsurfaces.
TEXTURE	Smallest discernible subsurfaces, usually defined by: (1) observing distance, or scale, and (2) grain size--a function of tone (gray shade) in aerial photos, or radiometric sensor count values per resolution unit in multispectral images.
COVERAGE	The set of all surfaces represented in an aerial photo.
AREA	Any subset of adjacent surfaces observed in a coverage.
ELEVATION	Height of a texture element relative to lowest texture element on the surface or in the area.
RELIEF	Difference between highest and lowest texture elements on a surface or in a coverage or area.
SLOPE	Angle of a surface relative to horizontal plane; change in horizontal distance between two texture elements on a surface divided by change in elevation across the same elements. First derivative of elevation.
SPACING	Horizontal distance between specified sets of linear texture elements (e.g., stream channels, ridge crests).
CURVATURE	Rate of change of angle between two surfaces (vertical plane) or between sets of linear texture elements (horizontal plane). Second derivative of elevation.

Table 10.6.

Determinants of the Observer's Perspective on a Terrain Surface

ANGLE	DEFINITION
ANGLE OF REGARD	Perspective from which a surface is viewed.
AERIAL ANGLE OF REGARD	Viewing a surface from an aerial perspective. Observer is above Earth's surface and gaze direction is downward in a near-vertical or vertical plane.
TERRESTRIAL ANGLE OF REGARD	Viewing a surface from a terrestrial perspective. Observer is on Earth's surface and gaze direction is in a horizontal or near-horizontal plane.

Table 10.7.

Sample Designs for "Slices" Used to Abstract and Describe Terrain Surfaces

SAMPLE DESIGN	DEFINITION
PLAN VIEW	Outline formed by intersection of a horizontal plane and Earth's surface. Plane is perpendicular to an aerial angle of regard and parallel to a terrestrial angle of regard.
LONGITUDINAL PROFILE	Outline formed by intersection of a vertical plane and Earth's surface, where plane is parallel to a terrestrial angle of regard.
TRANSVERSE CROSS (PROFILE)	Outline formed by intersection of a vertical plane and Earth's surface, where plane is perpendicular to a terrestrial angle of regard.

one knows where the terrain form begins and ends. Otherwise, how can "longest" be defined? Figure 10.3 is an easy case because the terrain form—actually a landform (a volcano)—has fairly distinct boundaries, that is, the stream to the right (east). However, the problem of delimiting terrain forms consistently frustrates the application of numerical analysis (Evans, 1972). One promising solution, dividing a continuous surface into discrete units (drainage basins) by digital techniques (Mark, 1988; Savitt et al., 1992), is not appropriate for all types of terrain.

Another important relativity, the assumption that the two profiles must lie at right angles, simply follows cartographic convention. Indeed, it is an empirical question; for many terrain forms an acute angle may be more appropriate for the relative position of the two profiles. Following the Gibsonian line of reasoning, just about any few slices (profiles) should suffice for a description of a given terrain form. Indeed, the more complex the form and the less discernible its boundaries, the less possible it is to establish "the" normal perspective and the less the observer's orientation should matter—although this too is an empirical issue (for a discussion of terrain homogeneity, or *stationarity*, see Pike & Rozema, 1975, pp. 505–506).

10.10   Measures of Topographic Form

The basic concepts defined thus far enable us to generate outlines that sample the terrain forms we perceive. The next step is to choose measures that describe the shapes thus outlined. Among the most common numerical attributes are elevation, relief (the difference between highest and lowest elevation in a specified area), and slope—

the first derivative of elevation (Evans, 1972; Neuenschwander, 1944; Zakrzewska, 1967). The definitions, or typical ranges for relief and slope values (e.g., Way, 1978), are given in Table 10.8.

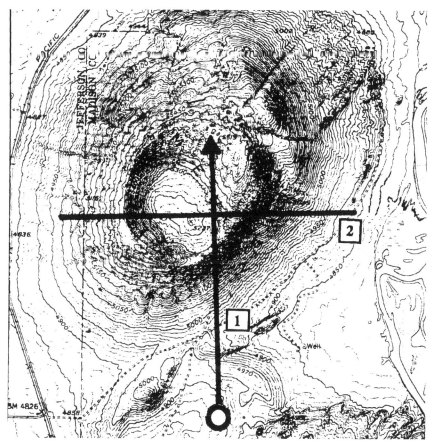

Figure 10.3. *How location of the observer (circle) influences positioning of planes for longitudinal (1) and transverse (2) profiles that sample a landform. Reproduced and modified from U.S Geological Survey Map N4245-W11152.5/7.5 (Menan, Buttes, Utah), 1951.*

These common measures alone do not fully describe the appearance of terrain and hence are insufficient to define topographic terms and descriptors. Descriptors need to account for the variability and arrangement of slopes, both within a particular form and across forms within an area. A simple measure of this is the Slope Change Index

(SCI) (Rinker & Corl, 1984; proposed earlier by Wood & Snell, 1960), an inverse measure of terrain-element spacing. To determine the SCI one counts the number of terrain segments between hilltops and valley bottoms that occur along a profile. Slopes must have a direction, or sign: "positive" slopes face one end of the profile, and "negative" slopes the opposite end. Summing both counts gives the SCI, an index of the number of terrain reversals. However, a steep mountain and a gentle hill could have the same SCI value (see Figure 10.4). Still more sensitive discriminants are needed.

Table 10.8.

Ranges for Relief and Slope

RELIEF
Low	0–100 m
Moderate	100–300 m
Strong	Over 300 m

SLOPE
Horizontal	0
Very Gentle	0–5%
Gentle	6–15%
Moderate	16–40%
Steep	41–60%
Very Steep	Over 60%

Many operational approaches treat the problem of uniquely capturing terrain form. For example, a hillslope profile may be divided into quartiles, defining the 25% nearest the summit as the upper quartile, the 25% nearest the base as the lower quartile, and the enclosed 50% as the middle two quartiles. Across these three—summit, side, and base—one could compute mean and variance for both elevation and slope, and using raw elevation data, compute the higher statistical moments. No one who has taken a course in psychological statistics can help but notice that some size–frequency distributions are reminiscent of profiles of hills. Elevation skew and kurtosis have in fact been applied to the description of actual terrain shapes (e.g., Evans, 1972; Tanner, 1960). Indeed, many investigators in the earth sciences have been successful in specifying terrain geometry in statistical terms (Hammond, 1964; Horton, 1945; Pike & Rozema, 1975; Strahler, 1964). With the advent of large digital data sets, terrain now can be characterized by statistics of slope curvature—the second derivative of elevation—and other complex measures (Dikau, 1990; Evans, 1980; Pike, 1988).

Some basic equivalences between terrain form and statistical specifications of elevation and slope are presented here. Figure 10.4 shows what each of several measures can and cannot discriminate (with due caution for the uncertainty of meaning of kurtosis). For example, a terrain form with high variance in either slope or elevation is more "rugged" than one with low values (a "smooth" form). A terrain form with high kurtosis of elevation, could be a pinnacle; one with low elevation kurtosis might be a mound. A terrain form highly skewed in slope along a sampling profile could be gentle on one side and steep on the other. Computing such values, for summits, sides, and bases, for opposing sides of a form with respect to both profiles, might generate numerical criteria that distinguish, say, a field of dunes from a glacial end, moraine from karst terrain. The computer algorithms and digital elevation data needed to automate an analysis of this kind are already available (Dikau, Brabb, & Mark, 1991; Evans, 1980; Pike, 1988).

No philosopher's stone exists for terrain analysis or geomorphology; no single "magic number" can distinguish all or most terrain forms (Evans, 1972; Pike & Rozema, 1975). Combining the parameters of spacing, elevation, slope, and other attributes, however, dramatically increases the discriminative power of terrain measures (e.g., Hammond, 1964; Wood & Snell, 1960). For example, slope variance and elevation skew could, together, distinguish certain types of rugged asymmetrical hills from smooth symmetrical ones. This concept has been developed most fully in the *geometric signature*—the set of measurements that describes topographic form well enough to distinguish geomorphically disparate landscapes (Pike, 1988). Moreover, the plan-view shape (Figure 10.3, Table 10.7) of such closed forms as volcanoes and impact craters may be effectively quantified by an index of circularity (e.g., Pike, 1974). Much work lies ahead, particularly in devising effective measures of properties and patterns in the spatial (horizontal) domain (Mark, 1988). For example, Pike's (1988) multivariate signature of landslide-prone terrain types in California, although successful, omitted some of the most diagnostic attributes of perceived terrain in plan view: elongation, parallelism, angularity, branching contiguity, degree of nesting, and connectivity.

10.11   Some Explicit Definitions

Even some of the simpler attributes described here are sufficient to operationally define the generic concept of a *form* more precisely than by merely declaring that a form is "any perceptually discriminable area or set of surfaces." A *form* is any area or set of surfaces described in terms of—for example—its relief, slope, spacing of slope reversals, slope variance and azimuth, and elevation skew and kurtosis (cf. Hammond, 1964; Pike, 1988; Pike & Rozema, 1975; Wood & Snell, 1960). What one perceiver sees as a separate form may not be what another perceiver

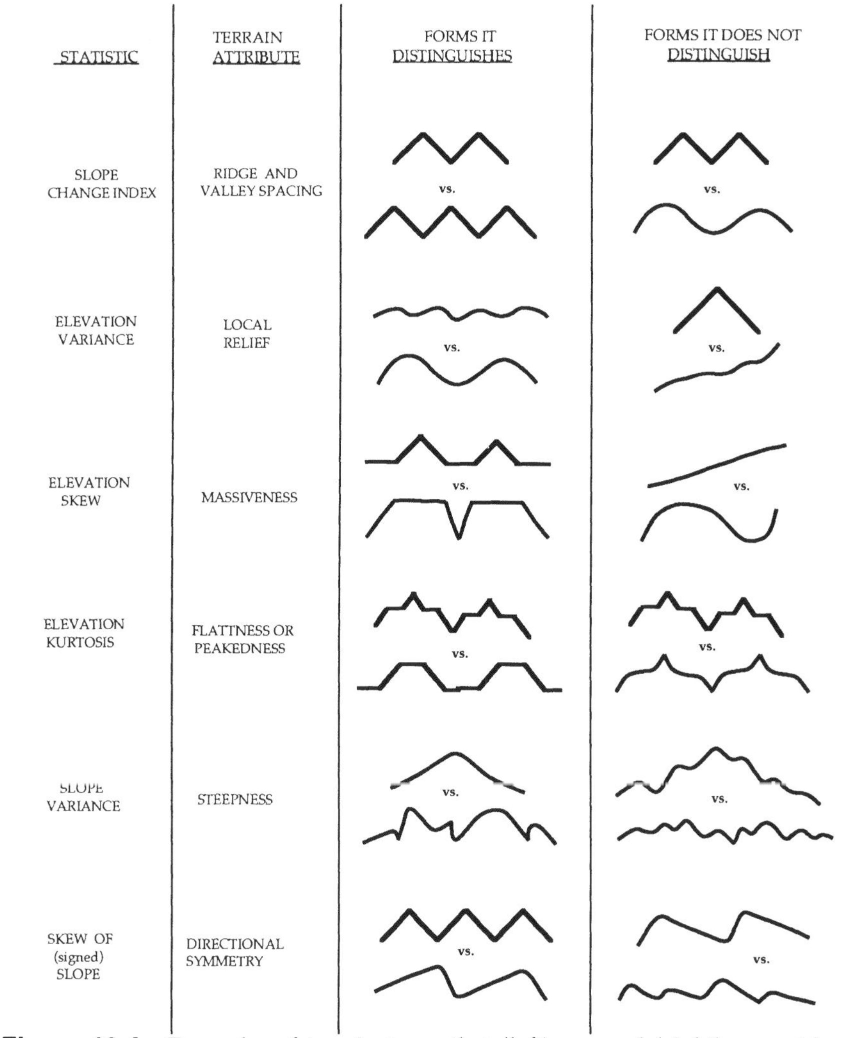

Figure 10.4. *Examples of terrain types that (left) can and (right) cannot be discriminated by single measures of profile shape.*

sees, nor need it correspond to what terrain analysts and geomorphologists call landforms. Quantifiable definitions of generic terms and descriptors thus provide further specification for each particular form.

Having defined terrain slices and some attributes of terrain geometry, we can begin to define terrain descriptors and terms. Only a few of the hundreds of terms in the TAD are defined here. Example descriptor definitions are given in Table 10.9, and example term definitions in Table 10.10.

The terms in Tables 10.9 and 10.10 appear alphabetically, although the proper ordering involves sequential dependencies. For example, a *horn* is a kind of mountain peak, and so the definition of a peak must be established first. A *peak* is a kind of summit, a *summit* is a kind of *elevated form*, and an elevated form is one kind of sloping surface. The sequential dependencies are required because all the various descriptors and terms must be consistent with one another. For example, a rugged plain would have to combine the definitions of *plain* and *rugged* in such a way as to preserve the basic meaning of the individual words: A *rugged plain* is a surface of significant horizontal extent that has low relief and gentle slope in broad scale (the "plain" part), and moderate to strong relief and steep to very steep slope in fine scale (the "rugged" part).

More detailed description of terrain forms at the fine scale (e.g., Table 10.9) would depend on both the character of the terrain and the context of the definition. Another example, a *knobby ridge*, might be: a summit that is a crest, e.g., long and narrow (the "ridge" part) and is composed of forms with narrow, convex summits and moderate to steeply sloping sides, and moderate to low variance in slope angle (the "knobby" part).

There are, no doubt, inconsistencies in these definitions—both individually and in combination (e.g., see Bates & Jackson, 1987)—and literally hundreds of such combinations have to be worked through. Furthermore, definitions have to be derived for landforms (e.g., coastal plain, drumlin, sinkhole, etc.), as well for forms within continuous topography described by the generic terms (e.g., hill, cliff, etc.).

10.12   Prospect

In this chapter we only really pointed to the problem—linking terrain terminology to terrain perception. Only after a considerable amount of computational work could one say, "Such-and-such terrain metric

Table 10.9.

Some Provisional Definitions of Terrain Descriptors, Based on Concepts in Tables 10.5, 10.6, 10.7, 10.8, and Figure 10.4.

DESCRIPTOR	DEFINITION
BLOCKY	Angular-textured elevated form or cluster of continuous forms of low relief, low slope variance, and steep or very steeply sloping sides.
BRANCHING	Division into two or more forms along profile, or division of a set of continuous repeated linear forms into two or more sets of continuous forms with parallel profiles.
BROAD	Profile is wider than its relief is high, and profile is wider than those of most similar forms within area (or photo coverage).
CONICAL	Axisymmetric plan view (elevated or inverted) with moderately to very steep slopes, low to moderate slope variance, and leptokurtic and skewed elevation
CONVEX	Profiles decrease in relief and slope toward center of planes of intersection (at elevation maximum) and increase in relief toward edge of at least one plane.
FLAT	Nonvarying slop, not necessarily horizontal.
HOG BACKED	Jagged (rugged or angular) profile, with repeated ascending and descending sides roughly equal in spacing and homogeneous in relief and orientation.
LARGE	Extent of profiles or plan view exceeds range of those of most similar forms in the area or the coverage.
LOBATE	Compound elevated forms with high slope variance and high SCI values. Component forms are continuous, highly skewed and moderately kurtotic in slope, and have convex summits.
ROLLING	Low relief, slightly elevated or inverted forms, with low SCI, low slope variance, and platykurtic elevation.
RUGGED	Moderate to strong relief, steep to very steep slopes, very high SCI, and moderate to high slope variance.
SCALLOPED	Ridge or base with repeated concave sides with concavity-maximizing cross-sections or profiles that are parallel.
SHALLOW	Relief of an inverted form is less than the range of relief in neighboring areas.
STAIR-STEPPED	Summits, sides, or bases whose plan views or profiles have surfaces alternating between low relief with very gentle to horizontal slope and moderate to very strong relief with moderate to very steep slope.

Table 10.10. *Some Provisional Definitions of Terrain Terms, Based on Concepts in Tables 10.5, 10.6, 10.7, 10.8 and Figure 10.4.*

TERM	DEFINITION
BUTTE	Isolated hill, smaller than a plateau, with steep sides and a nearly flat or horizontal summit. High elevation skewness.
CLIFF	Surface of low to strong relief, sloping greater than 50° or overhanging (beyond 90°).
DOME	Hill or mountain, of any size, with convex profiles and a circular to elliptical plan view. Elevation skewness may be high.
GORGE	Deep, very steeply sloping and narrow valley. Floor ("base" of an inverted form) commonly occupied by a river. Very low skewness of elevation.
PEAK	Mountain summit, small in plan view, with moderate to very steep sides (within the summit) and low elevation skewness.
PINNACLE	Narrow, isolated, leptokurtic form with very steep to vertically sloping sides, a base that is flared, and a peak or flat summit.
PLAIN	Broad, continuous surface of low relief, usually horizontal, flat, or gentle in slope, with only few small variations in relief and few or no prominent forms.
PLATEAU	Broad, raised horizontal or flat plain bordered on at least one side by escarpment or cliff typically 150 to 300 m in relief. High elevation skewness.
SUMMIT	Highest elevation on a profile; is generally a crest if form is elongate.
VALLEY	Depression enclosed by upland on at least two sides. Usually elongate, but breadth and relief variable. Floor usually occupied by a river.

values depicted as such-and-such patterns of grey tones in photographs represent the invariant information which specifies such-and-such a terrain form." The very attempt to systematically define generic topographic terms and descriptors, in ways that link perceptible attributes with topographic features, is something of a novelty, especially in psychology. However, the statistically based concepts advocated here and elsewhere (e.g., Strahler, 1964) already have been put into practice for actual forms over large areas—previously by manual methods (Hammond, 1964), but increasingly using digital elevation data (cf. Dikau et al., 1991; Evans, 1980; Pike & Thelin, 1989). A complementary tool for perceiving topography, mechanized relief shading, also has been implemented digitally (Pike, 1992; Thelin & Pike, 1991).

It has not escaped the notice of investigators in the earth sciences

(Dikau, 1990; Evans, 1972; Mark, 1988; Pike, 1988) that such concepts as we have presented here suggest entire programs of basic research into topography and its perception and characterization, perhaps involving both field measurements and theory-driven calculations. We look forward to more psychological research into the human factors involved in depicting terrain by multispectral remote-sensing data (e.g., see Hoffman & Conway, 1989). We also anticipate more ecologically focused work on the perception of terrain and topographic maps (Smith & Heinrichs, 1991; Savitt et al., 1992).

New satellite and Space Shuttle systems have created a severe "data overload" of archived images awaiting analysis (see Hoffman, 1984, 1991). Although some remote-sensing researchers have complained about the growing overload for years, the planned Earth Observation System of satellites will generate on the order of a terabyte of new multispectral image data per day! More trained experts are needed; many more. Thus, anything that we experimental psychologists can do now to reveal and codify the expertise of the terrain analyst can contribute to our understanding and preservation of the environment. This chapter was an exercise in what *we* mean by *ecological psychology*— applying experimental psychology to the solution of problems in environmental science. Perhaps the idea will become contagious!

Acknowledgments

The research reported here was supported by a grant from the Scientific Services Program of the U. S. Army Research Office and was conducted at the Engineer Topographic Laboratory (now the Topographic Engineering Center) of the U. S. Army Corps of Engineers, Ft. Belvoir, VA. The senior author would like to thank Robin Akerstrom for her help in preparing the final version of the manuscript.

10.13 References

Bates, R .L., & Jackson, J.A. (Eds.). (1987). *Glossary of geology* (3rd ed.). Alexandria, VA: American Geological Institute.

Carr, D. D., & Van Lopik, J. R. (1962). *Terrain quantification Phase I: Surface geometry measurements* (Rep. No. 63-208, U. S. Air Force Contract No. AF 19(628) 181 Project No. 7628 Task No. 762805). Cambridge, MA: Cambridge Research Laboratories.

Craik, K. R. (1972). Psychological factors in landscape appraisal. *Environment and Behavior, 4*, 255–266.

Curran, H. A., Justis, P. S., Young, D. M., & Garver, J. B. (1984). *Atlas of landforms* (3rd ed.). New York: Wiley.

DaSilva, J. A. (1985). Scales for perceived egocentric distance in a large open field. *American Journal of Psychology, 98*, 119–144.

Davis, F. W., Quattrochi, D. A., Ridd, M. K., Lam, N., Walsh, S. J., Michaelsen, J. C., Franklin, J., Stow, D. A., Johanssen, C. J., & Johnston, C. A. (1991). Environmental analysis using integrated GIS and remotely sensed data: Some research needs and priorities. *Photogrammetric Engineering and Remote Sensing, 57*, 689–697.

Dikau, R. (1990). Derivatives from detailed geoscientific maps using computer methods. *Zeitschrift für Geomorphologie*, Suppl. *80*, 45–55.

Dikau, R., Brabb, E. E., & Mark, R. K. (1991). *Landform classification of New Mexico by computer* (Open-file Rep. 91-634). Reston, VA: U.S. Geological Survey.

Evans, I. S. (1972). General geomorphometry, derivatives of altitude and descriptive statistics. In R. J. Chorley (Ed.), *Spatial analysis in geomorphology* (pp. 17–90). New York: Harper and Row.

Evans, I. S. (1980). An integrated system of terrain analysis and slope mapping: *Zeitschrift für Geomorphologie,* Suppl. *36*, 274–295.

Fairbridge, R. W. (Ed.). (1968). *The encyclopedia of geomorphology*. Stroudsburg, PA: Dowden, Hutchinson, and Ross.

Forman, R. T. T., & Gordon, M. (1986). *Landscape ecology*. New York: Wiley.

Frost, R. E. (1953). *A manual on airphoto interpretation of soils and rocks for engineering purposes* (Instructional manual). West Lafayette, IN: Department of Civil Engineering, Purdue University.

Gibson, E. J. (1969). *Principles of perceptual learning and development*. Englewood Cliffs, NJ: Prentice-Hall.

Gibson, J. J. (1950). *The perception of the visual world.* Cambridge, MA:

Riverside Press.

Gibson, J. J. (1966). *The senses considered as perceptual systems.* Boston: Houghton Mifflin.

Gibson, J. J. (1971). The information available in pictures. *Leonardo, 4,* 27–35.

Gibson, J. J. (1979). *The ecological approach to visual perception.* Boston: Houghton Mifflin.

Golledge, R. G. (1987). Environmental cognition. In D. Stokols & I. Altman (Eds.), *Handbook of environmental psychology* (pp. 131–174). New York: Wiley.

Haber, R. N. (1985). Toward a theory of the perceived spatial layout of scenes. *Computer Vision, Graphics, and Image Processing, 31,* 282–321.

Hammond, E. H. (1964). Analysis of properties in land form geography: an application to broad-scale land form mapping. *Annals of the Association of American Geographers, 54,* 11–19.

Hoffman, R. R. (1984). *Methodological preliminaries to the development of an expert system for aerial photo interpretation* (Rep. No. ETL-0342). Ft. Belvoir, VA: Engineer Topographic Laboratories, U. S. Army Corps of Engineers.

Hoffman, R. R. (1985). *What's a hill? An analysis of the meanings of generic topographic terms* (Final Rep., Contract No. DAAG-29-D-0100). Alexandria, VA: Scientific Services Program, U. S. Army Research Office.

Hoffman, R. R. (1987). The problem of extracting the knowledge of experts from the perspective of experimental psychology. *The AI Magazine, 8* (2), 53–66.

Hoffman, R. R. (1990). Remote perceiving: A step toward a unified science of remote sensing. *Geocarto International, 5,* 3–13.

Hoffman, R. R. (Ed.). (1992). *The psychology of expertise: Cognitive research and empirical AI.* New York: Springer-Verlag.

Hoffman, R. R., & Conway, J. A. (1989). Psychological factors in remote sensing. *Geocarto International, 4,* 3-21.

Horton, R. E. (1945). Erosional development of streams and their drainage basins, hydrophysical approach to quantitative morphology. *Geological Society of America Bulletin, 56,* (3), 275–370.

Johansson, G., & Borjesson, E. (1991). Toward a new theory of vision studies in wide-angle space perception. *Ecological Psychology, 1,* 301–331.

Lillesand, T. M., & Kiefer, R. W. (1979). *Remote sensing and image interpretation.* New York: Wiley.

Lobeck, A. K. (1939). *Geomorphology.* New York: McGraw-Hill.

Mark, D. M. (1988). Network models in geomorphology. In M. G. Anderson (Ed.) *Modelling geomoprhological systems* (pp. 73–97). New York: Wiley.

Mintzer, O., & Messmore, J. A. (1984). *Terrain analysis procedural guide* (Rep. No. ETL-0352). Ft. Belvoir, VA: Engineer Topographic Laboratories, U. S. Army Corps of Engineers.

Morisawa, M. (1985). Development of quantitative geomorphology. In E. T. Drake & W. M. Jordan (Eds.), *Geologists and ideas, A history of North American geology* (pp. 79–107). Boulder, CO: Geological Society of America.

Neisser, U. (1991, January). *Without perception, there is no knowledge: Implications for artificial intelligence* (Report). Atlanta: Emory Cognition Project, Department of Psychology, Emory University.

Neuenschwander, G. (1944). *Morphometrische Begriffe, Eine kritische Übersicht auf Grund der Literatur.* Inaugural Dissertation, Universität Zürich, Switzerland.

Parker, S. P. (Ed.). (1984). *The McGraw-Hill dictionary of earth sciences.* New York: McGraw-Hill.

Pike, R. J. (1974). Craters on earth, moon, and mars: Multivariate classification and mode of origin. *Earth and Planetary Science Letters, 22,* 245–255.

Pike, R. J. (1988). The geometric signature: Quantifying landslide-terrain types from digital elevation models. *Mathematical Geology, 20,* 491–511.

Pike, R .J. (1992). Machine visualization of synoptic topography by digital image processing. In D. A. Wiltshire (Ed.), *Selected papers in the applied computer sciences* (chap. B, Bulletin 2016). Reston, VA: U. S. Geological Survey.

Pike, R. J., & Rozema, W. J. (1975). Spectral analysis of landforms. *Annals of the Association of American Geographers, 65,* 499–516.

Pike, R. J., & Thelin, G. P. (1989). Cartographic analysis of U. S. topography from digital data. *Proceedings of Auto-Carto 9, The international symposium on automated cartography* (pp. 631–640). Baltimore: International Society for Photogrammetric Engineering and Remote Sensing.

Quattrochi, D. A., & Pelletier, R. E. (1991). *Quantitative methods in landscape ecology* (Rep.). Stennis Space Center, MS: National Aeronautics and Space Administration, John C. Stennis Space

322   HOFFMAN & PIKE

Center.
Reeves, J. (Ed.). (1975). Manual of remote sensing. Falls Church, VA: American Society of Photogrammetry and Remote Sensing.
Rinker, J. N., & Corl, P. A. (1984). Air photo analysis and feature extraction (Rep.). Engineer Topographic Laboratory, Ft. Belvoir, VA: U. S. Army Corps of Engineers.
Savitt, S. L., Henderson, T. C., & Colvin, T L. (1992). Feature extraction for localization. Proceedings of the DARPA Image Understanding Workshop (pp. 327–331). Arlington, VA: Defense Advanced Research Projects Agency.
Smith, K., & Heinrichs, M. (1991, September). How to figure out where you're at. Paper presented at the 25th Reunion of the Center for Research in Learning, Perception, and Cognition, University of Minnesota, Minneapolis, MN.
Strahler, A .N. (1964). Quantitative geomorphology of drainage basins and channel networks. In V. Chow (Ed.), Handbook of applied hydrology (pp. 4–39). New York: McGraw-Hill.
Stringer, P. (1975). The natural environment. In D. Canter, P. Stringer, I. Griffiths, P. Boyce, D. Walters, & C. Kenny (Eds.), Environmental interaction (pp. 281–319). New York: International Universities Press.
Tanner, W. F. (1960). Numerical comparison of geomorphic samples. Science, 131, 1525–1526.
Thelin, G. P., & Pike, R. J. (1991). Landforms of the conterminous United States — A digital shaded-relief portrayal (U.S. Geological Survey Miscellaneous Investigations Map I-2206, scale 1:3,500,000). Reston, VA: U. S. Geological Survey.
Wall, S. D., Farr, T. G., Muller, J.-P., Lewis, P., & Leberl, F. W. (1991). Measurement of surface microtopography. Photogrammetric Engineering & Remote Sensing, 57, 1075–1078.
Way, D. S. (1978). Terrain analysis. Stroudsburg, PA: Dowden, Hutchinson and Ross.
Weszka, J. A., Dyer, C. R., & Rosenfeld, A. (1976). A comparative study of texture measures for terrain classification. IEEE Transactions on Systems, Man, and Cybernetics, SMC-6, 269–285.
Wolfanger, L. A. (1941). Landform types (Michigan State College Agricultural Experiment Station Tech. Bulletin 175). East Lansing, MI: Michigan State College.
Wood, W. F., & Snell, J. B. (1960). A quantitative system for classifying landforms (Rep. No. EP-124). Natick, MA: U.S. Army Quartermaster Research and Engineering Command, U. S.

Army.

Zakrzewska, B. (1967). Trends and methods in land form geography. *Annals of the Association of American Geographers, 57*, 128–165.

Chapter 11

The Role of Mental Simulation in Problem Solving and Decision Making

Gary Klein and Beth W. Crandall

Klein Associates Inc.
Fairborn, OH

11.0 Introduction

Mental simulation is the process of consciously enacting a sequence of events. Mental simulation plays an important role in phenomena such as problem solving, judgment, decision making, and planning. In this chapter we describe how mental simulation can serve as an important source of power. We also identify some difficulties that can arise when people use mental simulation.

Our own thinking about mental simulation evolved out of a perspective on decision making that emphasizes perceptual-recognitional processes. We developed a Recognition-Primed Decision (RPD) model to account for problem solving and decision making outside of well-defined laboratory tasks. In the following section we describe the RPD model in order to put mental simulation into context. Then we describe similarities and differences between mental simulation and related concepts in cognitive psychology. Next we describe some of the major functions of mental simulation. We present a descriptive model of mental simulation and examine some of the vulnerabilities that arise from its use. Finally, we examine the linkage between mental simulation and ecological psychology.

11.1 The Recognition-Primed Decision Model

We began by studying command-and-control decision making, observing and obtaining protocols from urban fireground commanders (FGCs) in charge of allocating resources and directing personnel. We studied the decisions they made in handling nonroutine incidents during emergency events (Klein, Calderwood, & Clinton-Cirocco, 1986). Some examples of the types of decisions these commanders had to make included whether to initiate search and rescue, whether to initiate an offensive attack or concentrate on defensive precautions, and where to allocate resources.

The FGCs' accounts of their decision making did not fit into a decision-tree framework. The FGCs argued that they were not "making choices," "considering alternatives," or "assessing probabilities." They saw themselves as acting and reacting on the basis of prior experience; they were generating, monitoring, and modifying plans in response to the demands of the situations. We found no evidence for extensive option generation. Rarely were even two options concurrently evaluated, so that opportunities for comparisons between the utilities of outcomes were largely absent. We could see no way in which the concept of optimal choice might be applied. Moreover, it appeared that a search for an optimal choice could stall the FGCs long enough to lose control of the operation altogether. The FGCs were more interested in finding an action that was "workable," "timely," and "cost-effective."

Nonetheless, the FGCs were clearly encountering choice points during each incident. They were aware that alternative courses of action were possible, but insisted that they rarely deliberated about the advantages and disadvantages of the different options. Instead, the FGCs relied on their ability to recognize and appropriately classify a situation. Once they knew it was "that" type of case, they usually also knew the typical way of reacting to it. They would use available time to evaluate an option's feasibility before implementing it. Mental simulation might be used to "watch" the option being implemented, to search for flaws, and to discover what might go wrong. If problems were foreseen, then the option might be modified or rejected altogether and a next most typical reaction explored. This mental search would continue until a workable solution had been identified.

We described this strategy as a Recognition-Primed Decision (RPD) model of rapid decision making (e.g., Klein, 1989; Klein & Calderwood, 1991; Klein et al., 1986; Klein, Calderwood, & MacGregor, 1989). For many task environments, a recognitional strategy appears to be highly efficient. The proficient FGCs we studied used their experience to generate a workable option as the first to consider. If they had tried to generate a large set of options, and then had systematically evaluated these, it is likely that the fires would have gotten out of control before the FGCs could make any decisions.

The RPD model explains how people can make decisions without having to compare options. It fuses two processes: situation assessment and mental simulation. In brief, the RPD model asserts that people use situation assessment to generate a feasible course of action and use mental simulation to evaluate the course of action. Additionally, when there are several plausible hypotheses about the nature of the situation, mental simulation is also used to choose the best explanation.

The RPD model is presented in Figure 11.1. Level 1, the simplest case, is one in which the situation is recognized and the obvious reaction is implemented. In Level 2, mental simulation is used to diagnose or explain the current situation. In Level 3, mental simulation is used to explain the development of a situation, and/or mental simulation is used to uncover problems with a course of action prior to carrying it out, so that the option has to be strengthened or rejected.

The RPD model is characterized by the following features:

• Situational recognition allows the decision maker to classify the task as familiar versus unfamiliar or atypical.

• The recognition classified as familiar carries with it recognition of the following types of information: plausible goals, cues to monitor, expectancies about the unfolding of the situation, and typical reactions.

• Options for courses of action are generated serially, with a very typical course of action as the first one considered.

• Option evaluation is also performed serially, using mental simulation to test the adequacy of the option, to identify weaknesses of that option, and to find ways to overcome the weaknesses.

• Skilled decision makers are able to respond quickly by using experience to identify a plausible course of action as the first one considered rather than having to generate and evaluate a large set of options.

• Under time pressure, the decision maker is poised to act while evaluating a promising course of action, rather than being paralyzed while waiting to complete an evaluation of alternative options.

A key process for evaluation is the use of mental simulation to portray the way the option would be implemented in that specific environment, to allow the decision maker to develop expectancies about how the situation will evolve, to allow the decision maker to detect possible barriers in the specific environment, and to allow the decision maker to notice unexpected opportunities for useful interventions. Mental simulation also can help the decision maker improve options, alert the decision maker to important dynamics, and modify the assessment of the situation. Finally, mental simulation can be used to

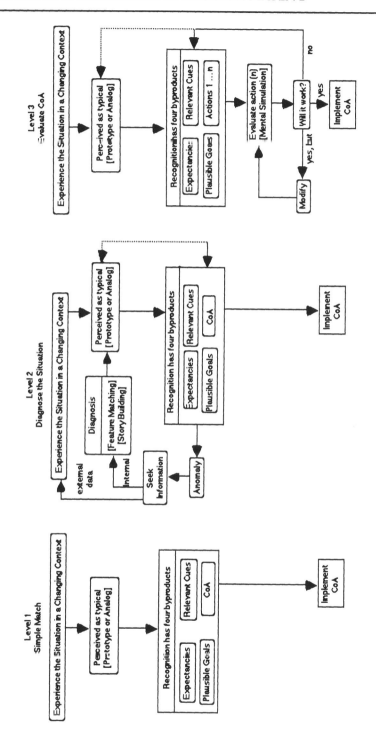

Figure 11.1. *The Recognition Primed-Decision model (RPD).*

explain anomalies and thereby contribute to situation assessment.

We do not propose that the RPD model is an alternative to analytical strategies. Klein (1989) described the conditions under which a recognitional strategy is more useful and those conditions favoring the use of analytical strategies such as Multi-Attribute Utility Analysis (MAUA).

In our studies (see Klein, 1989, for more detail), we found that 60% to 80% of decisions can be classified as recognitional, as long as the decision maker is reasonably experienced. These proportions were observed in different settings, ranging from fireground commanders to design engineers. As expected, with less experienced decision makers (e.g., novice tank platoon leaders), the proportion declined to 40%. Inter-coder reliabilities were high in these analyses, ranging from 87% to 94%. These data are consistent with the contingency decision-making framework of Beach and Mitchell (1978) and Payne (1976). In our research we applied ethnographic methods to try to learn more about the nature of nonanalytical strategies used in naturalistic settings and to try to estimate the likelihood of decision makers using analytical versus non-analytical strategies in the particular domains we studied. Clearly, there are times when MAUA methods are required, and we observed many such examples. However, even though we studied nonroutine decisions, we still found that people mostly used recognitional, rather than analytical, strategies.

We tested applications of the RPD model in a variety of tasks and domains, including fireground command, battle planning, critical care nursing, corporate information management, and aircrew coordination (e.g., Crandall & Calderwood, 1989; Crandall & Gamblian, 1991; Crandall & Klein, 1987; Klein et al., 1986; Thordsen, Galushka, Klein, Young, & Brezovic, 1990; Thordsen, Klein, & Wolf, 1990). These studies replicated the frequency of recognitional strategies and repeatedly demonstrated the importance of perceptual processes during decision making. The studies also helped us understand and confirm the importance of mental simulation in naturalistic decision making.

It should be clear by now that the RPD model synthesizes aspects of problem solving along with decision making. The subprocesses of problem solving are to define goals, generate alternative ways of reaching those goals, evaluate options, and implement the most promising alternative. These are all included in the RPD model, except that the model shows how people with experience can short-circuit the need to generate and evaluate sets of options. It is this activity, the evaluation of different options, that is the traditional focus of decision-making research. It is the part of the problem-solving process that calls for decision making. The RPD model explains how problem solving can proceed without directly contrasting different courses of action. It describes a strategy that people use under time pressure, but our results with design engineers show that it is frequently used even under adequate time allowances. Within the RPD model, mental simulation serves a variety of essential functions, including evaluating situation assessment, learning about the dynamics of the situation, evaluating courses of action, improving courses of action, and generating expectancies.

11.1.1   Time Pressure, Recognitional Decision Making, and Mental Simulation

The issues surrounding time-pressured decision making reflect a larger problem. Normative decision research has often centered around well-defined and static tasks for which an optimal decision could be identified. Under these conditions the performance of subjects has generally been suboptimal. However, it is unlikely that powerful analytical decision strategies could be applied to dynamic tasks with ill-defined goals and moderate to high levels of time pressure. In their predominant focus on human frailties, analytical decision models have not provided an account of how proficient decision makers can function effectively in time-pressured environments.

In one study of cognitive performance under time pressure, we measured the quality of chess moves from actual games played under routine conditions (e.g., approximately 150 sec per move) and under blitz conditions (e.g., approximately 6 sec per move) (Calderwood, Klein, & Crandall, 1988). Chess masters showed virtually no reduction in move quality as time pressure increased. Class B players showed a significant reduction, but move quality was still quite high under the time pressure. These data suggest that people can maintain fairly high levels of performance despite time pressure that should prevent the use of analytical strategies of decision making and problem solving.

In developing the RPD model, a primary goal has been to account for the impact of time stress on decision making. The RPD model was developed and refined using incident accounts drawn primarily from time-pressured settings. We estimated that in our initial study more than 85% of these difficult, nonroutine, and high-risk decisions took less than a minute, and in many cases the FGCs seemed to require only a few seconds. In contrast to the analytical strategies, recognitional decision strategies can be implemented under high degrees of time pressure, because these strategies describe decision makers' attempts to identify satisfactory, rather than optimal, options.

The aspect of recognitional decision making that we suspect *is* affected by time pressure is the mental simulation component. We expect that this usually occurs as a graceful degradation—a diminished forward search that is done with less care and thoroughness.

Therefore, we have sought to understand mental simulation in order to gain a clearer perspective on how time pressure affects decision making and problem solving. Mental simulation may be the key process in the RPD model that is directly affected by time stress. Most of the RPD components are perceptual, such as recognition of typicality and of changes in the relevance of cues. These processes would not be affected by time pressure (except in the mechanical case in which the time pressure is so great and the environment so dense with information that the decision maker is prevented from receiving all the messages). In contrast, mental simulation requires conscious examination and therefore is vulnerable to time pressure. The decision maker attempts to work out a scenario, either by moving forward from the current situation (to evaluate a course of action) or by starting in the past and moving up to the current situation (to understand how the current situation might have developed). This takes time, and it takes cognitive resources.

If time pressure is sufficiently great, it can limit the extent of mental simulation, thereby reducing the chance to play out an option far enough to detect possible flaws. It can prevent mental simulation altogether. Time pressure may also distort the way mental simulation is carried out. If this is so, then there is the added potential for degraded decision making and problem solving, particularly when experience levels are low.

11.2   Accounts of Mental Simulation

In this section, we describe previous research and models that pertain to mental simulation. Einhorn and Hogarth (1981) introduced the term *mental simulation* as an aspect of decision making. In discussing anchoring and adjustment strategies, Einhorn and Hogarth presented the idea "that adjustments are based on a mental simulation in which 'what might be,' or 'what might have been,' is combined with 'what is' (the anchor)" (p. 456). In other words, they suggested that decision makers imagine different configurations of events in order to apply a heuristic of starting with a known value and adjusting this value to fit a new situation. Although this was one of the first instances of the term *mental simulation*, there have been a number of accounts of the same process from a variety of perspectives, including:

> Simulation Heuristic
> Image Theory
> Progressive Deepening
> Story Model
> Mental Imagery
> Mental Models
> Computer Simulations
> Additional Models

In examining each of these concepts, we show similarities and differences compared to mental simulation.

11.2.1 The Simulation Heuristic

Kahneman and Tversky (1982) described a simulation heuristic that consisted of a decision maker "running" a mental model of a situation in order to determine how to react. Although our theoretical perspective on decision making is somewhat different from that of Kahneman and Tversky, our approach to mental simulation overlaps with their work.

Kahneman and Tversky (1982) listed five functions of the simulation heuristic: generate predictions, assess event probabilities, generate conditional probabilities, assess causality, and generate counterfactual assessments. Kahneman and Tversky also noted a number of properties of mental simulation. The simulation can be run with default conditions or specific settings. The target state can be set or left unspecified. It is more plausible to make changes that increase typicality than changes that are unlikely. An outcome is judged as unlikely if it can only be reached via "uphill" (that is, implausible) assumptions. The plausibility of a mental simulation is a function of factors such as availability, ease of retrieval, and memory for specific instances such as analogues. We have observed these phenomena in our research. We would note, however, that Kahneman and Tversky characterized mental simulation as a conscious, deliberative, highly analytic procedure.

Jungermann and Thüring (1987) described the simulation heuristic in a 4-step model. The steps are: activation of problem knowledge, constitution of mental model, simulation of mental model, and selection of inferences. Although this model is a good starting point, the steps are far too general to be of much use as a descriptive model.

11.2.2 Image Theory

Beach and Mitchell (1987; see also Beach, 1990) presented a model of decision making that is oriented around different types of images—images pertaining to individual values, images about the eventual outcome of the current situation, images of the desired outcome, and images of the effects of adopting different courses of action. The retrieval, use, and comparison of these various types of images are held to explain how decision makers function. Conceptually, image theory is much broader than the mental simulation construct we are investigating. Of greatest relevance for our purposes are the images regarding the trajectory of the current situation, the desired situation, and the anticipated situation. In its current form, image theory does not address the situational dynamics of how a course of action in a certain setting might lead to other side effects. In addition, image theory has not yet been linked with basic cognitive processes such as capacity limitations.

11.2.3 Progressive Deepening

De Groot (1965) described a phenomenon that he called progressive deepening to capture how chess players consider which move to make. De Groot observed that chess players attempt to search a decision tree more deeply, rather than more broadly. Their skill enables them to identify a small set of options worth considering, and then a small set of countermoves that might be attempted, and a few reactions to these, and so forth. Moreover, the chess players return to a given move and follow it out more deeply to see if there are any pitfalls or opportunities. This is a type of gaming that relies on mental simulation and that we have observed in interviews we have conducted. Still, there are minor differences worth noting. First, chess is a limited context game. The nature of the simulation is very constrained as compared to the task of a fireground commander imagining how a rescue might be made. Second, mental simulation does not have to be progressive. In many cases one passthrough is sufficient, although there are times when decision makers are innovative and report playing the course of action again and again.

11.2.4 The Story Model

Pennington and Hastie (1993) described jurors making decisions in terms of a story model. They saw the jurors as sifting through a wide variety of data and organizing these data into a coherent account of what probably happened. In other words, the jurors were trying to build a story that was the best match with the evidence and with their own sense of how people behave. We find this a very attractive notion because it applies to a wide variety of domains in which a person must assess a situation in the face of ambiguous and conflicting evidence.

The story model seems to be a type of mental simulation that provides a problem solver or decision maker with an explanation of how a situation has evolved. In our research on Navy Combat Information Centers (e.g., Kaempf, Wolf, Thordsen, & Klein, 1992), we found that key decisions are often of this type, for example, whether an unknown target has hostile intent. Here the story model is particularly apt. There are also differences that suggest that mental simulation is broader than constructing a story. First, the story model was designed for explaining how a situation developed, whereas mental simulation is also valuable for evaluating how a course of action will play out. Second, the model restricts the concept of a story to incidents involving human agents, capable of volition and intention. The story is about what people choose to do. In contrast, mental simulation might also apply to inanimate objects and forces; a fireground commander imagines how a fire might spread in order to judge where to send in crews.

11.2.5 Mental Imagery

Shepard and Metzler (1971) and Kosslyn and Pomerantz (1977) summarized the literature on mental imagery, a phenomenon that is related to mental simulation. However, mental imagery research has been concerned with tasks such as attempts to rotate complex shapes. It has not involved causal agents, the impact of several different components interacting, or the influence of contextual factors such as personality styles of agents involved in carrying out a plan.

11.2.6 Mental Models

The concept of mental models (e.g., Rouse & Morris, 1986) is obviously linked closely to mental simulation. During mental simulation, a decision maker cognitively constructs a model and sets it in motion to see what happens.

We would point out that the concept of mental models is much broader than that of mental simulation. Mental models include the memory representation of a task or domain. When someone is trained to use a system, it is common to say that person is building up a mental model of the system (e.g., Kieras & Bovair, 1984). And when information about the functioning of the system is provided, it is claimed that the manipulation provides a mental model of the system. As Rouse and Morris pointed out, the problem with this usage is that "mental model" comes to refer to everything that the person knows.

The concept of mental simulation is a subset of the concept of mental model. It refers to the information that people use when they construct and run action sequences. Because of limits in working memory, the action sequences cannot become too complex. People would have difficulty identifying a dozen objects as agents, and another dozen causal factors, and then tracing how the causal factors influence all the objects/agents. That is too much to keep track of. We hypothesize that some of the metacognitive skill in constructing a mental simulation is in selecting only a few components at an appropriate level of abstraction. Therefore, mental simulation is not concerned with everything a person knows (the mental model of the domain), but with the information set that is used to construct and run the action sequence.

In approaching mental models from the perspective of playing out a scenario or story, rather than building an organizational scheme, we are following the path chosen by other researchers. Thus, Williams, Hollan, and Stevens (1983) noted that "Running a mental model corresponds to modifying the parameters of the model by propagating information using the internal rules and specified topology. Running a mental model can also occur when autonomous objects change state" (p. 133). DeKleer and Brown (1983) carefully examined the differences between building the simulation, which they called *envisioning*, and setting the simulation into action, which they called *running*. These are central to their interest in what happens when a person activates a mental model in his or her mind's eye. Forbus (1983) presented an account of what happens when a person runs a mental model, in which the model is transitioned from one state to the next. Forbus used the term *action sequence* to refer to the successive states of the mental model as it is being run, and we adopted this terminology.

11.2.7 Computer Simulations

Computer simulations offer an interesting case to consider, particularly as a contrast to mental simulation. In constructing a computer simulation, a developer identifies a key set of variables, links these together using a set of relationships, and initializes the variables to begin the simulation run. The simulation proceeds mechanically to change states and enter transactions in accord with the parameters defined. This is essentially what happens during certain types of mental simulation. Still, there are some instructive differences. A computer simulation can accept a much larger set of variables and interactions than a decision maker could keep track of. This is one reason for the popularity of computer simulations, because they allow the decision maker to introduce a greater variety of factors into the equation. A second difference is that at many boundary points there will be no default values and computer simulations rely on Monte Carlo methods. A third difference stems from one of the major limitations of computer simulations—that they cannot capitalize on all of the rich domain-specific associations that people have, because it is so difficult to build this everyday knowledge into the simulation.

11.2.8 Additional Models and Phenomena

There are a number of other phenomena that are related to mental simulation. It would take us too far afield to review all of these, but we briefly mention some additional concepts that are linked to mental simulation. The Piagetian stage of *formal operations* is reflected in a person's ability to engage in hypothetical reasoning. This developmental stage lines up nicely with the concept of mental simulation, which is also a process of hypothetical reasoning. The concepts of *scripts* and *schemata* are also relevant here. The use of a script allows one to know and anticipate what is going to happen in a situation. In our view, the issues become more interesting when the script breaks down and the decision maker must use personal experience to go beyond it. For example, we would be less interested in a person reciting a restaurant script to imagine a future meal than in a person imagining what he or she might do if a specific restaurant caught fire. Another related concept is a decision tree that can trace out all the branches growing from each choice point or mode. This resembles a person playing out a scenario, the difference being that the decision makers we have studied do *not* trace each of the branches, and do not evolve a complex tree structure. Rather, they identify one or a few alternatives at each node and consider only these. Their experience seems to enable them to only consider plausible branches rather than having to generate all branches and then *prune* these back. Finally, the concept of *fantasy* (Lynn & Rhue, 1988) overlaps mental simulation. We mention these diverse concepts to show how varied a role mental simulation has played in different domains and theories in psychology.

We should also mention some applied work in the realm of training (e.g., Druckman & Bjork, 1991), in which the use of *mental imagery* for people such as pilots, divers, and archers has been studied and shown to have some value. The review conducted by Driskell and Mullen (1991) showed that mental simulation methods have been generally effective in providing training benefits. However, such imagery is too static to be of interest in studying decision making. The person rehearses the same image repeatedly (e.g., making a perfect tennis stroke rather than playing with different possibilities). Also, Sheikh (1983) did a considerable amount of work on mental imagery for psychotherapy and for education. However, this work also focused on static rather than dynamic images.

In summary, mental simulation is closely tied to many different topics in psychology and cognitive science. Nevertheless, it remains an interesting and important topic in its own right, one that has received surprisingly little direct attention.

11.3   Mental Simulation in Naturalistic Decision Making

How is mental simulation different from other forms of reasoning, or from rule-based accounts of behavior? Clearly, the phenomenon overlaps other forms of inference, and yet it appears to have its own underlying characteristics. For this reason, it is important to provide a more detailed account of mental simulation so that its unique characteristics can be better appreciated and its role in decision making better understood.

Our approach to the study of decision making has consistently emphasized the value of grounding research in the tasks and settings in which the activity occurs. It made sense to us to take a look at how people use mental simulation in their everyday lives and to work from there to develop a model that describes those activities.

We undertook an ethnographic study of mental simulation, designed to offer an initial, exploratory look at the role of mental simulation in planning, decision making, and problem solving. For the study, we developed a database of 102 mental simulation descriptions—incidents in which decision makers reported relying on some form of mental simulation for problem solving and decision making.

Within our sample of mental simulation incidents, we identified four primary functions served by the simulations: generate a course of action, inspect and evaluate a course of action, explain a phenomenon, and discover and explore models of a phenomenon. Category descriptions and examples follow.[1]

11.3.1 Generate a Course of Action

This category includes mental simulations wherein the decision maker attempts to envision a planned course of action. The mental simulation is used to create a plan, generate information, and develop expectancies about events that are likely to occur in the mentally simulated situation. The simulations are typically convergent, focusing on identifying action sequences and time bound. They typically take a current situation or event and project its development over time, from the present into the future. The simulations allow the decision maker to anticipate the "look and feel" of ensuing events and to adequately prepare for them. Once actually in the situation, the decision maker uses these simulation-based expectancies to evaluate his or her assessment of the situation. As the situation unfolds, discrepancies between what the decision maker expected and what he or she observed can serve as a flag that the original situation assessment was off base, or that critical components have changed and the situation assessment needs to be reevaluated.

Mental simulations that serve a planning/anticipation function appear to be particularly important when time limits are severe or risks high. An experienced decision maker knows that once in the situation there will not be time to analyze and evaluate, so he or she uses mental simulation to perform those functions ahead of time. The office

---

[1]Two coders independently categorized each item in the database into the four categories. Coding reliability was assessed using kappa, a chance-corrected index of agreement (Fleiss, 1981). Overall kappa across all four categories was .47, where .40 is deemed acceptable and .75 is deemed very good. The informal nature of the incident accounts made it hard to judge category membership in many cases.

manager for a small business described how she develops Dbase programs:

> *I don't begin until I have a clear picture in my mind of what 'path' a piece of data needs to follow to be sent to another spreadsheet or to be transformed. If I get stuck, I use the help functions to form these "maps." Before I have a full understanding of all the steps involved, the map isn't formed and I have a lot of trouble programming successfully. As I build the map, the correct "route" glows a little brighter than the other ones. If I get an error message, these maps make debugging easier, because I can see where things have gone wrong.*

A parts engineer at a manufacturing firm described how he generates cost estimates for a bid:

> *Once I have the part figured out pretty well, I mentally "run the job"—I do this by imagining one part going through the entire manufacturing process. At the same time I am mentally processing the part, I set a timeclock in my head and measure the time it takes for each phase of the manufacturing process. This lets me figure out how we would build the part and price out each phase of manufacturing for a single part. Once I know the processes and the amount of time it would take to produce one part to spec, I run the timeclock with 10 parts, then 200, and then whatever the lot size is for the actual order. By running the process with different size orders, I can see what the tradeoffs will be for repetitions, learning curve, scrap, and so forth.*

A surgical nurse reported:

> *Prior to any surgery in which I am assisting, I mentally step through the processes we will be doing. First, I imagine the preparation for the IV and endotracheal tubing, then I move on to general prep (non-sterile drapes, shaving, sterile drapes). Once I have stepped myself through prep, I walk through the surgery itself. I identify the surgical tools and other materials that will be needed and prioritize them. This lets me be prepared for quick access to the particular containers and materials I'll need. Once I'm done, I set up the surgical room in accord with this mental run-through.*

11.3.2 Inspect/Evaluate

This category includes mental simulations that allow a decision maker to evoke and assess the impact of a specific action or sequence of events, including the flaws or risks associated with it. The decision maker asks, "Will this work?" or "What could go wrong at this point?"

On a two-lane state highway, a driver passed several houses set close to the road. The driver recalled:

> *There was a toddler on one of the porches. As I passed the house, the child stepped off the porch and headed for the road. Moments before I had noticed a semi in my lane, about a quarter mile behind me. As the child moved towards the road, I could `see' the scene from behind the wheel of the semi. I knew the truckdriver could not stop the truck before reaching the point in the road the child was heading towards. The size of the truck and the speed at which it was moving meant that an attempt to swerve would send the truck out of control. As all this was playing out in my mind, I had pulled off the road, jumped out of the car, and was running back toward the child. I could see the truck coming at us. At that moment, the child's mother ran out of the house, into the road, and scooped the child up and out of harm's way.*

The parent of two young boys described her search for a new house:

> *I went to look at a house I had seen listed for sale. As I went through the house I imagined various activities in each room, and how we would `fit' there. I have a visual disability, so one set of images involved how the house would be for me to move around in. I imagined myself moving through the house, and found places that seemed like they would be a problem for me. I looked at the spiral staircase, thought about the kids and me using it, and how that might be difficult. The house only had two bedrooms, and I imagined the boys sharing a room. I thought over how I might divide the space up; then I could just see the two of them in the middle of the room fighting, instead of on their respective sides. Although there were many aspects of the house I really liked, I decided it had too many disadvantages.*

Evaluation simulations often involve a directed search for specific problems and an assessment of probable outcomes that includes an affective component (satisfaction vs. disappointment). The affective response evoked by the simulation guides the person's decision of whether to actually implement the course of action envisioned. The person uses experience to detect flaws and warning signs as the simulation unfolds. Analogies, rules of thumb, and so on, enter into the detection of problems. These flaws relate to other criteria suggested by Robinson and Hawpe (1986): consistency and plausibility. In addition to taking consistency and plausibility into account, the inspection will cover potential flaws in the course of action.

If the simulation satisfies these tests, the person can proceed. If it fails, the person may be able to modify it. In this way, the course of action is improved. In our sample, we found that the decision makers tried to find a workable scenario and often took the time to evaluate it for flaws. Sometimes they tried to optimize their choice, but there were cases in which they just tried to confirm that a course of action would not lead to trouble.

11.3.3 Explain

This category includes mental simulations conducted to help the decision maker understand how and why an event could have occurred. These mental simulations involve the search for a plausible explanation and are typically focused backward in time. The decision maker attempts to generate a feasible account of how an end state could have arisen or the factors that could account for an anomaly that has occurred. Inability to generate a plausible explanation can cause the decision maker to abandon the mental simulation altogether and to reexamine other aspects of his or her thinking. Some examples follow:

> The captain of a Mississippi tanker, the Pisces, made radio contact with the captain of another, the Trademaster, as the two ships entered opposite ends of an S-shaped stretch of the river. The captains agreed to pass port-to-port, as usual. As they came into visual range, the captain of the Trademaster spotted several slow-moving barges just ahead of him and realized he would have to go around them. He radioed the Pisces to switch to a starboard-to-starboard passage, and the Pisces agreed. However, the Pisces captain noticed barges approaching his vessel that would make it difficult for him to maintain his course on the right of the river. He radioed the Trademaster to shift back to a port-to-port passage. The Trademaster captain never received the message, and when the Pisces began to shift position, the Trademaster's captain decided there was a temporary problem that

accounted for the Pisces maneuver. Thus, his mental simulation accounted for the Pisces's behavior without raising any alarms. When the Pisces did not "correct" its course, the Trademaster edged to the left to give the other ship more room. Because the Pisces' captain believed they were going to pass port-to-port, he too edged over. The two ships eventually collided as each attempted to maintain the "agreed upon" course.[2]

During the Iran-Iraq war, a Tactical Action Officer (TAO) in charge of Air Defense for a Navy Battle Group was asked by one of the cruisers for permission to shoot down a threatening aircraft. The target was unidentified, was coming from the direction of an Iranian airport, did not generate IFF signals, was flying at a low altitude, and was heading straight toward the cruiser. Moreover, it failed to respond to radio warnings. In short, it fit the Rules of Engagement and was a legitimate target. Nevertheless, the Tactical Action Officer refused permission to fire. He noted that the unknown aircraft was flying at a very low speed and was far from the coast. He could not imagine a hostile Iranian fighter aircraft moving so slowly. He tried to imagine a light aircraft or helicopter on a suicide mission, but the aircraft's slow speed and erratic course made even this seem unlikely. Unable to formulate an explanation that involved a hostile aircraft, the Tactical Action Officer generated a mental simulation involving helicopters from other ships in the Battle Group that had become lost. He directed aircraft to the area for visual identification. His suspicions were correct — it was a British helicopter that had lost its way.

11.3.4 Model/Discover

This type of mental simulation is used to deepen the decision maker's understanding of some process or problem. The simulations often have a musing, expansive quality. The goal of these simulations is to create a hypothetical model or system and to "observe" it in action. These simulations often involve examination of causal factors and the interactions among them, and they tend to be multifaceted. The simulations typically proceed iteratively. Knowledge gained from each run of the simulation enhances subsequent simulations of the same problem or process. For example:

---

[2]Based on an account in *Normal Accidents* (Perrow, 1984).

Einstein conducted many thought experiments. In a famous one, he imagined that he was riding in an elevator. The elevator had a pinpoint opening, through which shone a beam of light. From his vantage point inside the elevator, Einstein pictured himself riding up and down at various speeds, watching changes in the light beam. In this way he was able to "observe" how increased velocity made the light seem to bend more sharply.

For a study of tank platoon training, we found that cadets and instructors were both sensitive to the same range of visual and perceptual cues, but that they used this information quite differently (Brezovic, Klein, & Thordsen, 1987). The instructors excelled at constructing a view of the exercise from the adversary's perspective. For example, the instructors envisioned the adversary forces on the other side of a hill, looking at access routes and setting up their own offensive positions. The instructors evaluated any maneuver or positioning for how it appeared to the adversary, and what the adversary's likely response would be.

We found 19 cases in which the simulation was used to generate a course of action or to generate predictions and estimates, 36 cases in which the course of action was evaluated or inspected for risk, 11 cases in which a model was derived, and 13 cases in which the simulation was used to explain a situation. There were an additional 23 instances that were judged not to meet the criteria for mental simulation, some because of insufficient information.

11.4   A Model of Mental Simulation

Based on our examination of these incidents, we derived a general account of mental simulation, as presented in Figure 11.2. A person uses mental simulation to accomplish some goal, so the first function is to understand the need motivating the activity. For a given need, the next step depicted in Figure 11.2 is to specify the components of the simulation. The person needs to determine whether the task is to interpolate between a known initial condition and a known terminal condition, to extrapolate forward from a known initial condition (as in predicting the future), or to extrapolate backward from a known terminal state, as in trying to explain a situation. The person needs to identify the causal factors that will be used in the simulation.

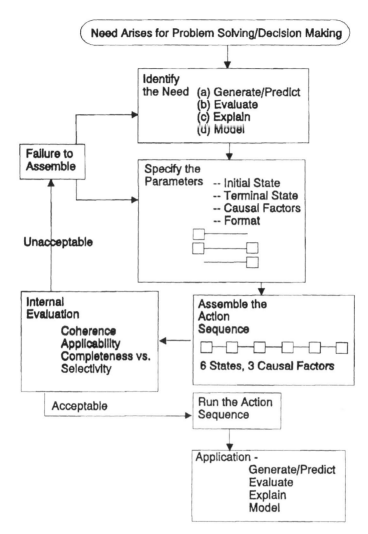

Figure 11.2. *A general model of mental simulation.*

The third step depicted in Figure 11.2 is to assemble the simulation. In our database, many simulations consisted of transitions from one state to another. The purpose of the simulation was to examine the nature of these transitions. It appears that people naturally limit the number of factors and transitional states they employ in a simulation. In our study, the median number of transitions was about 6, and the median of causal factors that varied was 3, with a range of 1 to 6. So we represented mental simulation as an action sequence with 6 progressions and 3 operators or causal factors. Sometimes the initial state is a given, and the person has to extrapolate to an end state; at other times both the initial and terminal states are known, and the task is to interpolate—to bridge the gap.

We depicted the transitions from one state to another as discrete, rather than continuous. Therefore, "configuring the simulation" in one's head involves selecting a small number of variables and manipulating these over several transitions—like successive frames in an animated sequence. Not all mental simulations are experienced as a series of discontinuous action states; many appear to run in a continuous action flow. But in these instances as well, the number of factors and the span of the simulation are constrained.

The internal coherence of the simulation is judged using some of the criteria listed by Robinson and Hawpe (1986) for evaluating stories: coherence, completeness/selectivity, and applicability. Pennington and Hastie (1993) suggested that stories are more acceptable as a function of completeness, consistency, plausibility, and uniqueness. In other words, the simulation has to make sense; it has to address the important features of the situation but not drown in excessive detail, and it has to apply to the goals. If the simulation fails these tests, it will be necessary to try again or to rethink the goals. If it succeeds, then the person runs the simulation and uses the information gained from it.

Figure 11.3 shows a straightforward elaboration of the general model as presented in Figure 11.2. Here the mental simulation is used to generate a course of action, as described earlier. The person builds the simulation by specifying the parameters, assembling them into an action sequence, checking whether the action sequence is appropriate (e.g., is coherent, applies to the need at hand, and is sufficiently complete), running the action sequence, and using the run either as the plan itself or to obtain predictions.

Figure 11.4 shows the use of a mental simulation to evaluate a course of action, once it is generated. The functions of generating and evaluating a course of action are closely related. The only difference is that Figure 11.4 elaborates some of the final steps. The running of the action sequence takes the form of reviewing each step to spot problem areas such as implausible continuations, inconsistencies, and pitfalls. These trouble spots may themselves be evaluated, using mental simulations. This is referred to as *micro-simulations* in Figure 11.4. Finally, the decision maker or problem solver may form a global, perhaps affective evaluation of whether the course of action seems

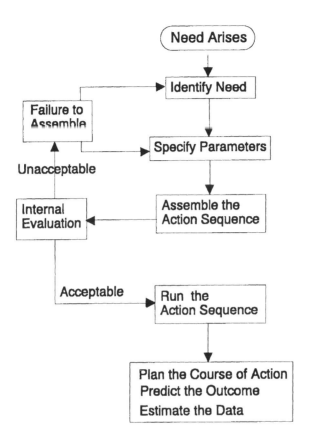

Figure 11.3. *Generate a course of action.*

workable.

Figures 11.5 and 11.6 show a different aspect of mental simulation—explaining and modeling. Here the focus is on situation assessment rather than on a course of action. Figure 11.5 covers the use of a mental simulation to build an account of how an event may have evolved. Thus, it starts from the past and ends in the present. The primary difference from Figure 11.3 is that here, the *failure* to build an acceptable mental simulation has significance. It is often taken as a sign that the explanation is not reasonable. For example, in the Iran-Iraq case presented earlier, the Tactical Action Officer used his failure to build a credible threat scenario as the basis for concluding that the track was not a threat.

Figure 11.6 covers the use of mental simulation to learn more about

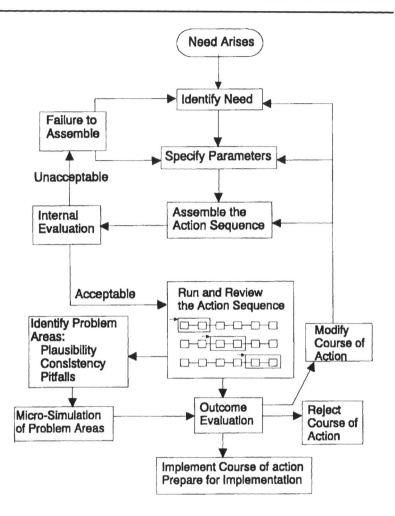

Figure 11.4. *Evaluate and inspect risks.*

the situation or the phenomenon. It expands on Figure 11.5 by showing that the evaluation can be used to revise and vary parameters in the simulation, so that the intent is to improve and elaborate the action sequence.

We must be careful about the way we treat these processes. There are many cartoons about complex sequences that depend on a stage marked "Miracle." There are also many computer simulations that include functions such as "Understand," which, in fact, do not understand anything but merely signify the need to someday figure out

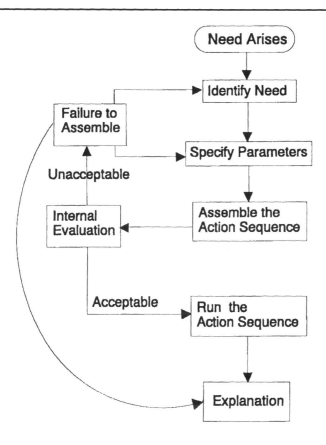

Figure 11.5. *Explain.*

what understanding is. None of the stages in Figures 11.3–11.6 seem to require miracles, but some of them may appear to describe a process while only naming that process. The descriptions in Figures 11.2–11.6 are attempts to move to increasingly detailed accounts.

11.5   Potential Flaws in Mental Simulation

The review of the 102 cases suggested several important properties of mental simulation, including ways in which it facilitates planning and decision making. However, we also found instances in which outcomes were not successful, and in because mental simulation was not entirely beneficial. The Trademaster/Pisces incident described earlier is an example because the mental simulation blinded the captain of the

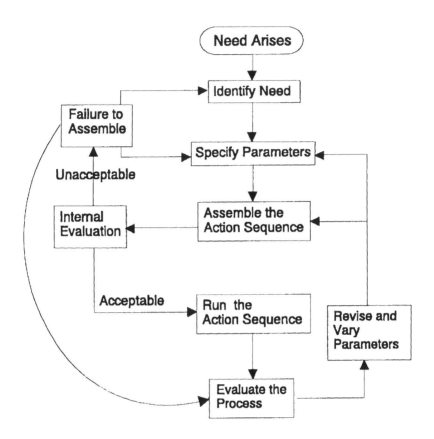

Figure 11.6. *Model.*

Trademaster to the fact that the Pisces had changed the plan. In this
section we discuss some vulnerabilities and sources of potential biases in
mental simulation. These include:

De minimus explanations
Commitment/overconfidence
Failure to decenter
Multiple factors
Person/situation factors

11.5.1 De minimus Explanations

Perrow (1984) distinguished *De minimus* from *De maximus* explanations. In a *De minimus* explanation, the attempt is to show that the situation is normal, and that anomalies can be explained away. Mental simulations are directed toward understanding how observed events fit within an assessment of the situation as usual, rather than focusing on the anomaly and running simulations that might account for its occurrence. Three-Mile Island falls into this category. Faced with a loss-of-coolant accident, the operators focused on one key indicator—an instrument. The instrument had sometimes given unreliable readings, and the operators realized that if they ignored it, everything else made sense. Unfortunately, the instrument was giving accurate readings, and it was their understanding of the events that was at fault. A related problem is that in which a person fails to notice subtle changes in conditions and persists in stereotyped reactions to a situation that no longer exists.

In other words, people can use mental simulations to explain away inconvenient or worrisome cues. This is a serious problem, and several of the failures noted in our database were due to this type of error. At the same time, people cannot be continually seeking *De maximus* explanations—playing what-if games forever. There is a delicate balance in sensing that it is time to abandon a situation assessment because expectancies have been violated or because conditions have changed markedly. Cohen (1988) referred to this shift as between the "low temperature" mode (explaining away apparently disconfirming data) and the "high temperature" mode (readily shifting hypotheses in the face of conflict). Cohen (1991) also tried to identify ways of helping people detect when they have gone too far in explaining away inconsistencies.

11.5.2 Commitment/ Overconfidence

As people assemble and run a mental simulation, they can become convinced of its plausibility and committed to its efficacy. They may be so persuaded that it is difficult for them to give proper credence to a different scenario (Anderson & Sechler, 1986), and they may be more inclined to bypass the evaluation phase that lets the person inspect a plan of action for defects. The advantages of mentally testing a plan for fatal flaws or unacceptable risks before it is implemented are obvious. On the other hand, it is inefficient to always conduct inspections, so the key is to judge when inspections are needed and when they can be bypassed. It is also the case that evaluation/inspection may be cut short because of time constraints (see Person/Situation Factors below). In our database, there were decision errors or poor outcomes because evaluation simulations were not carried out.

11.5.3 Failure to Decenter

There appear to be marked differences in people's abilities to mentally simulate an event and then observe it from several perspectives. Decentering allows a decision maker to view a scene from a different angle or a situation from another's point of view. Although we did not observe instances of failure to decenter in the cases we studied, we identify it as a possibility, especially in adversarial situations.

11.5.4 Multiple Factors

We noted that because of capacity limits people may restrict the scope of their mental simulations, allowing a limited number of causal factors and transitions within any given simulation. People seem to have difficulty simulating multiple causes or interactions among several factors. For example, software code inspectors described how their ability to imagine the way a program worked was severely restricted when the code included many subroutines that interacted. Similarly, when several malfunctions occur, people have difficulty disentangling them. This means that for certain complex environments, mental simulations may lead the decision maker astray by oversimplifying the scope of the problem.

11.5.5 Person/Situation Factors

We observed several noncognitive factors—aspects of the person or the setting—that can affect the quality or utility of mental simulations. These include experience level, cognitive style, and time constraints.

Experience Level. When people lack task experience or domain knowledge, they do not have the building blocks with which to assemble an adequate mental simulation. There may be too few causal factors employed or so many that the simulation becomes confusing. Constraints may be set inaccurately. The values used to initialize the simulation may be wrong. Part of the skill of running a simulation is metacognitive—being able to judge the level of cognitive complexity that one can handle without becoming confused. A skilled person who senses that things are getting too complex may move to a higher level of abstraction to keep the causal factors at a manageable level. In contrast, someone who is less skilled may try to juggle too many factors at once, may not see how to move to a more abstract level, and may be unable to generate a useful simulation.

For example, as part of our sample we asked four people, working independently, to mentally simulate the changes in the Polish economy that would result from the transition to a market economy. The transition was announced January 1, 1990, and the interviews were conducted during the following months. One informant, a Polish economist, had little trouble specifying key parameters (exchange rate, inflation rate, and unemployment) and constructing action sequences to generate predictions. In contrast, the other people interviewed, who had much less domain knowledge, could not identify many initial parameters. They only considered inflation and unemployment, and they did not know the current values of either. They could not assemble action sequences, or generate predictions, and so we judged that they had failed to construct a mental simulation, due to their inadequate experience level.

Although we also found real-life cases in which, due to lack of experience, people could not construct a simulation, none of these resulted in a failed decision. We expect that such cases do exist, however. The meaning of this error type is clear: When people do not have enough experience, they may construct incomplete or inaccurate scenarios. Worse than that, some novices we studied could not even find initial values for the simulation. Also they lacked the domain knowledge to estimate trajectories, to keep the number of paths from increasing exponentially.

Cognitive Style. The research literature on mental imagery suggests that there are strong individual differences in people's abilities to envision objects and then move them in space or modify some aspect of them (Sheikh, 1983). A total of 72% of the mental simulation accounts in our database contain explicit references to visual imagery, but 28% did not. We suspect that people's own abilities and the demands of the task interact to yield simulations that are more or less effective. For example, certain chefs may be able to simulate the taste and texture of a dish from the ingredients and cooking procedures they will use to create it, and sense the impact the dish will have in the context of accompanying foods. People who do not have this ability must rely on menu planners or trial and error in preparing a meal.

Although the topic of individual differences in cognition seems important, neither our review of the literature nor our review of the cases highlighted any critical dimensions. Nevertheless, we include the topic because it may trigger hypotheses for future research.

Time Constraints. We posited that mental simulation plays a central role in decision making by providing a means by which a person can quickly size up a situation, generate a feasible course of action, inspect the plan for risks or flaws, modify the plan if need be, and implement it. The combination of perceptual-recognitional processes and mental simulation is extremely powerful and allows an experienced decision maker to function effectively even under severe time constraints, but not always. We know that decision making can become compromised as time becomes more limited.

Our database contained instances of mental simulations under high and low time-pressured conditions. We categorized each instance of mental simulation according to the degree of time pressure. We then examined whether the decision maker was able to configure and run the simulation, and if so, whether the simulation proved beneficial to the outcome of the event.

Time pressure does not appear to affect decision making by interfering with the configuration or running of mental simulations. Under both low and high time pressure, people were able to conduct a mental simulation in almost all cases (93.2% and 90.3%, respectively). But simulations were much less likely to aid the decision maker under high rather than under low time pressure. When time pressure was low, 92% of simulations had a positive impact on the outcome of the event. When time pressure was high, only 61% of simulations benefited the outcome. The difference between these proportions was significant (z = 2.04; p < .05). We suspect that as time limitations become severe, people truncate the evaluation/inspection phase of mental simulations. They do not have the opportunity to check for flaws in their situation assessments or risks associated with a particular course of action. We must interpret these data with caution, however, because in some circumstances decision quality might reduce under time pressure, regardless of whether mental simulations were involved. Thus, if fewer time-pressured decisions are effective, there will be less chance for a mental simulation to be judged as benefiting the decision.

11.5.6 Other Potential Biases

There are also aspects of judgment and decision making that have been discussed in the research literature as sources of bias, but that appear to us to be less worrisome. These include:

Availability and Representativeness Heuristics. These heuristics were described by Kahneman and Tversky (1982) as having the potential for resulting in errors. But the whole idea of using analogues and prototypes is that one's past experience has demonstrated what is likely to occur in the future. Although the analogues and prototypes can result in poor decisions, the alternatives—either to avoid using analogues and prototypes or to carefully calibrate their applicability—seem even less desirable, especially under conditions such as time pressure.

Emphasis on Dramatic Events. Kahneman and Tversky (1982) posited that in constructing scenarios people overestimate the impact of dramatic events on causal transitions and underestimate the likelihood that events are produced by slow, incremental changes. We did not find the tendency to emphasize dramatic changes to be typical of the simulations in our database, and we are unsure of how serious or pervasive a problem it is. Perhaps the more grounded a simulation is in a specific context—task, event, or situation assessment—the less likely it is to ignore the typical, usual, and mundane, in favor of the dramatic event.

We find it interesting that decision makers are assumed to be biased toward predicting a future that is too normal while simultaneously placing undue emphasis on dramatic events. It is not clear to us which source of bias is assumed to prevail at any given point in time. What we have observed is that people rely on their experience with similar events and situations—the dramatic and the mundane—to guide their assessments, expectations, goals, and mental simulations.

11.6 Mental Simulation and Ecological Psychology

So far in this chapter we have focused on naturalistic decision making and the role of mental simulation within it. We have said little about ecological psychology or how the two perspectives mesh. There are many points of contact between the two paradigms. At the core of the

RPD model is the notion that perceptual events convey an understanding of a situation as typical and evoke a course of action in accord with that situation assessment. Decision makers do not have to figure out a course of action—they can see it as they perceive the situation. Further, the RPD model is keyed to the strengths of the decision makers, not their frailties, and this is also true of ecological psychology. The links with Gibsonian notions of direct perception, affordances, and an inherent resonance between the organism and its environment are evident.

It is worth noting that in developing the RPD model, we were not attempting to apply an ecological perspective to decision making. The RPD model grew out of our attempts to test analytic models of decision making in natural settings. Analytic decision models take as their focus the internal cognitive representations of events, and the processes decision makers use to manipulate and transform those mental representations in order to select from an array of options.

However, analytic approaches are difficult to use in environments marked by risk, time stress, and uncertain information. In attempting to account for problem solving and decision making in those settings, we found ourselves placing greater and greater emphasis on decision context—the nature of the situation and the role of perceptual processes in how the decision maker assesses and interprets an event. The RPD model asserts that skilled decision makers do not have to calculate choices but can know how to react to a situation based on the way they perceive it. The perception, or situation assessment, carries with it an appreciation of how to respond. In this sense, the RPD model is a source of convergent validity for the ecological perspective. Working from different research paradigms and issues, we developed accounts of human behavior that emphasize many of the same perceptual-recognitional processes.

In our view, one of the most important and powerful features of mental simulation is that it provides a means of planning and evaluating that is perceptually based, rather than requiring abstraction from an analysis of an event. This may be a point of convergence with an ecological approach. People are able to experience a situation "as if" it were occurring and use their perceptual-recognitional abilities to understand how they might respond to it. The affordances recognized in the simulated event are experienced and evaluated in terms of decisions and plans of action that "could be" initiated. Mental simulation allows the decision maker to assess the affordances provided by the particular imagined situation. Moreover, people are not fixated on the immediate stimulus field, but perceive the situation retrospectively to develop explanations and identify causal factors, and prospectively to develop plans, look for risks, anticipate problems, and generate expectations. One of the important insights of ecological psychology is to address the event, rather than the time cycle. A decision maker is typically entering events in the middle and can rely on mental simulation to reconstitute the origins and to anticipate outcomes in order to maintain the unity of the event. So we can again see the correspondence between mental simulation and ecological perspectives.

There are points of difference as well. The ecological perspective provides an account of a context-sensitive organism, sensing and directly reacting to its environment. According to this view, one need not invoke complex chains of mental operations to account for behavior. But what happens when the decision maker goes beyond the immediate perceptual event and the affordances of that particular situation? Ecological psychology has difficulty in accounting for such cognitively complex events as planning, anticipating problems, decision making, evaluating intended courses of action, or monitoring one's own decision processes. Although it is possible to describe such processes in the terminology of the ecological paradigm, it is not clear what is gained by doing so. Beyond fostering debates over preferred forms of language, we are not convinced that recasting cognitive phenomena in ecological terms provides an account of behavior that has greater explanatory power. The ecological approach to visual perception has shown itself to be valuable because of the phenomena it identified and explained. The ecological approach to perceptual–cognitive phenomena has yet to prove its utility.

The use of mental simulation to experience and scrutinize a variety of possible pasts and futures in order to respond effectively and efficiently represents a remarkable perceptual–cognitive skill. In our view, mental simulation and its function within the RPD model offers a challenge to ecological psychology. We must seek explanations for behavior that merge human perceptual capabilities with the complex cognitive events that characterize human goal setting, planning, problem solving, and decision making.

Acknowledgments

We would like to thank Julian Weitzenfeld and John Flach for their thoughtful reviews on an earlier draft of this chapter.

This research was funded through contract MDA903-89-C-0032 with the U.S. Army Research Institute for the Behavioral and Social Sciences.

The views, opinions, and findings contained in this chapter are those of
the authors and should not be construed as an official Department of the
Army position, policy, or decision.

11.7 References

Anderson, C. A., & Sechler, E. S. (1986). Effects of explanation and
    counterexplanation on the development and use of social
    theories. *Journal of Personality and Social Psychology, 50*(1), 24–34.
Beach, L. R. (1990). *Image theory: Decision making in personal and
    organizational contexts.* West Sussex, England: Wiley.
Beach, L. R., & Mitchell, T. R. (1978). A contingency model for the
    selection of decision strategies. *Academy of Management Review,
    3,* 439–449.
Beach, L. R., & Mitchell, T. R. (1987). Image theory: Principles, goals,
    and plans. *Acta Psychologica, 66,* 201–220.
Brezovic, C. P., Klein, G. A., & Thordsen, M. (1987). *Decision making in
    armored platoon command* (AD-A231775). Alexandria, VA:
    Defense Technical Information Center.
Calderwood, R., Klein, G. A., & Crandall, B. W. (1988). Time pressure,
    skill, and move quality in chess. *American Journal of Psychology,
    101,* 481–493.
Cohen, M. S. (1988). *Supporting metacognition: The significance of decision
    making biases for decision aiding* (Tech. Rep.). NASA. Washington,
    DC: Cognitive Technologies, Inc.
Cohen, M. S. (1991). Can decision analysis define the goals of training?
    Proceedings of the Human Factors Society 35th Annual Meeting, 2,
    1353–1357.
Crandall, B., & Calderwood, R. (1989). *Clinical assessment skills of
    experienced neonatal intensive care nurses* (Final Rep. prepared for
    the National Center for Nursing Research, National Institutes
    for Health under Contract No. 1 R43 NR01911 01). Yellow
    Springs, OH: Klein Associates.
Crandall, B., & Gamblian, V. (1991). *Guide to early sepsis assessment in the
    NICU* (Instruction manual prepared for the Ohio Department of
    Development under the Ohio SBIR Bridge Grant program).
    Fairborn, OH: Klein Associates.
Crandall, B., & Klein, G. A. (1987). *Key components of MIS performance*
    (Rep. prepared for Standard Register, Dayton, OH). Yellow
    Springs, OH: Klein Associates.
de Groot, A. D. (1965). *Thought and choice in chess.* New York: Mouton.

(Original work published 1946)

deKleer, J., & Brown, J. S. (1983). Assumptions and ambiguities in mechanistic mental models. In D. Gentner & A. L. Stevens (Eds.), *Mental models* (pp. 155–190). Hillsdale, NJ: Lawrence Erlbaum Associates.

Driskell, J., & Mullen, B. (1991, January). Presentation at In-Process Review for Army Research Institute, Springfield, OH.

Druckman, D., & Bjork, R. A. (Eds.). (1991). *In the mind's eye: Enhancing human performance.* Washington, DC: National Academy Press.

Einhorn, H. J., & Hogarth, R. M. (1981). Behavioral decision theory: Processes of judgment and choice. *Annual Review of Psychology, 32,* 53–88.

Fleiss, J. L. (1981). *Statistical methods for rates and proportions* (2nd ed.). New York: Wiley.

Forbus, K. D. (1983). Qualitative reasoning about space and motion. In D. Gentner & A. L. Stevens (Eds.), *Mental models* (pp. 53–72). Hillsdale, NJ: Lawrence Erlbaum Associates.

Jungermann, H., & Thüring, M. (1987). The use of causal knowledge of inferential reasoning. *NATO ASI Series, F35,* 131–146.

Kaempf, G., Wolf, S., Thordsen, M. L., & Klein, G. (1992). *Decision making in the AEGIS Combat Information Center.* Fairborn, OH: Klein Associates. (Prepared under Contract No. N66001-90-C-6023 for the Naval Command, Control and Ocean Surveillance Center, San Diego, CA).

Kahneman, D., & Tversky, A. (1982). The simulation heuristic. In D. Kahneman, P. Slovic, & A. Tversky (Eds.), *Judgment under uncertainty: Heuristics and biases* (pp. 201–208). Cambridge, MA: Cambridge University Press.

Kieras, D. E., & Bovair, S. (1984). The role of a mental model in learning to operate a device. *Cognitive Science, 8,* 255–274.

Klein, G. A. (1989). Recognition-primed decisions. In W.B. Rouse (Ed.), *Advances in man-machine system research* (Vol 5, pp. 47–92). Greenwich, CT: JAI Press.

Klein, G. A., & Calderwood, R. (1991). Decision models: Some lessons from the field. *IEEE Transactions on Systems, Man, and Cybernetics, 21*(5), 1018–1026.

Klein, G. A., Calderwood, R., & Clinton-Cirocco, A. (1986). Rapid decision making on the fire ground, *Proceedings of the 30th Annual Human Factors Society* (Vol. 1, pp. 576–580). Dayton, OH: Human Factors Society.

Klein, G. A., Calderwood, R., & MacGregor, D. (1989). Critical decision method for eliciting knowledge. *IEEE Transactions on Systems, Man, and Cybernetics, 19*(3), 462–472.

Kosslyn, S. M., & Pomerantz, J. R. (1977). Imagery, propositions, and the form of internal representations. *Cognitive Psychology, 9,* 52–76.

Lynn, S. J., & Rhue, J. W. (1988). Fantasy proneness: Hypnosis, developmental antecedents, and psychopathology. *American Psychologist, 43*(1), 35–44.

Payne, J. W. (1976). Task complexity and contingent processing in decision making: An information search and protocol analysis. *Organizational Behavior and Human Performance, 16,* 366–387.

Pennington, N., & Hastie, R. (1993). A theory of explanation-based decision making. In G. Klein, J. Orasanu, R. Calderwood, & C. E. Zsambok (Eds.), *Decision making in action: Models and methods* (pp. 188–201). Norwood, NJ: Ablex.

Perrow, C. (1984). *Normal accidents: Living with high-risk technologies.* New York: Basic Books.

Robinson, J. A., & Hawpe, L. (1986). Narrative thinking as a heuristic process. In T. Sarbin (Ed.), *Narrative psychology: The storied nature of human conduct* (pp. 111–125). New York: Praeger.

Rouse, W. B., & Morris, N. M. (1986). On looking into the black box: Prospects and limits on the search for mental models. *Psychological Bulletin, 100*(3), 349–363.

Shepard, R. N., & Metzler, J. (1971). Mental rotation of three-dimensional objects. *Science, 171,* 701–703.

Sheikh, A. A., (1983). *Imagery: Current theory, research, and application.* New York: Wiley.

Thordsen, M. L., Galushka, J., Klein, G. A., Young, S., & Brezovic, C. P. (1990). *A knowledge elicitation study of military planning* (Tech. Rep. 876). Alexandria, VA: U.S. Army Research Institute for the Behavioral and Social Sciences.

Thordsen, M. L., Klein, G. A., & Wolf, S. (1990). *Observing team coordination within Army rotary-wing aircraft crews* (Final Tech. Rep.). Yellow Springs, OH: Klein Associates Inc. (Prepared under contract MDA903-87-C-0523 for the U.S. Army Research Institute, Aviation Research and Development Activity, Ft. Rucker, AL).

Williams, M. D., Hollan, J. D., & Stevens, A. L. (1983). Human reasoning about a simple physical system. In D. Gentner & A. L. Stevens (Eds.), *Mental models* (pp. 131–153). Hillsdale, NJ: Lawrence Erlbaum Associates.

Chapter 12

A Situated Cognition Approach to Problem Solving

Michael F. Young

University of Connecticut

Michael D. McNeese

Armstrong Laboratory/Human Engineering Division

12.0   Introduction

Problem solving is a cognitive skill often considered the consummate achievement of a well-educated person. It is an explicit goal of America 2000 and related efforts to revitalize American schools (SCANS, 1991). It is also a component of successful collaborative group work. Problem solving has often been reduced to an activity of solving 1- or 2-step written mathematics word problems, in which case a componential information-processing analysis has been considered sufficient. However, what has been missing from information-processing approaches to problem solving has been the importance of the problem-solving context and particularly its interaction with the skills and intentions brought to the situation by the problem solver.

There are many attributes of real problems that are missed when only simplified word problems are studied, attributes that should be considered in a more complete analysis of problem solving. For example, real-world complex problem solving requires coordination of multiple cognitive processes, applied through multiple paths (Siegler & Jenkins, 1989), and it occurs within contexts that provide critical perceptual information regarding potential solutions. Real world problem solving is often interpersonal, ill structured, and involves interwoven problems, extended time frames, and several possible competing solutions (Meacham & Emont, 1989). Real-world problem solving also involves discovering problems and subproblems, detecting key attributes of the problem (Bransford, Sherwood, Vye, & Rieser, 1986), and the "generation" of relevant subproblems (Cognition and Technology Group at Vanderbilt, 1992). Successful problem solving often occurs in groups and requires the social construction of knowledge. For group problem solving, the group itself is a transactive organism (Wegner, 1987), wherein members may provide affordances for one another (e.g., members may act as external knowledge stores for other people to enhance the overall efficiency of the group). Most of these attributes can be considered part of the "situation" or context for problem solving, and they unfold through time as a problem solver interacts with the problem space.

Recent developments in educational psychology have led to a theory of situated cognition in which knowledge (and problem solving) is seen as always being contextualized rather than constructed from static representations in memory. If as this theory holds, context gives meaning to all representations, then meaning itself is always "interactive," in that the representational nature of the context arises from the interaction of the agent with the context. This can be contrasted with schema theories in which meaning is stored and retrieved from memory, not produced from each situated context (e.g., Clancey & Roschelle, 1992). In the situated cognition model, meaning is created on the fly, rather than being translated from something (representational or schematic) in the head. In short, from a situated cognition perspective, problem solving is a product of the interactions between neurological processes and environmental information perceived by the individual, much like the process of visual perception as described by Gibson (1966, 1979/1986).

In our analysis of complex realistic problem solving, we found several Gibsonian ideas useful. For example, we consider problem solving to be an interaction between an agent and a problem environment, attributes of each constraining the interaction. We consider problem solving to be a goal-directed, self-organizing, intentional process. And we believe that the development of problem-solving expertise involves perceptual attunement across a generator set of problems. In contrast to Gibson, we emphasize the social construction of problem solving in which other people are significant parts of the problem-solving environment. Also, we allow for a role of representations in the problem-solving process, representations created "on the fly" during the problem-solving process.

In this chapter we discuss a situated cognition approach to problem solving, specifically borrowing concepts from ecological psychology to characterize problem solving as a perception–action process. Rather than trivializing problem solving by using traditional 1- or 2-step mathematics word problems, we describe the nature of real-world problem solving in all its richness, complexity, and often ill-defined nature. Then we discuss one example of an instructionally designed problem-solving situation, the Jasper series, that we studied as a controlled situation for understanding complex problem solving. Finally, we describe some examples of how our view of problem solving has been enhanced by borrowing concepts from ecological psychology.

12.1   The Nature of Real-World Complex Problem Solving

In contrast to the word problems presented in most school situations and various performance problems used in human factors research, real-world problems have important perceptual and structural characteristics often not addressed by their simpler instructional cousins. Consider three examples of real-world problem-solving situations: surgery, piloting, and engineering design.

12.1.1   Surgery

Problem solving in nonemergency surgery begins weeks or months before the actual operation. The physician, through discussion with the patient and often with assistance from consulting physicians, explores laboratory and radiological reports, physical exams, patient histories, medical databases, requested test results, alternative treatments, specific details of the operation, possible side-effects and probable outcomes. In short, the physician's role requires the active integration of knowledge distributed across many sources. Wegner (1987) portrayed the physician–patient dialogue as one that meshes two levels of expertise. The patient is an expert on his or her symptoms and their manifestation, whereas the physician is an expert in diagnosing and classifying illnesses. Together they form what Wegner (1987) referred to as a "transactive retrieval system."

When the operation is undertaken, varied professionals must coordinate their skills, including nurses, anesthesiologists, assisting physicians, pathologists, medical students, and the surgeon. During the operation information must be exchanged about the patient's ongoing condition, the progress of the operation, unexpected or unique situations, and how to resolve problems. Such people-to-people exchanges can convey more than information, they can set a relaxed, or serious, or excited/emergency atmosphere and affect the nature of the problem solving that occurs. Because each operation is unique and despite all the planning possible, complications and the specifics of the operation require the continuous identification of problems and their solutions on multiple fronts (e.g., anesthesiology, vital signs, surgical and suture techniques, etc.). After the operation itself is completed (usually in a matter of hours), the patient's recovery must be monitored for days extending even to months.

Although the situation preceding surgery involves diagnosis, coordination, planning, and sometimes hypothesis testing, the situation during surgery is highly focused on perceptual cues defining the status of the patient. This situation is one of confirming or denying the continued state with respect to the next move to be taken. For example, Crandall and Calderwood's (1989) study of nurses in a neonatal intensive care unit showed that nurses working with "microbabies" (24–48 weeks after conception) were sensitive to subtle information that alerted them to infections and other problems before these problems were revealed by standard tests. The study indicated that much of the information nurses picked up was contrary to information used to diagnose full term-babies.

Additionally, the various activities identified by Crandall and Calderwood (1989) occurred within the situated institutional context of a hospital. Cicourel (1990) pointed out that this context fosters group-derived norms that channel people with particular titles and presumed competencies, responsibilities, and entitlements into certain physical spaces at particular times for particular tasks. This context subsequently affects levels of discourse and interaction among medical team members that may be divergent from established organizational decision making as established by official roles and duties. Cicourel suggested that interpersonal networks typically facilitated the day-to-day functioning inherent in the hospital context. He also noted that the hospital context relies on "high technology" that affects data analysis, communicating, and problem solving. One must observe that often the highest technology is directly interfaced with the human patient to form a medical human-machine system. For example, during surgery the patient is connected to a variety of monitors (e.g., EKG, EEG) and supplementary life-support technologies (e.g., oxygen mask). The technology often creates distributed problem-solving teams, monitoring the patient from different locations but needing to collaborate to manage treatment.

12.1.2   Piloting

Next we consider the nature of combat piloting. Mission planning begins days or weeks before an actual flight as targets are identified, classified, and prioritized. Commanding officers explore information (intelligence and reconnaissance data), alternative mission scenarios, potential losses and threats, and outline the specific details of a mission. Once a flight is undertaken, the pilot must coordinate his or her efforts with the other members of his crew (e.g., navigator and weapons officer), other pilots on the mission, and information sources in the air and on the ground (e.g. air traffic controllers)— hopefully only making minor adjustments to the mission plan. Incidental battle damage, weapon type, threat potential, predicted weather conditions, terrain features, force size, time of day, and equipment malfunction can all potentially create conditions that alter the original mission and tactics plan and create new problems to be solved as the mission unfolds (McNeese et al., 1990). As the pilot embarks on the actual mission, technology plays an important role— from air traffic control workstations and heads-up cockpit displays, to communications instruments that connect pilot to wingman. One of the greatest needs within piloting is the proper integration of pilot and aircraft. The information that appears on the displays, the tactile feedback from the throttle, and the noise level of the cockpit are all practical examples of influences of the human-machine system on the mission plan. Once the mission is completed, the outcomes (such as battle damage assessment) and debriefing can take days or weeks to complete.

Throughout the course of a combat mission, people-to-people exchanges are critical to the success of the mission as all members of the crew must coordinate their actions and share information about the status of the aircraft and the mission. McNeese et al. (1990) provided evidence that pilots themselves feel there is a distinct need for explicit communications between the pilot and the "backseater" (navigation and weapons officer). Pilots expressed this need in the context of providing input for possible problem-solving dialogue with an intelligent associate performing in the role of a backseater. The pilots also noted the social context of the communication between themselves and their backseaters. For example, they wanted the backseater to give information (as opposed to data); they wanted them to help identify problems, to not control conversation, and to give direct commentary. Wellens (1993) indicated that several studies of airline crew coordination suggest that the sharing of information among crew members is a critical factor in obtaining situated awareness and task performance. Embedded within the social and technological context of piloting is the potential effect of a number of social psychological variables (e.g., affiliation and self-perception, attraction, conformity, social facilitation, and group structure and communication; Wellens & McNeese, 1987) that can influence the construction of meaning.

Because each flight is unique and to some extent unpredictable, and despite all the planning possible, the actual situation develops as the pilot approaches the target. Various aspects of the mission are invariant, whereas other aspects change in the midst of the situation. In fact, this is often referred to by the aviation/human factors community as "situation awareness" (see Sarter & Woods, 1991). Endsley (1988, p. 790) specified *situation awareness* as "the perception of elements in the environment within a volume of time and space, the comprehension of their meaning, and the projection of their status in the near future." Wellens (1993) indicated this requires a shift in attention from the cockpit to the surrounding airspace without a loss of control of the aircraft or falling prey to the enemy. He argued that as automation impinges on the development of today's cockpit, the pilot must attend to complex navigation, weapons, control systems, and electronic displays and alarms while being sealed in a pressurized vehicle, and yet complete the mission. The concept of situation awareness highlights the primacy of perception operating within the social and problem-solving context of piloting.

The transformation from planning to perception that must be accomplished "on the fly" by pilots illustrates this primacy of perception in real-world problem solving. McNeese et al. (1990, p. 55) documented that "the concept of 'visual acquisition' needed to acquire a target has multiple connections to a 'preflight-planning' concept. For example, the use of the strategy of big-to-small relies on characteristics of 'most prominent features' identified in aerial photos, a major concept of preflight planning. Pilots also indicated that an adjustment to the initially presented maps during preflight planning must be accomplished via mental rotation during the course of visual acquisition."

12.1.3 Designing

As a third example of complex real-world problem solving, consider engineering design. Perhaps design conjures up worn images of a single individual at a drawing board working on the details of a product design. In todays engineering, business, and production environments, this is far from accurate. Design today encompasses groups of individuals (e.g., the designer, the production planner, the engineer, the programmer, the engineering psychologist, and the business manager) working together with new design technologies (e.g., computer-supported electronic conferencing, relational databases, computer-aided drafting machines, and hypermedia presentations) under a variety of constraints and imposed tradeoffs.

Design often begins with an in-depth understanding of the end user's needs. Greenberg (1991) indicated that participatory design is a new approach to design that highlights the role of the design team (comprised of developers and eventual system users) that cooperatively participates in the planning and decision making necessary to build a system that matches workers' needs. The inclusion of users as a primary component of design teams places emphasis on the development of a product or system in the context of the user's work domain.

The real-world problem solving encountered in design points to the reciprocation of abilities between people and machines within a natural work domain. Design problems are often ill-defined and require designers' perceptual-recognitional abilities. Recognition-based decision making in design (see Klein, 1987) points to the necessity that design is a process that occurs in a situated context. Gero (1990) described design activity from two contexts:

> the context within which the designer operates and the context produced by the developing design itself. The designer's perception of what the context is affects the implication of the context of design. The context shifts as the designer's perceptions change. Design activity can be now characterized as a goal-oriented, constrained, decision-making, exploration, and learning activity that operates within a context that depends on the designer's perception of the context. (p. 28)

Design also requires input from interdisciplinary perspectives. Boff (1987) identified cross-disciplinary chokepoints inherent in the problem of systems design that cause many layers of complexity. For example, he addressed the problem that much of the specialized information required from multiple domains/technologies, needed to impact design, cannot be accessed due to the fragmented nature of information in an organization. Within the context of time constraints, available resources, lack of understanding between different design team members, and design bias, design problem solving may quickly be directed down suboptimal pathways. Ancona and Caldwell (1990) suggested that product design teams engage in activities to determine resource availability, what the product can and should be, what other areas of the organization want the product to be, what technologies are available for building the product, what markets the product might serve, to gather distributed information, to determine other's expectations, to coordinate, to keep others informed, and to work intensively with other teams responsible for other aspects of product design (e.g., marketing). Inherent in all these tasks is the fact that design teams include social-based processes such as getting to know other team members, determining who has particular skills, and who can be trusted. From this analysis it is clear that due to organizational, situational, social, and personal biases, the design process is affected by social constraints as well as engineering constraints.

These three real-world, complex, problem-solving situations highlight 10 important characteristics of situated problem solving:

1. Real-world problem solving requires the coordination of multiple cognitive processes, applied through multiple paths (Siegler & Jenkins, 1989), dynamically and continuously unfolding. Examples include analysis, planning, problem identification, metacognitive monitoring, and problem solving while comparing multiple solutions to multiple subproblems.
2. Real-world problem solving occurs within complex contexts that provide critical perceptual cues and rich situational affordances (Rogoff & Lave, 1984); for example, the information, discourse, technology, and atmosphere of an operating room.
3. Real-world problem solving is interpersonal. Greeno, Smith, and Moore (1993) wrote, "the issue of [problem solving] is social in a fundamental way. Learning occurs as people engage in activities, and the meanings and significance of objects and information in the situation derive from their roles in the activities that people are engaged in" (p. 100). Examples include surgeons coordinating with anesthesiologists, nurses, and medical students; pilots coordinating with their crews and other pilots; or designers communicating with engineering psychologists and marketing experts.
4. Being interpersonal, real world group problem solving requires the social construction of knowledge (see Bereiter &

Scardamalia, 1989; Edwards & Middleton, 1986). More than simply communicating or coordinating within a group, problem solving requires that group members construct a shared perception of the problem and the solution, often mediated by technology.

5. Real-world problem solving is often ill-structured and requires generation of relevant subproblems (Cognition and Technology Group at Vanderbilt, 1992). Despite careful and extended planning, real situations vary widely from case to case and require continuous identification of problems, sub-problems to these problems, and solutions. When complex problem solving is done on the fly, problems that are detected must be conceptualized into manageable subproblems that afford specific actions: Planning is integrated throughout the problem-solving process.

6. Real-world problem solving involves the integration of distributed information, typically from various specialties and domains. For example, solving the problem of a successful combat mission involves intelligence information, weather information, piloting skills, and appropriate weapons selection and use, and so on.

7. Real-world problem solving takes place across extended time frames. Such problems cannot be solved in a few minutes or even in a few hours and are often completely beyond the time and space constraints of a single individual. They have a developmental history and future all contextualized in the ongoing situation.

8. Real-world problem solving involves several possible competing solutions (Meacham & Emont, 1989). Rather than a single correct solution, most real-world problems have multiple correct solutions as well as "almost workable" solutions. Alternatives must inevitably be planned, worked out in detail, compared, and subsequently selected or rejected (see Kugler, Shaw, Vicente, & Kinsella-Shaw, 1991, for a topological description of multiple goal paths).

9. Real-world problem solving involves discovering problems and noticing perceptual attributes of the problem, such as detecting relevant from irrelevant information (Bransford, et al., 1986). Problem detection and noticing that a problem affords a particular action or solution is a critical component of the problem solving process often missing in traditional training,

education, human factors, or artificial intelligence domains. When simplified, well-defined problems are substituted for realistic problems, students are denied an opportunity to acquire and practice this important perceptual skill.

10. Finally, real-world problem solving involves inherent values, intentions, and goals that often have personal and social significance (Johnson, Moen, & Thompson, 1988). For example, value issues may arise surrounding the implications of personal error during surgery and affect problem solving activities.

These are the characteristics that can be used to establish the ecological validity of an environment to be used to study complex realistic problem solving.

12.2   Situated Cognition and Problem Solving

Recent developments in cognitive and educational psychology have led to a view of situated cognition in which knowledge and thinking is seen as always being contextualized rather than constructed from static representations in memory (Brown, Collins, & Duguid, 1989). Hoffman & Deffenbacher (1992) discussed a similar concept by proposing a general world view, called "contextualism," that challenges traditional information processing theories and suggests alternatives based more on ecological psychology (see also Coulter, 1989, & Saxe, 1991). The view that learning must be "situated" has a rich history and is rooted in the psychology of Dewey (1938), Jenkins (1977), and Gibson (1966). Recently, this contention has been made stronger to include not only learning but all thinking, a theory of situated cognition (see, e.g., Brown & Campione, 1991; Brown et al., 1989; Greeno, 1989; Greeno, et al., 1993; Lave, 1988). This approach emphasizes the study of cognitive phenomena that relate to actual experience and activity, rely on the ecological validity of research, and focus on natural settings.

Situated cognition can be contrasted with schema theories in which meaning is stored and retrieved from memory, not interactively created from each situated context (e.g., Clancey & Roschelle, 1992). In the situated cognition model, the processes of perceiving and acting create meaning on the fly, rather than reading it back from something (representational or schematic) in the head. Consider the three examples of real-world problem solving given earlier: the surgeon during an operation, a pilot acquiring the target, and a design team developing a new idea. The knowledge in these situations is not static and not solely inside the individual(s) involved. The heart monitors, airplane avionics, and computer designs all provide information while continually signaling problems and changing conditions. Various sources point to the significance of a perceptual component to expertise in these domains and others, for example, aerial photo reconnaissance, identification of rocks and landforms, chess mastery, firefighting, dermatology, clinical psychology, and judging livestock (see Bransford & Stein, 1984; De Groot, 1965; Hoffman & Deffenbacher, 1992; Klein, 1987). Using the principles of ecological psychology, this list can be extended to a more general class of problems. For current purposes then, problem solving must be viewed as a product of the interactions between neurological processes and environmental information perceived by the individual: much like the process of visual perception as described by Gibson (1966, 1979/1986).

If, as the theory of situated cognition contends, all problem solving is situated, then there are always two components to successful problem solving: the agent and the context. Problem solving must be viewed as the relationship between the actor and the environment (mutually determined effectivites and affordances). It would be misleading at best to assert that the properties of the problem, the problem space, or the problem-solving environment merely influence problem solving. Rather, it is the relationship between the agent and the problem that is problem solving. It would not be meaningful to characterize the problem solving of an individual apart from the context in which that problem solving occurs. Context broadly includes people, machines, design artifacts, environments, and other objects and agents that may interact to establish ecological problem-solving relationships—including the problem and solution spaces that constitute the information field for problem solving. This is what Pea (1988a) described as "distributed knowledge." A situated cognitive analysis of problem solving must describe the abilities that each person brings to the table, the relevant attributes of the environment including dimensions of the problem that afford certain actions, and how they interact.

Greeno et al. (1993) addressed this issue in attempting to characterize transfer from a learning situation to a novel situation by using the situated cognition model. They wrote that an activity like problem solving,

jointly depends on properties of things and materials in the situation and on characteristics of the person or group. Following Gibson

(1979/1986) and Shaw et al. (1982)... affordances and abilities are relative to each other: a situation can afford an activity for an agent who has appropriate abilities, and an agent can have an ability for an activity in a situation that has appropriate affordances. (p. 101)

Thus, from the perspective of situated cognition, problem solving must be viewed as the interaction of an agent with a specific problem-solving situation.

Vicente (1995) explained the role of context by making reference to an allegory of the ant and the beach (Simon, 1981): "Viewed as a geometric figure, the ant's path is irregular, complex, and hard to describe. But its complexity is really a complexity in the surface of the beach, not a complexity of the ant" (p. 64). The ant's behavior is constrained by the landscape of the beach more than by internal organismic forces. When the ant is placed on a different part of the beach or when external disturbances (e.g., wind, earth movement) occur, different navigation actions are required to reach the same goal. Different ants may have different strategies for navigation that are used dependent on the type of disturbance experienced. Vicente (1995) suggested that understanding such a situated activity requires the determination of (a) regularities in the ways the landscape affects the ant, and simultaneously (b) the psychological invariance of the ant across different tasks.

Such analysis can be extended into problem solving, especially the transfer of knowledge from one situation to another (Greeno et al., 1993). The regularities in this case would be those between problem situations and the problem-solving activities taken by the agent. The invariants would include the neurological properties of the problem solver and general problem-solving heuristics. In a real-world setting, much of the regulation is in the perceived contrasts/comparisons of events and the metacognitive strategies that are invariant across settings. For situated cognition that has social/cultural components, the regulation includes other people who provide mutual affordances for each other (e.g., cognitive apprenticeship, Brown et al., 1989; reciprocal teaching, Brown & Palinscar, 1988; distributed intelligence, Pea, 1988a; external memory, Wegner, 1987).

Inherently, a situated learning environment also employs aspects of a particular cultural practice that impact the observations and interactions of an expert with an environment. Various sociocultural factors constrain recognition and utilization of affordances. For example, in the school setting students often discover that leaning their chair back off its back legs affords them a much more ergonomic position and decreased back strain. However, the sociocultural norms typically reinforced by the local expert (the teacher) prohibit such activity. Cultural practice can often be most salient during transfer of problem solving from the learned context to a new context. Misconceptions, misapplied analogies, and "malrules" (Brown & Burton, 1978) can be the result of attempts to apply actions that were correct in the learned culture to a new sociocultural situation with similar affordances.

Certain practices that are legitimate in one setting may be taboo in another; for example, winking is acceptable in American culture but considered rude in some Asian cultures, or on a microlevel, nurse–doctor interactions may be different in the operating room than in other hospital or office settings. The situations create different "allowable" responses for nurses, technicians, and doctors. Denison and Sutton (1990) reported that in observing operating rooms, they were surprised to find loud rock music, joking, and storytelling among the surgical team during a coronary bypass procedure. They attributed this to the fact that joking reduced tension and alleviated boredom that could negatively impact work. But they also observed that surgeons constrained the activities of nurses by not allowing spontaneous joking and laughing until after cues were given. When nurses interacted with doctors outside the operating room, the authority initiated in the operating room was not observed. This would pose problems for novice nurses who had yet to experience the sociocultural nature of the operating room. When their behavior from outside was transferred to the operating room, they immediately experienced sanctions that determined what behaviors were accepted in this new setting.

If we are to design a context in which to study problem solving using a situated cognition approach, then that context must engage the learner in activity and interaction with a complex realistic problem. Only if the environment affords the problem-solving actions that students would normally engage in in a real-world setting, can it then be said that the environment can serve as a situated learning context. The "Jasper Series" (see Acknowledgments) that we discuss next provides an example of such a situated learning context delivered to learners through multimedia technology.

12.3   Jasper as an Example of a Constrained Realistic Context

To understand more about the nature of situated problem solving, the nearly unlimited number of variables present in real-world problem solving must be controlled. Without some constraints on the problem space and activities possible in that environment (affordances), the inherently nonlinear multivariate nature of the context would be too complex for any existing psychometric models or research designs (assuming one could muster the sample sizes needed to test such models). Our solution to this problem was to create a multistep problem, with a limited number of solution paths possible, and to constrain the problem by presenting it on videodisc. Problem solutions were restricted to materials, values, and agents shown in the video: Solutions had to be provable based on information from the video. With the problem space and solution space thus constrained, we were able to begin to ask questions about the abilities and perceptions of agents (such as prior domain knowledge and group processes) that contribute to situated problem solving.

Research on situated learning suggested a need to develop in students a knowledge of real situations that encompasses more than merely engaging in mathematical calculations or isolating scientific facts (Cognition and Technology Group at Vanderbilt, 1990). The student's role in solving Jasper is to be an active problem solver. Successful problem solving in this context requires problem-solving intelligence, knowledge of how to access and retrieve relevant data, when and how to rely on alternative data sources, how to operationalize needed information as mathematical computations, and how to manage and intelligently take advantage of information throughout the problem-solving environment; in short, the ability to use what Pea (1988a, 1988b) called "distributed" intelligence.

Briefly, instruction using the Jasper macrocontext involves viewing a series of 15-min stories in which the major character, Jasper Woodbury, encounters a problem such as the discovery of a wounded eagle far out in the woods. All of the data required to obtain a quantitative solution to the rescue of the eagle was embedded in the story. Students are challenged to list all of the things they must consider to develop a workable rescue plan (e.g., time, payload of the rescue plane, fuel, etc.). Then, they are asked to generate and document their solutions. Throughout this time, the videodisc is made available for students to retrieve relevant facts and information on request, often accessing the disc themselves using a Hypercard® interface. The first two Jasper problems are essentially realistic distance-rate-time problems in the context of trip planning and emergency rescue.

Each problem in the 6-disc series presents a complex multistep problem that middle school students typically require more than a week of traditional 40-min classes to solve, either in small groups, individually, or as a class. The random-access capability of the videodisc makes the complexity of these problems manageable, because quantitative facts as well as story events can be quickly and easily reviewed. Even though the mathematics required to solve the Jasper physics distance-rate-time problems is important in a real-world sense, the computations themselves are not complex, and students are often challenged more by the process of dealing with the multistep nature of the problem than by the mathematics itself (for a discussion of such difficulties, see Campione, Brown, & Connell, 1988). Practice dealing with complexity can develop in students an appreciation of the need to plan, to retrieve relevant information when needed, and an appreciation that not all mathematics problems can be solved quickly, even if the required computations are not in themselves complex. This is a key component of changing students' attitudes about the nature of mathematics (Lesh, 1985).

The Jasper series has been used as an experimental context for investigating problem solving in mathematics and science, for teaching thinking skills to a wide variety of students (fourth graders to graduate students, at risk of academic failure to gifted and talented students), and for looking at the nature of collaborative groups (Barron, 1992; Cognition and Technology Group at Vanderbilt, 1990; McNeese, 1992). The broad results of these studies indicate that the Jasper series is a powerful context for learning. Further studies are underway to determine the specific attributes of the environment that are active in affording learning, their unique contributions to the learning process, and their transfer across related and unrelated contexts. For example, Barron (1992) is looking to discover the nature of cooperative groups that yield the most effective collaborative problem solving by using the computer to generate every possible pseudogroup from a pool of students. Young and Kulikowich (1992) looked at the creativity, reading comprehension, note taking, and attitudinal aspects of Jasper problem solving. A more elaborate study of attitude change toward mathematics and problem solving is being undertaken by Van Haneghan and his colleagues (Van Haneghan, 1992). All these research efforts point to the importance of understanding learning in complex realistic contexts, that is, situated learning.

The Jasper series is unique as an experimental stimulus for studying realistic situated problem solving and collaborative group work. Although each problem maintains the 10 attributes of real-world problem solving cited earlier, the problems are also constrained by restricting problem solvers to only the information provided in the video. Further, each presentation of the problem information can be reproduced exactly, making experimental replication and comparative groups possible across time and space. These attributes also have importance for instruction and particularly for reliable assessment.

Young and Kulikowich (1992) began to develop an automated system that integrates instruction on Jasper problem solving with assessment of Jasper problem solving. Their Hypercard®-based Jasper Planning Assistant (JPA) supplements traditional think-aloud protocol data with latencies for various problem-solving activities: planning, questioning, calculating, information-finding, and reviewing of facts (Kulikowich & Young, 1991). Although serving as a problem solving assistant, JPA is also capable of assessing attitudes, knowledge of the story structure, and confidence in solution. This research has also employed conventional paper-and-pencil measures of reading comprehension to demonstrate transfer from Jasper problem solving in mathematics to comprehension of reading passages with analogous content.

McNeese (1992) is pursuing research that assesses the effects of generative and metacognitive strategies on group and individual problem solving in the Jasper context. The basic approach is to look beyond the general success of collaborative groups versus individual learning and to address the generative and socially constructed nature of plans and solutions created on the fly within the Jasper context. The basic question was "What are the conditions in group collaboration that lead to a group member's use of knowledge as an individual?" Better individual transfer is expected from students who participated in the collaborative learning setting than from those who participated in the individual learning setting. Social construction of knowledge within the situated context of real world problem solving generates and utilizes affordances beyond situations that contain only a lone problem solver. Characteristics described by others include collective induction (Laughlin, 1989), metacognition (Brown, Bransford, Ferrara, & Campione, 1983) and distributed intelligence (Pea, 1988a). In this study, the Jasper context was thought to provide groups with multiple affordances for problem solving (see Cognition and Technology Group at Vanderbilt, 1990). However, superior performance by the collaborative group could also be explained by individual performance; that is, the group simply adopts the solution of its best member. The study uses pseudogroups (see McGrath, 1984) to assess this possibility.

This research also seeks to determine to what extent the social construction of knowledge through group problem solving facilitates recall and transfer. The number of shared transactions is taken to indicate the extent to which groups are more actively engaged in generating and recognizing affordances. Groups with few shared transactions typically show a strong dominant leader as revealed by several measures: time spent talking in the group, statement type (e.g., the number of plans suggested), or the quality of their statements (few statements that are highly valued by other members of the group). Advantages of the additional perceptions that accompany the shared transaction of socially constructed knowledge are expected to show on an isomorphic transfer task (Repsaj—that's Jasper spelled backwards). Greeno et al. (1993) described how such transfer could occur across situated learning contexts as students detect the invariants across the two problems using perception gained through social interaction.

There were three phases to this study: acquisition, transfer, and recall. During acquisition, students solved the Jasper problem in one of two learning settings: individually or in groups. Then each student was individually asked to solve the Repsaj problem (an isomorph of the Jasper problem containing the same structural relationships and storyline, but a different surface structure). This allowed the comparison of group-to-individual and individual-to-individual transfer to determine the extent to which collaboration (compared to individual acquisition) facilitated near transfer. All students returned again after 3 days to complete memory tests concerning the materials presented in the Jasper and Repsaj problems.

Protocol analysis was employed to analyze each student's problem-solving think-aloud transcript. A description of the Jasper problem space developed at Vanderbilt (see Goldman, Vye, Williams, Rewey, & Pellegrino, 1991) served as a basis for our protocol analysis. Statements were categorized into one of five arguments: goal, current state, means, outcome, and uncodable. The Vanderbilt problem-space description allowed each transcript to be assigned a score representing the extent to which the student explored various elements of the problem, including plans, problem-space elements, subgoals, and constraints necessary to solve the problem. Each element was scored as only mentioned, attempted to solve, or correctly solved. In addition to the protocol analysis, several other quantitative measures were taken, including problem-solving latency, percentage of items recognized, talk-aloud efficiency, and number of initiatives.

Results suggest that problem solving in the Jasper context results in spontaneous transfer in both individual and collaborative settings. The Jasper context definitely affords problem solving and metacognitive activities in a natural way, without the need for structured implementation of cooperative learning procedures. It has been observed that when people collaborate to jointly solve problems, there are social affordances that act to delimit or constrain generative activities. This can be positive when it serves to reduce misinformation and misconceptions. Yet, groups may also afford "loafing" (Latane, Williams, & Harkins, 1979) among members, which can have negative consequences on the transfer of knowledge.

Research in the Jasper context suggests that realistic problem solving and collaborative interactions can be developed and studied using a well-defined and constrained problem space presented on videodisc and enhanced with computer technology. This system has provided teachers with a new way to engage students in mathematics, science, and even integrated study across the curriculum on issues of river ecology, endangered species, and flight. But as a tool for research, it has only begun to shed light on the nature of situated learning and cooperative groups. This research will continue. For now, however, we would like to think beyond the implications of Jasper for teaching and for psychological research and consider how concepts from ecological psychology have impacted the design and analysis of the Jasper problem.

Collins, Brown, and Newman (1989) suggested that "learning in culture" supplies models of expertise, but these external models of expert process and reasoning must be observed and enacted by a learner in such a realistic environment containing all the complex information and affordances for action (including other people) available in the real setting. As such, the problem-solving interactions afforded by the Jasper problem can be characterized as the social construction of knowledge.

12.4   Problem Solving and Socially Constructed Knowledge

Bereiter and Scardamalia (1989) described the interaction that takes place among people during group problem solving as "socially constructed knowledge." Edwards and Middleton (1986) referred to joint remembering as an adaptive social function that serves as a transactional negotiative quality of human communication, whereby groups remind each other of things, influence each other's recall, and establish a mutuality of shared experience. In describing the social organization of cognitive apprenticeship, Collins et al. (1989) suggested that:

> *Apprentices learn skills in the context of their application to realistic problems, within a culture focused on and defined by expert practice. They continually see the skills they are learning being used in a way that clearly conveys how they are integrated into patterns of expertise and their efficacy and value within the subculture. And by advancing in skill, apprentices are increasing their participation in the community, becoming expert practitioners in their own right. (p. 486)*

As described in our three introductory examples, socially constructed knowledge is essential to most real-world problem solving. If we are to successfully develop the problem-solving abilities of others, then instruction must allow students to "take control of their own minds" and be generative rather than passive in the learning process (e.g., Slamecka & Graf, 1978; Wittrock, 1974). Traditionally, much of the research in cooperative learning and group problem solving (see McNeese, 1989, for a review) involved various manifestations of group process and learning, yet failed to take an ecological perspective that accommodates the social construction of knowledge.

Research looking at metacognitive actions and story recall within collaborative groups emphasized the social construction of knowledge (Collins et al., 1989; Dansereau, 1988; Palinscar & Brown, 1984; Stephenson, Branstatter, & Wagner, 1983). Of particular relevance are studies that focused on how collaboration influences subsequent transfer of each individual's knowledge to future problems. Dansereau (1988) showed how various metacognitive strategies, reciprocated in cooperative learning groups, impacted an individual's ability to complete similar problems. Larson et al., (1985) showed that different types of metacognitive activities experienced in groups contributed to acquisition and transfer of knowledge in different ways. Also, the collective induction research (e.g., Laughlin, 1989) and cooperative learning research (Gabbert, Johnson, & Johnson, 1986) specifically employed forms of group-to-individual transfer assessments of collaboration. Although many of these studies gave credence to the social nature of problem solving and learning, they failed to utilize problems and situated contexts that were ecologically valid, using our list of real-world attributes.

Work in collaborative groups is key to enabling students and co-workers to socially construct knowledge. It is critical that our view of problem solving be expanded to include the need for students not only to utilize their own knowledge structures (schemas or representations), but also to perceive meaning through social interactions and shared experiences. This issue is more than deciding if students should learn problem solving individually or in groups. Rather, it is an insistence that problem solving be learned in situations that afford collaborative construction of meaning and rely on shared knowledge, whether working in groups or individually with technology providing surrogate collaborator(s).

There are current programs designed to understand group problem solving and decision making as an overall information-processing system related to traditional theories of memory and cognitive psychology (Duffy, 1993; Hinsz, 1990; Wegner, 1987). Although these are refreshing attempts to see group process from a cognitive perspective, they run the risk of retaining some of the constructive, memorial aspects of static individual models of cognitive process. We prefer a perspective that emphasizes the situated nature of cognition along with a social collaborative view that emphasizes the perceptual aspects of learning and the affordances provided by the environment. In identifying the importance of collaboration as a new form of affordance structure for problem solving, Stephenson et al. (1983) pointed out the transactive nature of interactions between individuals and their working groups:

> (1) Whereas individuals may be discouraged by the daunting prospect of re-telling a long, complex story, the more realistic possibilities of successful performance by the group may lead to an enhancement of individual effort. (2) Individuals tend to be appropriately more doubtful about erroneous parts of their testimony and more confident about those parts which are correct. They may be expected to insist in the group on the retention of that which they are (rightly) confident about, whilst being prepared to yield (appropriately enough) to the more confident person when "doubtful." (3) Social validation or confirmation of individual recall by another may give the individual confidence that he or she is "on the right track," and lead to subsequent recall of material that would otherwise remain forgotten. (4) Material

"forgotten" by one individual but recalled by another may prompt the recall of additional "forgotten" information by the former (p. 177)

The group affords its members more opportunities to generate activities (e.g., elaborate ideas, monitor errors, and plan remedial actions) beyond what they could do individually. However, the social construction of knowledge within groups may necessarily be related to the level of synergy experienced by a group (McNeese, 1992). For example, groups that are dominated by one individual may afford what Latane et al. (1979) described as social loafing and thereby contribute to "two heads being less than one." The point here is that as we adopt an ecological perspective of situated learning, we must acknowledge the nonlinear complexity inherent in how groups engage in problem solving, and be prepared to describe multiple attractors in the information fields that constitute problem-solving environments.

12.4.1 Ecological Explanations of Situated Learning

We found it useful to view complex situated problem solving from the perspective of ecological psychology. From this perspective we acknowledge the primacy of the interaction between the skills and abilities brought to the situation by the problem solver (effectivities) and the affordances for action provided by the problem environment or problem space — a symmetry of acausal interactions (Shaw et al., 1982). This relationship is captured in the perceiving-acting cycle that temporally unfolds through the problem-solving process. We strive to describe both the abilities and intentions that each person brings to the interaction as well as all the information available in the environment, including dimensions of the problem and solution spaces that afford certain actions. When perception is emphasized over memory, it is the information picked up from the environment that is perceived and acted on in light of each agent's goals and intentions that becomes central to a full analysis of situated problem solving, not simply the actions or solutions of problem solving.

Kugler et al. (1991) described goal-directed activity, such as problem solving, as an interaction of attractor sets, specifically the attractor sets supplied by a complex realistic context (affordances) and the attractor processes (effectivities) by which we achieve the goal states set up by problem-solving intentions. Their analysis suggested that it is the information available from a situation that guides and constrains problem solving: "The behavior of inanimate systems is lawfully determined by a force field, whereas, the behavior of animate systems is lawfully specified by an information field" (p. 408). In their analysis of self-organization and intentional systems (such as people), they gave mathematic substance to Gibson's (1979/1986) principle of "organism-environment mutuality." Applying these ideas to situated cognition suggests that problem solving is an interaction between the problem-solving skills, goals, and intentions of the individual problem solver and the activities and manipulations that a particular problem situation affords. In their terms:

> Fields that have hidden degrees of freedom (internal fields) are said to be compactified. The relationship between local and global fields is promoted here to express the relationship between an environment and the organisms acting in that environment. The compacting of an external field by internal field properties expresses exactly the contributory role perceptual/cognitive variables play, along with physical variables, in codetermining the observed behavior of the organism. This, we propose, is what it means to say that an organism, as a perceptually attuned intentional system, is informationally as well as forcefully coupled to its environment. (p. 422)

In taking an ecological perspective of complex problem solving, we acknowledge that problem solving is individualistic in an essential way. Each person's interaction with the environment, specifically a problem space, is unique. The information field is constantly changing ("You never step in the same river twice"— Socrates), and the problem solver's ability to detect relevant information in the environment and act on it is also continually modified (perceptual learning). As a result, the interaction that is problem solving evolves— subproblems are discovered, new plans and goals are constructed, and in the terms of intentional dynamics, equilibrium points are created and annihilated in the problem-solution state space.

An ecological perspective of situated learning enhances many conceptions of cognitive processes that have until now been characterized only partially by information-processing and schema-theoretic models. The three examples we address are (a) planning as an essential element of complex problem solving, (b) learning as it occurs in traditional classrooms, and (c) transfer of learning (thought by many to be the hallmark of successful education).

12.4.1   Planning

Traditional models of complex problem solving characterize planning as one step in a linear (iterative) process. For example, planning is the second step in Polya's (1957) model and the third step in Bransford and Stein's (1984) IDEAL problem-solving model, identified as "Explore possible strategies." In this model, the problem must first be Identified and Defined/represented. The steps after planning include Acting on the strategy and Looking back (evaluating/monitoring) at the activities. In the componential subtheory of Sternberg's (1985) triachic theory of intelligence, planning is the one step characterized as finding "what things to do and how to do them." In these models, planning comes into problem solving as a linear/serial process. Planning is initiated and completed within a limited period of time and is not designed to be revised or revisited unless the whole process recycles.

As another traditional example, Mayer (1985) described planning as a metacognitive skill in mathematical problem solving. The model had basically two parts (problem representation and problem solution)— planning was part of the problem solution. Problem representation consisted of translation (linguistically parsing various words in a word problem) and integration (determining the problem "type" by selecting the appropriate problem schema). Planning was the first part of solving the problem once represented. Planning referred to selecting the solution strategy that would be the easiest, most efficient or preferred based on some other heuristic (e.g., selecting "counting on" vs. "counting up"). The model concluded with execution of the selected strategy, which required fluency.

In both of these traditional models of problem solving, planning is considered to be a cognitive process or skill, taking place all within a single individual's head, and executed from beginning to end, resulting in a plan (e.g., the pilot should fly from the city to the rescue site, then to the station to refuel and meet the ambulance for the drive to the hospital). But an ecological perspective suggests a very different model. Planning must be considered to arise from the interaction of the agent and the environment. As such, planning must be considered to be dynamic and continuous throughout the problem-solving process, unfolding across time as "compactified fields" are detected and explored, and active (to a greater or lesser extent) until the problem is resolved. This drove our design of the JPA assessment system to continually allow for and monitor planning throughout Jasper problem solving. Planning must also be viewed to be constrained by the information field, which in this case includes the problem space (actions that can be taken to solve the problem) and information distributed in an "intelligent" environment, primarily technology and people. This leads our analysis of planning in problem solving to concentrate on the social construction of planning, in contrast to the traditional conception of planning as taking place in the head of a single individual. It also leads us to consider each student's goals and intentions for problem solving and to take efforts to have them adopt the realistic goals inherent in the Jasper problems (e.g., rescuing an eagle to save its life).

12.4.2   Codetermined Classroom Interactions

Next we consider the explanation of learning as it occurs in the most common educational situation, a grade-school classroom. From a traditional information-processing perspective, learning occurs in a classroom context as information is transmitted from an informed source (usually the teacher) to the learner. Consider a recent Jasper activity at a suburban northeastern middle school that was teaching with the Jasper problems for the first time. In this case, the classroom contained 3 adults and 24 children. One adult was an educational researcher, experienced with Jasper and interested in studying the nature of complex problem solving. The second adult was a volunteer teacher, who had assisted with Jasper once before, but was teaching it for the first time. The third adult was the regular classroom teacher, with limited interest or enthusiasm for technology.

As students began to work with the Jasper videodisc and problem, information about mathematics and the problem-solving process was conveyed from the teacher to the students consistent with the information-processing model. But learning in this situated context was much richer than could be captured by the information-processing analysis. An ecological perspective required us to consider the unique perceptions of each agent in the environment, governed by the principle that knowledge is codetermined by the goals and intentions of each individual interacting with the properties of the environment. This led us to consider that the student activities were perceived very differently by the three adults who were detecting different affordances of the situation, and the learning that occurred was very different for each person involved.

Consider one classroom interaction as an example. Early on in the introduction of the Jasper problem, the volunteer teacher needed help in managing student access to the videodisc (using a hand-held controller) and enlisted the help of the regular teacher. Our analysis detected three unique perceptions of this activity. The regular teacher perceived being the "keeper of the controller" as a nonthreatening way to participate in the technology-rich environment; the volunteer teacher perceived this as classroom management of student access to the videodisc to facilitate student problem solving; and the researcher perceived this as the ideal use of the situation for teacher training through cognitive apprenticeship.

A second example can be taken from this same classroom activity. One team of students working on its solution to the Jasper problem first considered several possible solutions, narrowed their choices to 2, split the group in half, worked out both solutions, and decided to present the faster one as their group's solution to the class. When subsequently asked what they had learned from this activity, one student answered that he learned how groups could produce better solutions than a single individual. A second student answered that she learned that not all math problems can be solved in a day. A third student's response was that he had learned a lot about boats and hoped they would work on more boat problems (this latter response reflects a misconception based on common novice tendencies to classify on surface structure rather than on deep structure; see Chi, Feltovich & Glaser, 1980). Of course, not all the students were this insightful about their learning, but these responses illustrate the importance of considering learning to be codetermined through the interaction of problem context with each student's individual goals and intentions.

12.4.3 Transfer of Learning

In their analysis of transfer of situated learning, Greeno et al. (1993) drew on concepts from ecological psychology when they discussed how the identification of similar elements across transfer situations could be better described as attunement to affordances, than by constructed comparisons, element by element. Specifically, they suggested that problem solvers become attuned to the constraints of particular situations, and perceptions of the same or similar constraints lead to transfer across situations.

We rely on another concept borrowed from ecological psychology to enhance this description of transfer of learning. Using Jasper as an example, we showed that transfer from a single Jasper problem-solving experience can transfer across academic subject domains, from mathematics to reading comprehension (Young & Kulikowich, 1992).

However, this study showed transfer of only analogous content (information about boats and rivers), and no transfer across content (to horses and racing). Analyzing transfer from an ecological perspective leads us to consider that detection of invariance across a "generator set" of situations would be required for transfer to occur across content areas. Efforts are currently underway to look for such transfer after students have experienced several Jasper problems (a generator set of situations for detecting the invariance of higher cognitive skills such as planning and information finding), and teachers have made an effort to externalize and discuss the invariance with students. This leads us to characterize teaching to transfer in ecological terms as a social interaction designed to tune students' perceptions to the important invariants across problem-solving situations.

12.5   Conclusion

Real-world problems have many attributes that make them different in many important ways from problems used traditionally to teach problem solving. The theory of situated cognition, which highlights the perceptual aspects of expertise, provides a foundation for understanding real-world problem solving. Situated cognition, using concepts from ecological psychology, suggests that "the action" is not solely inside a student's head nor is it solely to be found in the environment. Knowledge and intelligent behavior are best characterized as a relationship between the abilities of an individual (effectivities) and the nature of situations (affordances of the environment). Accepting this view means taking the strong position that all learning and problem solving is situated. And, as in most real-world activities, a prominent feature of situations is often other people (co-workers, team members, and cooperative group partners). If knowledge is viewed as a relationship between agent and environment, and environments often contain other people, then knowledge is often socially constructed—based on the shared affordances provided by multiple agents in a situation. To understand situated learning and cognition, we must construct realistic problem-solving contexts that are more constrained and replicable (in contrast to the real world); problems that actively engage students in a way consistent with the 10 characteristics of real-world problem solving we identified.

The Jasper videodisc problem-solving context represents an example of such a problem-solving environment. Research on Jasper suggests that situated learning can occur in such environments, providing a mechanism to study real-world problem solving in an ecologically valid situation. We adopted the situated cognition approach and a related analysis of learning from an ecological perspective, which led us to a more complete description of important educational issues. We found it useful to consider problem solving as an interaction between the goals and intentions of each individual problem solver and the complex nonlinear properties of the problem-solving environment, including the problem space, the solution space, and "distributed intelligence" contained in technology and other people. We also found concepts from ecological psychology (e.g., a "generator set" of situations) useful for our analysis of higher level thinking skills such as planning, classroom problem-solving activities, and the transfer of problem solving across situations and subject domains.

In adopting this perspective, we found that concepts from schema theory and information processing are not as inaccurate as they are incomplete. For example, we found it useful to think that students have "representations" of problems that guide their progress through the solution space. However, we agree with Greeno et al. (1993) that these "action schemas" can be created on the fly from the information (constraints) contained in the problem context, rather than retrieved, in tact, from memory. This allows for a more dynamic, continuous description of important cognitive processes (our example given is planning) that seems to more completely capture the nature of complex realistic problem solving.

Acknowledgments

The "Jasper Series," developed by the Cognition and Technology Group at Vanderbilt, is marketed by Optical Data Corporation. The series is being used for ongoing research efforts at the University of Connecticut, Armstrong Laboratory, and elsewhere with permission and cooperation from the Learning Technology Center, Peabody College, Vanderbilt University, John Bransford and Susan Goldman, co-directors, Box 45, Peabody, Nashville, TN 37203.

12.6   References

Anconca, D. G., & Caldwell, D. F. (1990). Information technology and work groups: The case of new product teams. In J. Galegher, R. E. Kraut, & C. Egido (Eds.), *Intelligent teamwork* (pp. 173–190).

Hillsdale, NJ: Lawrence Erlbaum Associates.

Barron, B. (1992). *A pseudo-group analysis of Jasper problem solving.* Unpublished doctoral dissertation, Vanderbilt University, Nashville, TN.

Bereiter, C., & Scardamalia, M. (1989). Intentional learning as a goal of instruction. In L. B. Resnick (Ed.), *Knowing, learning, and instruction: Essays in honor of Robert Glaser* (pp. 361–392). Hillsdale, NJ: Lawrence Erlbaum Associates.

Boff, K. R. (1987). The tower of Babel revisited: On cross-disciplinary chokepoints in system design. In W. B. Rouse & K. R. Boff (Eds.), *System design: Behavioral perspectives on designers, tools, and organizations* (pp. 1–13). New York: Elsevier.

Bransford, J. D., Sherwood, R. D., Vye, N. J., & Rieser, J. (1986). Teaching thinking and problem solving. *American Psychologist, 41*(10), 1078–1089.

Bransford, J. D., & Stein, B. S. (1984). *The IDEAL problem solver.* New York: Freeman.

Brown, A. L., Bransford, J. D., Ferrara, R. A., & Campione, J.C. (1983). Learning, remembering, and understanding. In J. H. Flavell & E. M. Markman (Eds.), *Handbook of child psychology: Vol. 3: Cognitive development* (4th ed., pp. 393–451). New York: Wiley.

Brown, A. L., & Campione, J. C. (1991). *Communities of learning and thinking, or a context by any other name* (Tech. Rep.). Berkeley: School of Education, University of California.

Brown, A. L., & Palincsar, A. S. (1988). Guided, cooperative learning and individual knowledge acquisition. In L. Resnick (Ed.) *Knowing, learning, and instruction: Essays in honor of Robert Glaser.* Hillsdale, NJ: Lawrence Erlbaum Associates.

Brown, J. S., & Burton, R. R. (1978). Diagnostic models for procedural bugs in basic mathematical skills. *Cognitive Science, 2,* 155–191.

Brown, J. S., Collins, A., & Duguid, P. (1989, January–February). Situated cognition and the culture of learning. *Educational Researcher,* pp. 32–42.

Campione, J. C., Brown, A. L., & Connell, M. L. (1988). Metacognition: On the importance of understanding what you are doing. In R. I. Charles &. E. A. Silver (Ed.), *The teaching and assessing of mathematical problem solving* (pp. 93–114). Reston, VA: National Council of Teachers of Mathematics.

Chi, M. T. H., Feltovich, P., & Glaser, R. (1980). Categorization and representation of physics problems by experts and novices. *Cognitive Science, 5,* 121–152.

Cicourel, A. V. (1990). The integration of distributed knowledge in collaborative medical diagnosis. In J. Galegher, R. E. Kraut, & C. Egido (Eds.), *Intelligent teamwork* (pp. 221–242). Hillsdale, NJ: Lawrence Erlbaum Associates.

Clancey, W. J., & Roschelle, J. (1992). Situated cognition: How representations are created and given meaning. *Educational Psychologist, 27*(4), 435–453.

Cognition and Technology Group at Vanderbilt. (1990). Anchored instruction and its relationship to situated cognition. *Educational Researcher, 19*(6), 2–10.

Cognition and Technology Group at Vanderbilt. (1992). The Jasper experiment: An exploration of issues in learning and instructional design. *Educational Technology Research and Development, 40,* 65–80.

Collins, A., Brown, J. S., & Newman, S. E. (1989). Cognitive apprenticeship: Teaching the crafts of reading, writing, and mathematics. In J. Larkin (Ed.), *Knowing, learning, and instruction: Essays in honor of Robert Glaser* (pp. 453–494). Hillsdale, NJ: Lawrence Erlbaum Associates.

Coulter, J. (1989). Mind in action. Atlantic Highlands, NJ: Humanities Press.

Crandall, B., & Calderwood, R. (1989). *Clinical assessment skills of experienced neonatal intensive care nurses* (Final Rep. prepared for the National Center for Nursing Associates, NIH). Yellow Springs, OH: Klein Associates.

Dansereau, D. F. (1988). Cooperative learning strategies. In C. L. Weinstein, E. T. Goetz, & P. A. Alexander (Eds.), *Learning and study strategies: Issues in assessment, instruction, and evaluation* (pp. 103–120). San Diego: Academic Press.

De Groot, A. D. (1965). *Thought and choice in chess.* The Hague: Mouton.

Denison, D. R., & Sutton, R. I. (1990). Operating room nurses. In J. R. Hackman (Ed.), Groups that work (and those that don't): Creating conditions for effective teamwork (pp. 293–308). San Francisco: Josey-Bass.

Dewey, J. (1938). *Experience and education.* New York: Collier Macmillan.

Duffy, L. (1993). Team decision making biases: An information processing perspective. In G. A. Klein, J. Oransano, & R. Calderwood (Eds.), *Decision making in practice. Models and methods* (pp. 346–359). Norwood, NJ: Ablex.

Edwards, D., & Middleton, D. (1986). Joint remembering: Constructing an account of shared experience through conversational discourse. *Discourse Processes, 9*, 423–459.

Endsley, M. R. (1988). Situation awareness global assessment techniques (SAGAT). *Proceedings of the IEEE Aerospace and Electronics Conference, 3*, 789–795.

Gabbert, B., Johnson, D. W., & Johnson, R. T. (1986). Group-to-individual transfer, process gain, and the acquisition of cognitive reasoning strategies in cooperative learning groups. *Journal of Psychology, 120*(3), 265–278.

Gero, J. S. (1990). Design prototypes: A knowledge representation schema for design. *AI Magazine, 11*(4), 26–36.

Gibson, J. J. (1966). *The senses considered as perceptual systems.* Boston: Houghton Mifflin.

Gibson, J. J. (1986). The ecological approach to visual perception, Hillsdale, NJ: Lawrence Erlbaum Associates. (Original work published 1979)

Goldman, S. R., Vye, N. J., Williams, S., Rewey, K., & Pellegrino, J. W. (1991, April). *Problem space analyses of the Jasper problems and student's attempts to solve them.* Paper presented at the annual meeting of American Educational Research Association, Chicago, IL.

Greenberg, S. (1991). Computer-supported cooperative work and groupware: An introduction to the special issues. *International Journal of Man-Machine Studies, 34*, 133–141.

Greeno, J. G. (1989). A perspective on thinking. *American Psychologist, 44*, 134–141.

Greeno, J. G., Smith, D. R., & Moore, J. L. (1993) Transfer of situated learning. In D. Detterman & R. Sternberg (Eds.), *Transfer on Trial: Intelligence, cognition, and instruction* (pp. 99–167). Norwood, NJ: Ablex.

Hinsz, V. B. (1990). *A conceptual framework for a research program on groups as information processors* (Tech. Rep.). Wright-Patterson AFB, OH: Logistics and Human Factors Division, AF Human Resources Laboratory.

Hoffman, R. R., & Deffenbacher, K. A. (1992). A brief history of cognitive psychology. *Applied Cognitive Psychology, 6*(1), 1–48..

Jenkins, J. J. (1977). Remember that old theory of memory? Well forget it! In R. E. Shaw & J. D. Bransford (Eds.), *Perceiving, acting, and knowing* (pp. 413–429). Hillsdale, NJ: Lawrence Erlbaum Associates.

Johnson, P. E., Moen, J. B., & Thompson, W. B. (1988). Garden path errors in diagnostic reasoning. In L. Bolc & M. J. Coombs (Eds.),

Expert system applications (pp. 395–427). Berlin: Springer-Verlag.

Klein, G. (1987). Analystical versus recognitional approaches to design decision making. In W. B. Rouse & K. R. Boff (Eds.), *System design: Behavioral perspectives on designers, tools, and organizations* (pp. 175–186). New York: Elsevier.

Kugler, P. N., Shaw, R. E., Vicente, K. J., & Kinsella-Shaw, J. (1991). The role of attractors in the self-organization of intentional systems. In R. R. Hoffman & D. S. Palermo (Eds.), Cognition and the symbolic processes (pp. 367–431). Hillsdale, NJ: Lawrence Erlbaum Associates.

Kulikowich, J. M., & Young, M. F. (1991, October). *Complementing verbal report data.* Paper presented to the annual meeting of the Northeast Educational Research Association, Ellenville, NY.

Larson, C. O., Dansereau, D. F., O'Donnell, A.M., Hythecker, V. I., Lambiotte, J. G., & Rocklin, T. R. (1985). Verbal ability and cooperative learning: Transfer of effects. *Journal of Reading Behavior, 16,* 289–295.

Latane, B., Williams, K., & Harkins, S. (1979). Many hands make light the work: The causes and consequences of social loafing. *Journal of Personality and Social Psychology, 37,* 822–832.

Laughlin, P. R. (1989). *A theory of collective induction: Final report.* Arlington, VA: Office of Naval Research.

Lave, J. (1988). *Cognition in practice: Mind, mathematics and culture in everyday life.* Cambridge, UK: Cambridge University Press.

Lesh, R. (1985). Processes, abilities needed to use mathematics in everyday situations. *Education and Urban Society, 17,* 330–336.

Mayer, R. E. (1985). Mathematical ability. In R. J. Sternberg (Ed.), *Human abilities: An information-processing approach* (pp. 127–150), New York: W. H. Freeman.

McGrath, J. E. (1984). *Groups: Interaction and performance.* Englewood Cliffs, NJ: Prentice-Hall.

McNeese, M.D. (1989). *Explorations in cooperative systems: Thinking collectively to learn, learning individually to think* (Tech. Rep. AAMRL-TR-90-004). Wright-Patterson AFB, OH: Armstrong Aerospace Medical Research Laboratory.

McNeese, M. D. (1992). *Analogical transfer in situated, cooperative learning.* Unpublished doctoral dissertation, Vanderbilt University, Nashville, TN.

McNeese, M. D., Zaff, B. S., Peio, K. J., Snyder, D. E., Duncan, J. C., & McFarren, M. R. (1990). *An advanced knowledge and design acquisition methodology: Application for the pilot's associate* (Tech. Rep. AAMRL-TR-90-060). Wright-Patterson AFB, OH: Armstrong Aerospace Medical Research Laboratory.

Meacham, J. A., & Emont N. C. (1989). Interpersonal biases of everyday problem solving. In J. D. Sinnott (Ed.), *Everyday problem solving: Theory and applications* (pp. 7–23). New York: Praeger.

Palincsar, A. S., & Brown, A. L. (1984). Reciprocal teaching of comprehension-fostering and comprehension-monitoring activities. *Cognition and Instruction, 1*, 117–175.

Pea, R. D. (1988a, August). *Distributed intelligence in learning and reasoning processes.* Paper presented at the meeting of the Cognitive Science Society, Montreal.

Pea, R. D. (1988b). Putting knowledge to use. In R. S. Nickerson & P. P. Zodhiates (Eds.), *Technology in education: Looking toward 2020* (pp. 109–212). Hillsdale, NJ: Lawrence Erlbaum Associates.

Polya, G. (1957). *How to solve it!* (2nd ed.). Princeton, NJ: Princeton University Press.

Rogoff, B., & Lave, J. (1984). *Everyday cognition: Its development in social context.* Cambridge, MA: Harvard University Press.

Sarter, N. B., & Woods, D. D (1991). Situation awareness: A critical but ill-defined phenomenon. *International Journal of Aviation Psychology, 1*, 45–47.

Saxe, G. B. (1991). *Culture and cognitive development: Studies in mathematical understanding.* Hillsdale, NJ: Lawrence Erlbaum Associates.

Secretary's Commission on Achieving Necessary Skills (SCANS). (1991). *What work requires of schools: A SCANS report for America 2000 (Executive Summary).* Washington, DC: U.S. Department of Labor.

Shaw, R., Turvey, M. T., & Mace, W. (1982). Ecological psychology: The consequence of a commitment to realism. In W. B. Weimer & D. S. Palermo (Eds.), *Cognition and the symbolic processes* (Vol. 2, pp. 159–226). Hillsdale, NJ: Lawrence Erlbaum Associates.

Siegler, R. S., & Jenkins, E. (1989). *How children discover new strategies.* Hillsdale, NJ: Lawrence Erlbaum Associates.

Simon, H. A. (1981). *Sciences of the artificial* (2nd ed.), Cambridge, MA: MIT Press.

Slamecka, N. J., & Graf, P. (1978). The generation effect: Delineation of a phenomenon. *Journal of Experimental Psychology: Human Learning and Memory, 4*(6), 592–606.

Stephenson, G. M., Brandstatter, H., & Wagner, W. (1983). An experimental study of social performance and delay on the testimonial validity of story recall. *European Journal of Social Psychology, 13,* 175–191.

Sternberg, R. (1985). *Beyond IQ: A triarchic theory of human intelligence.* New York: Cambridge University Press.

Van Haneghan, J. (personal communication, June 5, 1992). Telephone and e mail messages describing development of an attitudes towards mathematics scale for Jasper problem solving.

Vicente, K. J. (1995). A few implications of an ecological approach to human factors. P. A. Hancock, J. M. Flach, J. K. Caird, & K. J. Vicente (Eds.), *Local applications of the ecological approach to human-machine systems.* Hillsdale, NJ: Lawrence Erlbaum Associates.

Wegner, D. M. (1987). Transactive memory: A contemporary analysis of group mind. In B. Mullen & G. R. Goethals (Eds.), *Theories of group behavior* (pp. 185–208). New York: Springer-Verlag.

Wellens, A. R. (1993). Group situation awareness and distributed decision making: From military to civilian applications. In J. Castillian, Jr. (Ed.), *Individual and group decision making: Current issues* (pp. 267–291), Hillsdale, NJ: Lawrence Erlbaum Associates.

Wellens, A. R., & McNeese, M. D. (1987). A research agenda for the social psychology of intelligent machines. *Proceedings of the IEEE National Aerospace and Electronics Conference, 4,* 944–950.

Wittrock, M. C. (1974). Learning as a generative process. *Educational Psychologist, 11*(2), 87–95.

Young, M. F., & Kulikowich, J. M. (1992, May). *Improving reading comprehension through real world situated problem solving.* Paper presented to American Educational Research Association annual meeting, San Francisco.

Chapter 13

Designing Team Workstations:
The Choreography of Teamwork

Leon D. Segal

Western Aerospace Laboratories

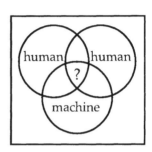

13.0   Introduction

When I started my research work at NASA-Ames Research Center, I was shown around the building and introduced to the people with whom I would be working. Some of my new colleagues had read, or heard about, my work: "Oh, you're the one who does that nonverbal stuff." It was then that I first became aware of the trap I had meticulously constructed then proceeded to fall into: Through the process of building an argument for the importance of including nonverbal behavior in the study of human communication and team workstation design, I had been put into a category that, taken out of context, has no functional meaning. Yes, I am indeed interested in "that nonverbal stuff," but that is only one facet of the complex design problems that stimulate my thinking and direct my work.

The parsing of communication into the dichotomy of "verbal" and "nonverbal" is an artifact of the human attempt to scientifically study the flow of information between two or more organisms. Although it does represent one of many classification schemes for describing human behavior, we must not confuse its usefulness with the fact that it is merely a tool for representing events, rather than the actual structure of events in "reality." I believe, and have tried to convey this in the past, that the study of communication is the study of a *context* created by the communicators and the environment upon which they act and within which they interact. I also believe that all investigations should have a practical purpose; my own theoretical and experimental work has always been directed and grounded by one objective—design. This chapter, therefore, begins with a discussion of different perspectives on human communication, then looks at their application to a specific problem in the design of complex systems—namely, the challenge of designing for team coordination and communication.

When designing a multioperator workstation, what behavioral issues should designers consider? Should a team simply be considered as an aggregate of single operators? Should there be a qualitative change in design as a result of a quantitative change in team size, that is, how does one address the synergetic aspects of combining two or more operators? Does the design of a workstation affect teamwork? If so, which aspects of teamwork should the design enhance, which should it sustain, which should it attempt to change, and which should it inhibit? As the designer of complex systems—and as a student of such systems—I believe these to be crucial questions that must be asked along the design process.

The concept of design as choreography sits at the center of the applied section of this chapter. My purpose is to vivify in the designer's mind the understanding that the creation of any novel artifact will inevitably result in the emergence of novel human perceptions and actions. Human behavior, however, is by nature nondeterministic; one can never determine with certainty what specific behaviors an individual, or team, may demonstrate. Designers, therefore, must attempt to define effective boundaries around the potential states that the team may occupy while operating the system, with the aim of creating an event space that they consider safe, productive, and humane.

This chapter focuses on team cooperation as seen from the perspective of ecological psychology, or rather, through the eyes of someone who sees in the ecological perspective a useful approach for the observation, description, and analysis of human behavior. I do not attempt to explain the beliefs, assumptions, and assertions that underlie the ecological approach; I am confident these fundamentals have been thoroughly explained elsewhere in this rich and heterogeneous volume. Readers will notice that most of the examples and illustrations provided throughout the text come from the world of aviation; having flown for many years, and since my research focuses on aircrew performance and cockpit design, I am most comfortable making the connection between my theories and this particular domain. It is important to remember, however, that the concepts discussed may be applied to a wide variety of team-machine systems, ranging from nuclear powerplants, air traffic control centers, and assembly lines, to television studios, office spaces, and large kitchens. I urge readers to search through their own experience for analogous situations with which they may feel comfortable to imagine, simulate, and evaluate the following ideas.

13.1   Two Heads Are Better Than One—But Not Simpler

In sports, being a good team player involves more than simply being proficient at the technical skills of the game and standing on the same court with other players. In basketball, for example, some players are considered good individual players, whereas others are considered good team players. Because basketball is fundamentally a team sport, and at the same time, one in which players often find themselves facing one-on-one situations, the combination of these two attributes—good individual and team skills—is what made great players such as Magic Johnson, Larry Bird, or Michael Jordan.

When teams are assembled—whether in sports, business, industry, or any other area of cooperative (*co-operative*) human activity—the context created by the aggregation of operators has unique attributes. Above all, it defines two tasks, two categories of interactions, pertaining to two activities that operators need to perform; two different—yet inevitably integrated—classes of interaction. One class of interaction emerges from the interface *between the team and the task environment* in which they operate; in this category, we look at the team's overall goals and how team actions affect the machine and the environment. The other class of interaction emerges from the *interface between individual team members*; here, we focus on how the flow of information within the team affects the coordination and cooperation between individual members.

In the particular context of team-machine interaction, aggregating operators creates a context that includes two categories of interactions: that which we usually call "control," and that which is called "communication." The first category includes the exchange of information between operators and machine through displays and controls; the second category includes the exchange of information between operators, utilizing all modalities of perception and action. Since the transfer of information from the human to the machine will in general be in the form of actions on the system (Rasmussen, 1986), an alternative perspective on these two categories may lean on the distinction between "action" and "information."[1] From this point of view, we may distinguish between the interactions between human and machine, interactions that occur through forces, and the interactions between humans, in which the mass term is absent, in which descriptions are kinematic (space/time), and/or geometric, and/or spectral—that is, interactions that are informational rather than forceful (Kugler & Turvey, 1987). Note that whatever definition one chooses to assign to these two categories, the boundary between them remains extremely fuzzy: If I say to you "Pass me the salt," am I communicating or controlling? When you sound your car's horn to warn a pedestrian, are you controlling (i.e., making them alter their behavior) or communicating? Cannot the action of pushing a button be seen as the transfer of information from operator to machine? This fuzzy region that lies between communication and control—the area in which information and action overlap—is the focal point of this chapter.

Further, implicit in all such multitask situations is the "metatask" of selecting which particular interaction - and what information - one needs to attend to at every moment. The extent to which the two categories of interaction are indeed distinguishable, or afford integration, depends on the design of the system, as well as on the level of expertise that individual operators bring into the team. From the design perspective, the two sources of information may be either separated or integrated physically, temporally, or semantically. The view developed in the following discussion outlines the critical role played by designers of team-machine systems in determining the particular relationship between the two tasks of control and communication. While acknowledging the added dimension of operator training, experience, and skill level, discussion of learning and expertise goes beyond the scope of this chapter.

For the most part, the fields of human factors and ergonomics have directed their attention to the study of the interaction between a single human operator and a machine; the resultant design guidelines focus primarily on interface considerations for single-operator workstations. What about team communication? Studies of team behavior have, by and large, been segregated from studies of interface design. Within the human factors paradigm, the most studied aspect of team coordination has been speech communication, that is, the audible, verbal interaction between operators (for a taste of publications ranging from theoretical to

---

[1] I am indebted to John Flach for this perspective.

experimental, see Barber, 1983; Costley, Johnson, & Lawson, 1989; Foushee & Manos, 1981; Gibbs & Muller, 1990; Kanki, Greaud, & Irwin, 1989; Kreckel, 1983; Larson & LaFasto, 1989). Although these, and other studies, have indeed uncovered important structures and patterns in cooperative human communication, they tend to assume that a record of the spoken information provides a sufficiently broad basis for the analysis of team coordination. Do team members limit their perception of teamwork to the auditory information they pick up or to the words they themselves utter? Should we focus on any one particular source—or modality—of information when studying team cooperation? Assuming that the team-machine context creates an overlap between control and communication, or action and information, one must further assume that there is more to crew communication that meets the ear. A look at the essential relationship between perception and redundancy will serve to illuminate this point.

13.2   Perception and Redundancy: A Multimedia Perspective

Perception and redundancy are intimately linked; perceptual systems—that is, living organisms—learn to capitalize on the redundancy of information available in their environment. Further, organisms learn to capitalize on redundancy itself as a source of higher level information ("higher" in relation to individual events), on the patterning or predictability of particular events within a larger aggregate of events (Bateson, 1972; Birdwhistell, 1970). For a good illustration of this point, we look to nature, which always provides wonderful examples of the ecology of behavior.

A small European bird, the indigo bunting, migrates at night, relying primarily on the pattern of stars in the sky to orient itself. Although it is sensitive to the earth's magnetic field, when the sky is visible, the indigo bunting will use the visual information even if it conflicts with the sensory information provided by the magnetic field. In a fascinating study, Emlen (1975) demonstrated that the bunting orients itself relative to that particular point in the night sky around which all stars rotate; thus, the bird inevitably directs itself in relation to the north, or polar, star. By placing the birds in a planetarium and rotating the night sky around stars other than the north star, Emlen produced a tendency in the birds to orient in the direction of the artificial polar star; even if the point of singularity, the fixed point, was occluded by clouds, the birds were able to accurately orient themselves toward it by referring to the angular motion of the stars that were visible. In this scenario, redundancy exists within the same modality—that is, all relevant information is detected by the visual system. The redundancy of information available in the visually rotating field of stars, and the organism's ability to capitalize on that redundancy, are key to understanding the indigo bunting's behavior—the dynamic organization of its perception and action.

The ability of pigeons to navigate accurately during day or night is well documented and has been exploited by man for many centuries. Experimental studies suggest that pigeons are able to use the sun's position in the sky as a directional guide (Keeton, 1974); on overcast days, or at night, they use their sensitivity to the earth's magnetic field to do the same (Wiltschko, Nohr, & Wiltschko, 1981). Interestingly, experimental disruptions of the bird's detection of the magnetic field (using a miniature electric coil fitted on its head, along with a battery strapped to its back—quite a hideous sight) have no effect whatsoever when the sun is visible; they do, however, cause disorientation when the sun is occluded by clouds. Thus, it seems that the information provided by the sun's position preempts the information provided by the magnetic field (Walcott, 1972). Further complexity was introduced by Kiepenheuer (1985), who demonstrated that pigeons can also use olfaction as a means for orientation. These findings demonstrate not only the pigeon's ability to sense different forms of directional information, but also its ability to differentiate between the different information fields and select the particular form that is most appropriate for any given context.

Like the indigo bunting and the pigeon, the human perceptual system too thrives on redundancy. For example, in the case of locomotion and navigation, information about the direction of motion is available redundantly in the entire optic flow field, in accelerations perceived by the vestibular system, in Doppler-type effects produced by sounds bouncing off stationary objects, in kinesthetic feedback from limbs and muscles, and in shifts perceived by the olfactory system. Although certain perceptual modalities are relied on more than others—for example, visual information usually plays a larger role in human navigation than does olfaction—there are situations in which the prioritization of information sources may change, or at least, situations in which the relationship between the information provided by two different modalities provides information at a higher level of analysis. For example, certain kinds of electrical fires may start with a strong smell of burning plastic, without any visual cues such as smoke or flames; the nose detects information that, along with *the lack* of visual information, suggests a particular type of event.

In communicating, humans rely on a broad spectrum of information provided by both words and actions, as well as by the environment within which the interaction takes place. Humans—just like the other living organisms described earlier—learn to capitalize on all forms of information that the situation presents (Hutchins, 1990; Klein, 1989; Lave, 1988; Rochlin, LaPorte, & Roberts, 1987). True, in certain cases, language and related symbol systems (e.g., American Sign Language, Morse Code) may be used to convey information that can only be presented symbolically, such as the description of a plan, the diagnosis of a situation, or discussing yesterday's dinner. At most times, however, particularly when people are actively involved in a physical task and interacting within the same physical environment, what people say is not independent on what they do; team activity and team communication are inevitably confounded.

13.3   Consequential Communication: Team Cooperation Within the Same Workspace

A distinction must be made between situations in which operators share the task of control, but do so from different stations, and those situations in which operators control the system while sharing one multioperator control station. In the first situation, operators in individual, segregated —and often remote—stations coordinate their activities *through* the system (the "machine"), using auditory and visual channels provided by the system to transmit and receive information; radios, telephones, and screens displaying verbal information, as well as feedback pertaining to changes in the system, provide individual operators with the information essential for crew coordination. Individual operators interact directly with the system only and indirectly with the task environment, which includes other operators.

When operators share the same workstation, however, they are exposed to an additional, entirely different, class of information; when sharing a workstation, every team member has access not only to environmental and system information, but also to the information provided *directly* by the presence and "co-operation" of team members. In this situation, the physical layout of the system creates a context in which each operator's activities can be directly perceived by other operators and thus can directly inform and affect team coordination. This shared situational context (Nickerson, 1981)—or physical co-presence (Gibbs & Muller, 1990)—has been noted to affect verbal communication in social settings and to structure team coordination tactics in goal-oriented task-performance activities (Chapanis, Ochsman, Parrish, & Weeks, 1972; Hutchins, 1989; Rochlin et al., 1987).

I call this unique class of information exchange *consequential communication*. Notice that consequential communication is a by-product of one operator's goal-driven activity; it is detected by the observer, rather than intentionally broadcast by the sender. Although the classic definition of *communication* suggests some intentionality on part of the sender—"The act of imparting, conferring or delivering, from one to another" (Webster, 1983)—consequential communication emerges from the purposeful interaction between human and environment, rather than purposeful interaction between humans. Here, the actor is more of an emitter than a sender, and the observer, more of a perceiver than a receiver.[2] The perceiver is responsible for the categorization of the relevant information as "communication"—that is, whether the event is to be categorized as an act of communication depends on the perceptual skills and directed attention of the perceiver.

In the language of ecological psychology, one might talk of the affordances of the team environment, what such an environment means to a perceiver (Michaels & Carello, 1981). If affordances are the acts or behaviors permitted by objects, places, and events, we might say that team coordination relies not only on speech, but on the entire spectrum of observable operator behaviors, system states, and environmental feedback, and on the affordances these specify for all operators. Team coordination—in this context—relies not only on intentional exchange of information, but also on each operator's ability to actively detect and employ the information inherent in another operator's activities.

13.4   Context: "A nod is as good as a wink to a blind bat" (Monty Python)

Without context, words and actions have no meaning at all (Bateson, 1979). Utterance and situation are bound up inextricably with each other, and the context of situation is indispensable for the understanding of the words (Malinowski, 1923). Perception and action take place

---

[2] I am indebted to John Grinder for this distinction.

within a particular context, one that determines the meanings and affordances of events and objects. The aggregate of perceivable information provides the context within which every particular bit affords a particular response for a particular observer. The same event (is it really "the same"?) may reveal significantly different affordances in different contexts: Throwing means one thing when the pitcher tries to evade the batter, another when he throws the ball to the player on first base; the sound \'no⁻\ gets its meaning from the context within which it appears ("no" vs. "know"); the meaning of a car's flashing turn signal is dependent on the context provided by the road and the surrounding traffic; a honk of the horn means one thing when it occurs as soon as the traffic light turns green, another when it alerts a pedestrian trying to cross the street, yet another when it is used to call someone to a waiting car. Perception of lower level affordances (of the event/action) may be constrained by the perception of higher order, value-related affordances defined by the perception of context (Vicente & Rasmussen, 1990).

The relationship between context and the interpretation of information is a pivotal point in the study of team communication and coordination. The physical form of the workstation, the task structure, and the environment with which the team-machine system interacts define a hierarchical set of nested contexts and thus constrain the range of information that may be perceived and understood by the operators. It is incumbent on the designers of complex systems to remain aware of the role they play in defining the context within which the different issues described here—activities of "control" and "communication," individual operator's perception of redundant information, and the emergence of consequential communication—take place.

Having viewed the communication-in-context subject from several perspectives, the discussion is now directed toward the "real world": How do operations in current designs reflect those aspects of team coordination previously described? The next section looks at several events from the aviation domain that will highlight some of these teamwork issues, as manifested by aircrews operating within the confines of the modern cockpit.

13.5   From Birds to Pilots: Stories from the Real World

The following accounts are taken from NASA's Aviation Safety Reporting System (ASRS, 1991), a database that holds reports submitted voluntarily by aircrew following incidents that they consider irregular. The reports are subjective accounts of what transpired in the cockpit during the incident, as narrated by operators who were actively involved in the control process. While reading through the brief reports, try to reconstruct the situation and imagine the sequence of events unfolding. Keep in mind the concepts discussed earlier; in particular, observe the interaction between the different sources of information—an interaction that is synthesized by the perceiving operator. (The excerpts have been slightly edited to allow for better reading.)

Example 1:

> *The warning horn indicating cabin altitude exceeding 10,000' came on while cruising at 35,000'. Before I could silence the horn and analyze the problem... the First Officer commenced and executed an emergency descent drill: throttles to idle, speed-brakes out, and transponder frequency to 7700 (emergency signal)—exactly by the book. The aircraft was well out of 35,000' before I could comprehend that an emergency descent was in progress. No discussion, just down we go! What happened to crew coordination? I have no answer. (ASRS #137152)*

Remarks: Note the shared context provided by the auditory information from the warning horn, a context within which the Captain (CA) was able to make sense of what the First Officer (FO) was doing (even though he did not necessarily agree with the actions). Notice the level of detail at which he was able to perceive the FO's actions; he did not need the FO to tell him *what* he was doing—that was obvious from seeing the actions; nor did he need to be told *why* it was being done— that was obvious from hearing the horn. The only thing missing, from the perspective of the reporting operator, was a discussion of whether that particular procedure, rather than any other, was appropriate.

Example 2:

> *As we added power for takeoff... the Second Officer (SO) pointed out a light aircraft immediately ahead of us at about 1000'.... I almost decided to abort the takeoff, but a second later the light aircraft began a right turn, so I continued the takeoff. Because of the disturbance and worry about the light aircraft, I did not notice or keep track of the rapidly accelerating airspeed, and missed the FO's callout of takeoff speeds. Meanwhile, because of the increasing speed, the airplane began*

to pitch up and the FO, sensing that we were tail-heavy, began to trim the stabilizer nose down.... I was unaware that the FO had changed the stabilizer setting, and therefore misunderstood when the takeoff warning horn sounded (it sounds automatically when the trim is beyond a certain setting). I began to reduce power to abort the takeoff, but both FO and SO correctly advised me that it was too late to abort, and that the trim had been changed. Takeoff was continued with no further problems. (ASRS #96238)

Remarks: Notice the different sources of information to which operators were directing their attention: The CA was engaged with looking at the light aircraft and missed the FO's calls for takeoff speed; the FO was engaged with the change in pitch and missed the connection between that and the increasing speed; the CA was unaware that the FO had changed the stabilizer setting and thus misinterpreted the warning horn. Notice that because both FO and SO knew that the setting was changed, they understood that the horn was warning about the change in stabilizer setting (trim out of the green band) and, realizing that the CA was unaware of their high speed and that the trim had been changed, correctly advised him to continue the takeoff.

Example 3:

On instrument approach to landing. The aircraft was well above the glide slope. At approximately 700' above ground level, the First Officer advised of the need for a missed approach procedure. Go around was called, power was applied by Captain, gear and flaps were raised by First Officer; no verbal commands were given, and no acknowledgments of actions were made. The aircraft immediately pitched up to excessive nose high attitude, well in excess of 35 degrees. The airspeed bled off to approximately 80 knots. At this point, the Captain was unable to exert enough forward control pressure and yelled for First Officer to help. Both pilots had full forward control pressure with the stick shaker sounding (stall warning). Neither pilots were able to take their hands off the controls. The Captain twice yelled for flaps up; after second call the Flight Engineer raised flaps in response to Captain's command. The nose of the aircraft began gradually lowering to less excessive pitch. After critical stage, the climb and approach checklist were then read again, and an uneventful approach and landing was made. (ASRS #79837)

Remarks: What can we learn from this account (other than to take the train)? Once the go around had been announced, that is, clear intention verbally stated, the actual control actions were performed silently; there was so much redundant information that indicated that engine power had been applied—for example, throttles forward, instrument readings, increase in engine sound, acceleration felt by the "seat of the pants"—that there was no need for the CA to say "power applied." There was, however, *not* enough information about the true flap position—although the FO reached for, and manipulated, the flap control, he apparently moved it to the wrong position. Throughout the entire event, the aircraft's extreme and unusual attitude provided all three crew members with a context within which actions were perceived and interpreted. Note the reference to the stick shaker, which provided both tactile and auditory information that warned of an eminent stall. Now, in contrast to this very dynamic situation, notice how informative a *lack* of action may be.

Example 4:

> *While all of this was happening, I noticed the Captain was unresponsive. I know that he was a diet controlled diabetic and told him to eat. He did, and became responsive in a few minutes. (ASRS #84100)*

Remarks: Within this particular control context, observing a *lack* of activity prompted the FO to draw certain conclusions and take corrective action to remedy this potentially disastrous situation. Thus, although actions provided the context for interpretation of events in the first three examples, *inaction* was the key piece of information in this account.

The more accounts one reads (this particular search identified 424 similar cases), the more one is convinced that in the context of a team workstation such as an aircraft cockpit, it is almost impossible for an individual operator to do anything that does *not* constitute intercrew communication. Actions performed by one operator, or expected actions that are not performed, mean something to other operators. Actions have consequences not only for the machine, but also for other operators who are present when the actions are performed. The study of team coordination, and the design of team-machine interface, must therefore include considerations for consequential communication.

13.6   Team Engagement State Space (TESS): A Descriptive Tool

When studying a situation such as described earlier, that is, when humans act and perceive in the context of two different categories of interaction, it is often useful to envision a two-dimensional space, one that contains all the states that describe the mutually constraining relationship between the two. The Team Engagement State Space, or TESS (Figure 13.1), defines a two dimensional space that can be used to describe the team's engagement in the two different tasks, and their direction of attention to the two different sources of information offered by the team context. It presents communication as one dimension of information, and one class of interactions; it is this class of information that, as discussed earlier, makes the difference between single-operator and multioperator contexts. The other dimension, control, refers to the class of information that is specifically related to the interface between operators and system, that is, the vertical axis describes the operators' engagement in the task of interacting with, and controlling, the system. Note that on both axes, the scale of 0–1 does not represent any specific, quantifiable measurement of information or behavior; its purpose is to illustrate, in an abstract way, the relative extent to which operators are engaged in the particular dimension described.

Imagine an aircraft parked at an airport, 1 hour before takeoff; while the cabin crew go about their duties, the pilots spend several minutes discussing the flight plan. In this situation, their involvement with the system is minimal, though their engagement with each other is high (point "d"). Now think of that same crew performing the first portion of their preflight checklist: Each operator is in charge of checking a different group of subsystems and paperwork; each is strongly engaged with the system, with very little exchange of information going on between them (point "e"). Finally, picture the situation during the takeoff roll itself: The captain is controlling the aircraft, while the first officer announces the speeds and sets the throttles to takeoff thrust; as soon as the aircraft is airborne, the captain calls out: "Positive climb, gear up," upon which the first officer reaches for the control that raises the gear. In this situation, both operators are engaged in both control and communication (point "f").

Beyond the description of specific instances, the figure can be used to describe different regions of operation that correspond to different types of team actions and interactions. As illustrated in the figure, the entire state space can be divided roughly into three regions, as described in the following paragraphs. Notice that from an investigative point of view, if a particular team tends to operate within a certain region, three primary applied questions can be asked: What characteristics, or qualities, of *the team* make it operate in this region? What elements of *the task* constrain the team to operate in this region? What properties of *the system* and *the environment* constrain the team to operate in this region?

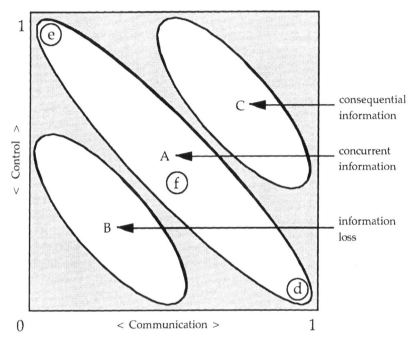

Figure 13.1. *Team Engagement State Space (TESS).*

Region A describes situations in which team coordination and systemic output are two classes of information that are concurrently available, yet independent of each other. Operators need to divide their attention between two tasks: that of attending to the system, and that of attending to the crew. Thus, the team's attention shifts from one source to the other, trading off one class of information for the other according to their perception of system state, environmental constraints, and task status. For example, in Example 2 in the previous section, the Captain was engaged in the task of monitoring the light aircraft and thus failed to perceive the First Officer's callout of airspeed for takeoff (see point "e" in the TESS, Figure 13.1).

Region B is where engagement in the two interactions is confounded by a conflicting task or some compounding and irrelevant source of information. For example, if an operator needs to leave his or her station and walk to an adjacent station in order to communicate with another (as was the case in older aircraft with large cockpits), his or her attention is by necessity diverted by the need to pick up information relevant for locomotion. Alternatively, pilots flying across the Pacific might intentionally divert their attention to activities other than the two described in the model, such as reading or listening to music, in order to break the monotony of long-haul flights. Thus, in the context of particular systems and specific situations, operators are unable, or unwilling, to share their attention *only* between system control and crew coordination. Two ASRS reports illustrate the nature of activities in this region: "As I reached over to the instrument, the passenger immediately behind the copilot's seat dropped the coffee pot, attracting both my attention, and the captain's, away from the panel" (#127627); "The flight attendant impatiently tapped the captain on the shoulder because he had not taken his beverage. He then took his beverage, tasted it, and realized we both had the other's beverage. We exchanged beverages and as the captain put his down, he saw the red altitude alert light flashing (meaning we passed our designated altitude)" (#62152).

Region C describes states in which information pertaining to systemic output and information pertaining to team coordination are integrated, whether by design or by task context. We might think of the pilot who has his or her hands on the controls while the other pilot flies the aircraft, perceiving his or her behavior through the movements of the controls; or the pilot who leans over to see where on the map the navigator's finger is pointing, using the physical interaction between navigator and map to detect specific, task related information. This region describes situations in which engagement in the control task provides operators with information that is relevant to team coordination; control activity creates the opportunity for consequential communication to emerge, thus making verbal communication redundant, as illustrated by the wealth of information described by the narrator in Example 1 in the previous section.

13.7   Design as Choreography

As a result of this physical nature of human-machine interaction, actions have the pivotal role of providing the most essential information—they tell one operator what the other is actually doing; beyond the verbal descriptions of intentions and plans, the "bottom line" is what an operator physically does in his or her interaction with the machine. It is, after all, called human machine inter-*action*, not human-machine inter-*speech* or inter-*thought*. Although one needs to acknowledge the rapid development of speech-recognition technology and voice-activated controls, it seems that the bulk of control activity, particularly continuous control tasks, will remain physical. This being the case, we must look at the intricate, and wonderful, relationship between the design of an environment and the behavior of living systems within that environment.

A useful description of the environment must reflect an environment's mutually constraining relationship to the organism (Turvey & Carello, 1981). The physical form and functionality of humans and machines constrains and shapes the nature of their interaction. Design guidelines suggest that the system be "reachable," "seeable," "controllable"—dimensions that are defined by the human's ability to perceive and act. At the same time, training programs are designed so that operators learn to interact with the particular forms of stimuli defined by the existing technology's ability to generate, display, receive, and process information. Once the system is given a physical form, interactions between humans and machines are constrained to *take place* in specific locations in the workstations. These points of interaction between operators and machine—displays in which information flows from machine to operators, and controls used by operators to enter information into the machine—define points in space through which all activities in the workstation flow, locations that become the focal point of the operator's control-related behavior.

Imagine a large control room: While operators are free to perform endless side tasks or peripheral activities between task-oriented actions—one controller might choose to stretch her legs by walking around the room, another might want to prepare and drink a cup of coffee—when the task itself is performed, the interaction always returns to specific locations in the workstation. The operator who got up to stretch must return to a certain location if she wants to press a particular switch; the one drinking coffee must direct his gaze at a specific point in space—the location of a particular display—in order to perform the monitoring task. As the operators' knowledge of the system grows, as he or she become familiar with the task's demands and the system's constraints, the operator learns *where* displays and controls are located in space, and *when* each display or control becomes relevant in the process of control.

Most importantly, in the context of the current discussion of team coordination, operators learn to act on the meaning of actions when they are carried out by other operators: "Because I know what happens when I do X, I know what to expect when I see you do X." The coordination of activities between team members depends on the degree to which individual teammates share the way in which they perceive the characteristics of the situation: "If you turn that dial to the right, I will reach for that lever and raise it two notches." Shared perception of affordances becomes the means for the coordination of control activities. The level of expertise in the task domain, along with the operator's familiarity with the system, determine the detail and accuracy with which such affordances may be perceived by each operator.

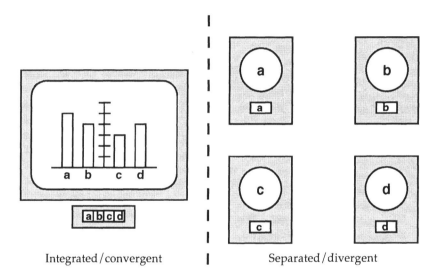

Figure 13.2. *Two approaches to display/control layout.*

The two different designs depicted in Figure 13.2 serve to illustrate the impact of display/control layout on operator behavior. Notice that although both layouts address the same interface problem—four controls for input, four displays for feedback—the difference in their configuration affords different styles of human-machine interaction. Imagine you were watching an operator interact with these interfaces: how would you perceive the effects of machine design on operator behavior?

The physical form of the workstation defines the constraints that *shape* operators' physical behavior; spatial layout of the control environment constrains operators to perform a specific set of movements. The specific location of displays constrains them to direct their gaze and focus on particular points in space; the location of specific controls constrains them to reach for those locations; the particular type of control constrains them to perform particular actions, for example, push, twist, flip, pull. Using the sample layouts described in Figure 13.2, imagine observing an operator monitor and control variables "a" and "d": Notice the different movements of eyes, hand, and head dictated by the two different layouts. If you were to supervise their performance, which layout would you prefer? Would that preference change if you were to operate the system yourself?

The machine's operating procedures define and constrain the *temporal* organization of operators' behaviors; procedures that describe specific sequences of inputs impose on the operator constraints that dictate a particular pattern of actions over time. For example, a procedure might determine that control "a" must never be manipulated before "b"; control "d" must always precede "c," which is usually followed by "b." Such procedures, coupled with the system's response time, define a pattern of actions over time that is particular to every design, as well as to every context of operation. From this perspective, the designer who defines the machine's logic of operation, the sequence of "if–then" statements that govern the system, is at the same time building a temporal sequence to which operators will conform in their interaction with the system. Again, use Figure 13.2 to imagine the difference between the patterns of actions emerging from interactions with the two different designs; note the differential impact of system layout on consequential communication.

Consider the effect of each configuration on team performance: How would each configuration affect the location of a team in the TESS described earlier (Figure 13.1)? Using these simple layouts as examples, one can see that as the displays and controls converge to a centralized location—a design approach that often enhances single operator performance—the team pays the price of losing valuable cooperative information. Conversely, the separation of displays and controls— which often imposes a higher workload on the single operator— enhances teamwork by providing information that may support consequential communication. Design and team performance are intimately linked; differences in design result in different types of individual activity, and, subsequently, different forms of teamwork performance.

Throughout the design process, the designer must be aware of these two principles: *the system's physical form shapes operator behavior, while its operating procedures organize that behavior temporally* (Segal, 1990). Designers must see themselves as choreographers; ideas that emerge from the designer's drafting table will define a set of actions unfolding over time, an operating "dance" that will be performed by operators whenever they interact with that system. It is the designer's responsibility that the dance performed allow for the smooth flow of information between team members and result in performance that is effective, productive, and safe.

13.8   Approaching Design Projects

> *Design as a problem-solving activity can never, by definition, yield the one right answer: it will always produce an infinite number of answers, some "righter" and some "wronger." The "rightness" of any design solution will depend on the meaning with which we invest the arrangement. (Papanek, 1985, p. 5)*

Unfortunately, the people responsible for the design of an environment are not always aware of its strong effect on the interactions that take place within that environment (Burgoon, Butler, & Woodall, 1989). One must ask: How do we design team workstations that take the concept of consequential communication into consideration? Focusing on the above mentioned choreography of the "operating dance," it seems easier to approach the problem by asserting what technology should *not* do, by looking at two extreme team-machine interface situations. Imagine, for example, the task of a copilot monitoring a pilot who is controlling certain aircraft parameters via a keyboard; it is virtually impossible to know what specifically that pilot is doing until it has already been done, that is, the input is recognized only once the aircraft's systems have reacted to that input. Alternatively, the picture of an excessively spread-out control station— and an operator reaching deliberately for each and every switch along the procedure—can be extremely distracting and overloading.

As mentioned in the previous section, there exists an inherent conflict between designing to support individuals and designing for teams. For example, although individual operators may benefit from the integration of different parameters into one display (Wickens, 1992), it is precisely this type of design that may annihilate critical nonverbal, consequential communication, from the team context. Similarly, although configurable displays afford the individual operator optimal control over the personal organization of information, they may lead to undesirable situations in which one operator is unable to read another's particular (and, perhaps, peculiar) display format. In this respect, designers may consider the need to compensate for consequential information that is lost through the convergence and flexibility of displays and controls, perhaps by providing enhanced feedback from the system indicating what specific actions each operator is performing.

In general, the design of critical displays and controls must not be so ambiguous as to obscure the actions of the operator interacting with them. By the same token, the design must not attribute too much importance to activities that need not be shared by operators. It is important to note, though, that there is no context-free solution for this design paradigm. Although it may be agreed on that, for example, highly compartmentalized workspaces tend to minimize cooperation, whereas open ones promote it (Burgoon et al., 1989), it is still incumbent on the designer to determine when, in the context of a particular machine and task, cooperation should be promoted and when it should be discouraged. For every operational context, it is important to decide not only what information is essential for interteam communication, but also what interaction information is better left invisible to other team members. This critical classification of information must serve as the foundation for the definition of specific criteria that will subsequently be applied to a specific design program.

There is no doubt that the ultimate solution will involve a tradeoff between the considerations for single-operator and team performance. As long as the investigation of these two remains divided between different disciplines—human factors and ergonomics designers focusing on individual human performance and single-operator interfaces, whereas sociologists, industrial psychologists, and anthropologists look at team cooperation and communication—the design of team workstations will continue to reflect the disintegrated nature of our understanding of team-machine interactions. We need to step back and include both perspectives in a new, expanded field of regard; the knowledge underlying team workstation design should emerge from the synergy of single-operator and team-dynamics research. The continuing challenge for designers is to use technology intelligently in order to engage the operators with each other, and simultaneously, with the machine that they control.

13.9   Summary

Although people do learn to cooperate under a variety of conditions, we must not rely on training as the sole means of creating productive and safe team dynamics. Beyond training, no one has more sweeping influence on performance than the designer (Senge, 1990). Designing a team workstation entails much more than simply aggregating several individual stations. The physical environment that surrounds the team defines a context that, along with the operating procedures, choreographs a particular cooperative team "dance." The designer has the power to create a dance that is not only natural and easy for any single operator to perform, but one that may also support the interaction between operators, creating a context in which they share both the perception of—and responsibility for—system state and task performance.

There are no recipe-type, context-free, design solutions. There are, however, several key issues that need to be considered throughout the design process. How does the design support single-operator performance? How does it enhance, or obstruct, the flow of information between team members? What is the subsequent impact on overall team performance? What design solutions can one provide for recovering essential information that may get lost, while covering redundant information that may be distracting? Technology must be utilized to shape an environment within which the synergetic nature of teams would be supported and enhanced, allowing for the smooth flow of information between individual operators and for truly ecological interactions between the team and the machine.

Dedication

This chapter is dedicated to my brother, David Gaster—a gifted teacher, an eager student, and an extraordinary pilot. As I write these lines, David is preparing to take off for a wondrous journey; may he fly high, fast, and far.

13.10   References

Aviation Safety Reporting System (ASRS). (1991). *Non-verbal flight crew coordination incidents* (Search Request No. 2277 & 2337). Mountainview, CA: ASRS Office, Battelle.

Barber, B. (1983). *The logic and limits of trust.* New Brunswick, NJ: Rutgers University Press.

Bateson, G. (1972). *Steps to an ecology of mind.* New York: Random House.

Bateson, G. (1979). *Mind and nature: A necessary unity.* New York: E.P. Dutton.

Birdwhistell, R. L. (1970). *Kinesics and context.* Philadelphia: University of Pennsylvania Press.

Burgoon, J. K., Butler, D. B., & Woodall, W. G. (1989). *Nonverbal communication: The unspoken dialogue.* New York: Harper & Row.

Chapanis, A., Ochsman, R. B., Parrish, R. N., & Weeks, G. D. (1972). Studies in interactive communication: I. The effects of four communication modes on the behavior of teams during cooperative problem-solving. *Human Factors, 14,* 487–510.

Costley, J., Johnson, D., & Lawson, D. (1989). A comparison of cockpit communication B737–B757. In R. S. Jensen (Ed.), *Proceedings of the Fifth International Symposium on Aviation Psychology* (pp. 413–418). Columbus: Ohio State University.

Emlen, S.T. (1975). The stellar-orientation system of a migratory bird. *Scientific American, 223,* 102–111.

Foushee, H. C., & Manos, D. (1981). Within cockpit communication patterns and flight performance. In C. E. Billings & E. Cheany (Eds.), Information transfer problems in the aviation system (NASA TP 1875, pp. 63–72).

Gibbs, R., & Muller, R. A. G. (1990). Conversation as coordinated, cooperative interaction. In S. P. Robertson, W. Zachary, & J. B. Black (Eds.), *Cognition, computing, and cooperation* (pp. 95–114). Norwood, NJ: Ablex.

Hutchins, E. (1989). The technology of team navigation. In J. Galegher, R. Kraut, & C. Edigo (Eds.), *Intellectual teamwork: social and technical bases of cooperative work* (pp. 191–220). Hillsdale, NJ: Lawrence Erlbaum Associates.

Kanki, B. G., Greaud, V. A., & Irwin, C. M. (1989). Communication variations and aircrew performance. In R. S. Jensen (Ed.), *Proceedings of the Fifth International Symposium on Aviation Psychology* (pp. 419–424). Columbus: Ohio State University.

Keeton, W. T. (1974). The mystery of pigeon homing. *Scientific American, 231,* 96–107.

Kiepenheuer, J. (1985). Can pigeons be fooled about the actual release site position be presenting them information from another site?

Behavioral Ecology and Sociobiology, 18, 75–82.

Klein, G.A. (1989). Recognition-primed decisions. *Advances in Man-Machine Systems Research, 5*, 47–92.

Kreckel, M. (1983). *Communicative acts and shared knowledge in natural discourse*. New York: Academic Press.

Kugler, P. N., & Turvey, M. T. (1987). *Information, natural law, and the self-assembly of rhythmic movement*. Hillsdale, NJ: Lawrence Erlbaum Associates.

Larson, C. E., & LaFasto, F. M. J. (1989). *Teamwork: What must go right / what can go wrong*. Newbury Park, CA: Sage.

Lave, J. (1988). *Cognition in practice*. Cambridge, UK: Cambridge University Press.

Malinowski, B. (1923). The problem of meaning in primitive languages. In C. K. Ogden & I. A. Richards (Eds.), *The meaning of meaning*. London: Routledge and Kegan Paul.

Michaels, C. F., & Carello, C. (1981). *Direct perception*. Englewood Cliffs, NJ: Prentice Hall.

Nickerson, R.S. (1981). Some characteristics of conversations. In B. Shackel (Ed.), *Man-computer interaction: Human factors aspects of computers and people* (pp. 53–64). Alphen aan den Rijn, The Netherlands: Sijthoff and Noordhoff.

Papanek, V. (1985). *Design for the real world: Human ecology and social change*. Chicago: Academy Chicago Publishers.

Rochlin, G. I., LaPorte, T. R., & Roberts K. H. (1987). The self designing, high reliability organization: Aircraft carrier flight operations at sea. *Naval War College Review, Autumn*, pp. 76–90.

Rasmussen, J. (1986). *Information processing and human machine interaction*. New York: Elsevier.

Segal, L. D. (1990). Effects of Aircraft Cockpit Design on Crew Communication. In E. J. Lovesey (Ed.), *Contemporary ergonomics* (pp. 247–252). London: Taylor & Francis.

Senge, P. M. (1990). *The fifth discipline: The art and practice of the learning organization*. New York: Doubleday/Currency.

Turvey, M. T., & Carello, C. (1981). Cognition: The view from ecological realism. *Cognition, 10*, 313–321.

Vicente K. J., & Rasmussen, J. (1990). The ecology of human-machine systems II: Mediating "direct perception" in complex work domains. *Ecological Psychology, 2*(3), 207–249.

Walcott, C. (1972). Bird navigation. *Natural History, 81*, 32–43.

Webster's New Universal Unabridged Dictionary. (1983). New York: Simon and Schuster, p. 367.

Wickens, C. D. (1992). *Engineering psychology and human performance* (2nd ed.). New York: HarperCollins.

Wiltschko, R., Nohr, D., & Wiltschko, W. (1981). Pigeons with a deficient sun compass use the magnetic compass. *Science, 214,* 343–345.

Chapter 14

Risk Management and the Evolution of Instability in Large-Scale, Industrial Systems

Peter N. Kugler

Department of Computer Science and
Center for Brain Research and Informational Sciences
Radford University

Gavan Lintern

Aviation Research Laboratory and Beckman Institute
University of Illinois

14.0   Risk Management

Of the challenges facing our global society, possibly none is more important than the requirement to control large-scale, dynamic systems such as those found in transportation, industry, economics, and politics. Systems that are both complex (many interacting components) and nonlinear (system response is a nonmonotonic function of inputs) can be exceedingly sensitive to minor events. Seemingly, our ability to design and build systems has outstripped our ability to control them (Perrow, 1984).

Furthermore, although the behavior of our own creations may surprise us and leave us at a loss for effective action, natural disasters in the form of drought, flood, and fire also challenge our ability to control and to manage. The complexity of these systems frequently leads to outcomes that have not been possible to predict. Small and seemingly trivial events that usually have no substantial impact on system behavior can occasionally cascade through the system to precipitate a major change of state. The evolution of these often surprising effects can be attributed to nonlinearities that permit minor perturbations to reverberate through a system with substantial amplification.

The general problem is one of interfacing human controllers to a nonlinear environment. This might be characterized as a problem of interfacing an animate "controller" dynamic to a nonlinear, inanimate "plant" dynamic, or that of interfacing a single controller with a large set of subsystems that interact nonlinearly. In essence, the issue raised is how a biological agent as an actor, perceiver, and controller interacts with a physical environment in which the dynamic has small nonlinear regions interspersed between large regions that approximate linearity. To gain some insight we pursue the issue of how the natural design of biological systems has solved the problem of negotiating with nonlinear environments. Is this problem one for which there is a common solution in nature? In pursuing further insights, we are influenced by the failure of classical physics to offer appropriate models for understanding nonlinear interactions and are motivated to analyze the principles that have emerged from the relatively new science of self-organization in animate and inanimate systems (Davies, 1988; Glieck, 1987; Prigogine & Stengers, 1984).

14.1   Nonlinearity

A nonlinear system is one in which monotonic adjustment of a control parameter can lead to a qualitative change in state. The transitions in a substance from solid to liquid and liquid to gas during heating provides an example. The transitions are evidence of the nonlinearities where temperature constitutes the control parameter. The development of convection rolls in a liquid when heated from below (the Rayleigh-Benard convection) offers another example of a nonlinear system. The convection pattern emerges via a nonlinear transition (known technically as a *bifurcation*) under the influence of a temperature gradient in the liquid acting as a control parameter.

Nonlinear systems have generic characteristics that are evident in these two simple examples. Firstly, there are large linear ranges; the nonlinearities comprise a relatively small part of the total system. Linear analyses will often give the appearance of accounting for most of the behavior of a nonlinear system, and it is a common strategy to relegate what is not accounted for by the linear analysis to noise or remnant. The tendency is to characterize this noise or remnant as unimportant and to ignore it. In contrast, a nonlinear perspective will take the view that the nonlinearities present the challenge to understanding. They, in fact, constitute what makes the system worthy of investigation.

Secondly, some characteristics of the form of organization that emerge after a state transition cannot be predicted. For example, convection rolls that develop in a fluid heated from below may roll in a clockwise or a counterclockwise direction. The selection of the direction is made at the time of the nonlinear transition. Specifically, with fluid molecules becoming more agitated as the fluid is heated, a chance dominance of molecular motion in one direction at the moment of the nonlinear transition will initiate a miniroll in that direction, and that direction of motion will quickly progress through (enslave) the remainder of the fluid. This is the bifurcation: One of the multiple potentialities (branches) is selected and subsequently sustained by a self-reinforcing (autocatalytic) process.

That is not to say that this or any other nonlinear systems are entirely unpredictable. Some of the global properties on either side of the nonlinear transition are quite regular, and the value of the control parameter that precipitates a transition is also regular. For example, the general topological features of convection rolls and the value of temperature gradient that causes them to emerge are regular. On the other hand, the direction of the rolling motion is not predictable. In essence, a nonlinear transition will produce a form of organization for which some macroproperties can be anticipated and others cannot.

14.2   Self-Organization

14.2.1   The Science of Isolated (Linear) Systems

Within the framework of thermodynamic law, an isolated system is one that does not exchange energy or matter with its surround. It is one that conserves energy and matter or energy/matter in accordance with the first law of thermodynamics. The organization of distributional characteristics of an isolated system are accounted for by the second law of thermodynamics. Ordered forms of energy are inevitably broken down into disordered forms to approach a state of maximum entropy (Atkins, 1984). This is the equilibrium condition. Once equilibrium has been achieved, an isolated system can never evolve into a more ordered state. It would, for example, be counter to the second law for a hot spot to spontaneously emerge in a piece of metal isolated from its surround, or for ink molecules dispersed in water to spontaneously come together to form an inkdrop.

14.2.2 A Physical Science for Open (Nonlinear) Systems

An open system is one that exchanges energy and matter with its surround. The imperatives of the second law, as exemplified in a dispersing inkdrop or the decay of a hot spot in a piece of metal, have puzzled many who observe increasing order in nature; the evolution of biological species being the most striking example. In recent years a new branch of physics, consistent with the imperatives of the second law of thermodynamics, emerged. It focuses specifically on the role that nonlinearities play in the emergence of order within open (and closed[1]) systems (Davies, 1989; Yates, 1987).

The theory of mechanism associated with emergence of order within open (and closed) systems requires the introduction of a new inventory of exotic physical and mathematical ideas to accommodate nonlinearity: constructs such as chaos, strange attractors, catastrophes, fractals, turbulence, universality, and topological dynamics. These new constructs have been accompanied by a new set of predictions for testing models of nonlinear phenomena. Whereas traditional mechanical and thermodynamic models invoke predictions about forces and entropy, the new models invoke predictions of transitional instabilities involving qualitative shifts in topological modes.

In addition, traditional physical analyses focus on linear (symmetry-preserving) transformations, whereas the new analyses focus on nonlinear (symmetry-breaking) transformations. The new models identify generic ways in which behavioral trajectories qualitatively unfold toward states of increasing order, that is, the forms and functions they embody as they self-organize. Accordingly, this new physics is sometimes referred to as a physics of self-organizing systems. A fundamental contribution to understanding the role of nonlinearities in (self-) organizing an open system has been made in Prigogine's (1980) theory of dissipative structures.

---

[1]A closed system exchanges energy (but not matter) with its surround.

14.2.3 The Theory of Dissipative Structures

> *Classical thermodynamics was associated... with the forgetting of*
> *initial conditions and the destruction of structure. We have seen,*
> *however, that there is another macroscopic region in which, within the*
> *framework of thermodynamics, structure may spontaneously appear.*
> *(Prigogine, 1980, p. 150)*

Some of the most influential early investigations of open systems were made by Prigogine and his colleagues, starting in the mid-1940s and extending for more than three decades (Babloyantz, 1986; Glansdorff & Prigogine, 1971; Nicolis & Prigogine, 1977, 1985; Prigogine, 1947, 1962, 1967, 1980). These investigations involved models of chemical systems in which competitions developed between the flows that coupled various components of the reaction mechanism. It had previously been assumed that the time evolution of a system's states was along a linear relaxational trajectory that moved the system inexorably through states of decreasing order toward a final state of maximum disorder (entropy). The trajectory followed the strictly destructive path prescribed by the second law of thermodynamics in an isolated system. It was further assumed that the consequences of the second law were the same for both isolated and open systems.

In contrast to the existing assumptions, Prigogine and his colleagues found that the linear relaxation dynamic could break down when some chemical systems were displaced far from equilibrium and was replaced by a nonlinear dynamic that drove such a system locally further from equilibrium. In this far-from-equilibrium region, a new thermodynamic path (branch) yielded constructive effects as a by-product of the second law's dissipative processes. A system drawn far from equilibrium could suddenly develop intrinsic force structures that drove it further from equilibrium, that is, the system started to self-organize. Recognizing the central role played by the dissipative processes in the self-organizing process, Prigogine termed these open systems *dissipative structures*. In 1977, Prigogine was awarded the Nobel Prize in Chemistry for this work.

14.2.4   Self-Organization of Chemical Clocks and Chemical Patterns

One of the best understood examples of a dissipative structure is the Belousov–Zhabotinski chemical reaction first reported by the Russian chemist Belousov in 1958 (see Chance, Pye, Ghosh, & Hess, 1973; Tyson, 1976; Winfree, 1974; Zhabotinski, 1974). While studying an intermediate transition phase of a complex reaction system involving four coupled flasks linking a water-soluble chemical medium, Belousov noticed that as one of the couplings between the flasks was varied, the reaction medium began to cycle through an evolutionary sequence in which it changed color periodically from red to blue and back. This open chemical system organized the machinery of a clock from dynamics intrinsic to the molecular interactions. Under slightly different variations in coupling the same chemical reaction self-organized into a spatiotemporal structure of red and blue stripes. Under other coupling, conditions, a thin layer of the reaction mixture gave rise to red and blue concentric curves that gradually unfolded around several centers in the medium.

The tendency of the Belousov–Zhabotinski reaction to self-organize opposed the evolutionary tendencies predicted by traditional reaction theories. For centuries chemists had been accustomed to mixing various reactive ingredients and observing relaxational dynamics that tended with time to approach equilibrium. For a reaction in which all reacting chemicals were soluble in water and the final products of reaction were colored a homogeneous red, it had been assumed that the evolutionary dynamic would inevitably reveal a gradual relaxational coloring of the medium onto a final red solution, a state that would then remain permanently motionless and uniform. Within the context of what was known at the time, discovery of chemical periodicities that would self organize as a result of variations in coupling was startling.

14.2.5 Intrinsic Competitions Between Linear and Nonlinear Forces

An account that could integrate the new self-organizing dynamic with the standard relaxation dynamic required the introduction of a nonlinear force term to compete with the linear relaxation force term. The proposed nonlinear forces are present under both isolated and open conditions, but are revealed only in conditions far from equilibrium. At and near equilibrium the linear forces dominate. Nonlinear forces are amplified if the system is driven far from equilibrium as is possible only in an open system. When flows of energy and matter amplify fluctuations to create far-from-equilibrium conditions, the latent nonlinear dynamic will seize control of the system and drive it toward states of greater organization (Prigogine & Stengers, 1984). The identification of a critical relationship between nonlinearity, open systems, and self-organization was a major discovery. In recent years the study of this relationship has become a prominent topic on the scientific agenda of the physical, biological, and social sciences.

14.3   Biological Self-Organization

An historical challenge for the physically minded scientist has been the removal of vitalism from explanatory accounts of biological systems (Davies, 1988; Prigogine & Stengers, 1984). The notion that there is something needed beyond natural law to account for life seemed unsatisfactory and unscientific. The natural response has been to turn to theories of mechanism that view living organisms as complex machines, the functioning of which can be understood in terms of ordinary physical laws. This classical reductionist approach has, however, failed to provide a satisfactory account. Although we remain committed to a physical science of natural laws, we believe that a focus on open, nonlinear systems promises to throw some light on the emergence of order in biological entities.[2]

The description of nonlinear transitions in the Rayleigh–Benard convection and in the chemical clocks of the Belousov–Zhabotinski reaction is provided here as an introduction to abstract principles that cut across the full range of physical systems, including those of biology. Indeed, the primary thrust of this chapter is to comment on human organization in complex industrial systems; a biological system of significance to us all. In doing so, we pursue the argument that principles of self-organization are nature's generic solutions to organization imposed on the smallest to the largest entities and are influential in the (self-) organization of living as well as nonliving systems. Human social organizations are no less subject to principles of self-organization than are other natural thermodynamic systems. Enroute to completion of this argument we present a model of nest construction by social insects to illustrate some principles of self-organization in a complex, biological system.

---

[2]It would be premature to suggest that the new insights now developed will provide an ultimate account of biological order. We remain open to the possibility that radically new laws beyond those now being formulated for open systems will be required. Any new principles or mechanisms, no matter how radical, will nevertheless take their place within the body of natural physical law.

14.3.1 Self-Organization in Social Termites

The insects of interest here are African termites who periodically cooperate to build nests that stand more than 15 ft high, weigh more than 10 tons, and persist in excess of 300 years.[3] This remarkable construction activity is made even more remarkable by the fact that termites work independently of each other and do not follow a structural plan. Nor, as noted by Prigogine and Stengers (1984), could it be said that these insects have a shared consciousness. Individual insects are locally controlled by pheromone (molecular) distributions that arise from materials excreted by the termites themselves and strewn by them around the building site, at first randomly and then in increasingly more regular ways. The pheromone-laden excretions fulfill two functions. They serve as the building material for the nest and also establish a perceptual field that constrains the patterning of the insect activity. The insect activity, in turn, determines the novel architectural structures that emerge from this dynamically improvised plan.

14.3.1.1   Unconstrained flight: An equilibrium mode.

Each spring the termites develop a sensitivity to a pheromone secretion in their waste. Once an insect deposits its waste it temporarily loses sensitivity to the pheromone, but atmospheric diffusion of the pheromone creates a gradient field that will orient other nearby insects who enter the field within the limits of their own perceptual threshold. The pheromone gradient establishes an information field relative to the insect that points to the deposit site. The recent deposit, which lies at the center of the diffusion field, can technically be referred to as a singularity, that is, a point in a field where the gradient goes to zero. Each deposit temporarily defines the spatial location of a singularity relative to the pheromone diffusion field, and thus also for the information field.

---

[3]The following details concerning nest construction by social termites are derived from naturalistic observations by Grasse (1959) and Bruinsma (1977), from a thermodynamic treatment advanced by Deneubourge and Leuthold (1977), and from an extended information analysis by Kugler and Turvey (1987).

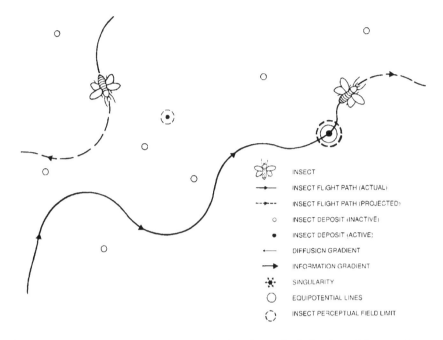

Figure 14.1. *During the random flight phase the behavior of insects is at equilibrium. There is a low probability that an insect will intercept the field of a recent deposit within the limit of its perceptual threshold. Insect flight paths remain independent of each other. (Adapted from Information, Natural Law, and the Self-Assembly of Rhythmic Movement (p. 69), by P.N. Kugler and M.T. Turvey, 1987, Hillsdale, NJ: Lawrence Erlbaum Associates. Reprinted by permission.*

The pheromone diffuses in accordance with Fick's law[4] which relates the rate of diffusion to the gradient of the field. As time passes the amount of pheromone at the singularity decreases (a dissipative process), scaling the field gradient accordingly. If only a few insects participate, depositing is so infrequent that the pheromone gradient of any recent deposit converges on global equilibrium[5] before another insect can be influenced by it (Figure 14.1). At global equilibrium there are no gradients and therefore no singularities.

---

[4] At Fick's law is that the rate of transport is linearly proportional to density.

[5] At equilibrium the global dynamic is identical to the local dynamic.

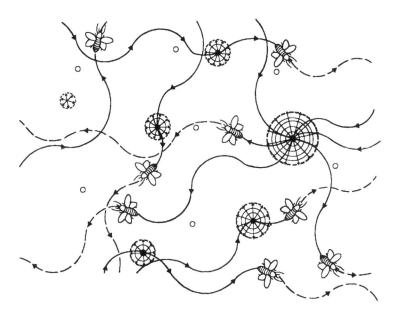

Figure 14.2. *The development of preferred sites marks a transition to coordinated activity of the insect population. As the size and number of preferred sites increase, insect trajectories become more tightly coordinated. (Adapted from <u>Information, Natural Law, and the Self-Assembly of Rhythmic Movement</u> (p. 72), by P.N. Kugler and M.T. Turvey, 1987, Hillsdale, NJ: Lawrence Erlbaum Associates. Reprinted by permission).*

14.3.1.2   Pillar construction: A near-equilibrium mode.

As more insects participate there is an increasing likelihood that one will intercept the pheromone gradient of an active site within the limit of that insect's perceptual threshold. Beyond a critical number of participating insects, the equilibrium condition of the flight pattern breaks down, and a few preferred deposit sites (singularities) emerge (Figure 14.2). These fixed points have symmetric attractor properties that influence insects similarly in all directions. Increases in the rate of depositing on preferred sites increases the size of the gradient field that attracts the insects, which in turn increases the rate of depositing, and so on. As the size of a deposit site grows, long-range coordination patterns begin to develop among the flight paths of individual insects, and an increasing number of them begin to orient their motion to the pheromone field. The result is an autocatalytic reaction resulting in an amplification of material deposits at points of highest pheromone concentration. As the autocatalytic reaction continues, pillars develop from the waste deposits (Figure 14.3). These pillars are constructed at the locations of singularities with only the tops of pillars remaining as active deposit sites.

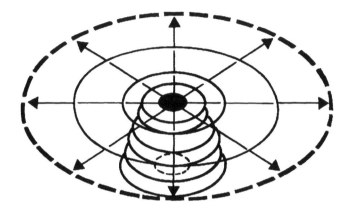

Figure 14.3. *Pillar construction results from an autocatalytic process in which the field effects of recent deposits accumulate and increase the probability that other insects will be stimulated to deposit at that site. (Adapted from from Information, Natural Law, and the Self-Assembly of Rhythmic Movement (p. 73), by P.N. Kugler and M.T. Turvey, 1987, Hillsdale, NJ: Lawrence Erlbaum Associates. Reprinted by permission.)*

14.3.1.3   Arch construction: A far-from-equilibrium mode.

As the size of the active gradient regions enlarge, competitions begin to develop between gradients generated by neighboring singularities. This competition occurs when the active portions of gradient fields begin to overlap. Saddlepoints[6] that establish a boundary between two gradient

---

[6]A saddlepoint has fixed point properties that attract in some directions and repel in others. In the nest construction system, a saddlepoint represents a virtual singularity midway between the singularities of two interacting fields. Insects who enter the field between the two pillars will have their flight trajectories first biased toward the saddlepoint. Once they reach the proximity of the saddlepoint their trajectories will be biased away from it.

fields emerge out of this interactive competition. As the system is displaced further from equilibrium, competitions begin to develop between neighboring singularities so that the linear dynamic (associated with relaxation to a singularity) breaks down. In this far-from-equilibrium region, multiple singularities begin to compete for local control over insect trajectories. The linear dynamic is replaced by a nonlinear dynamic.

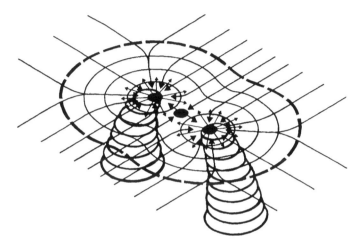

Figure 14.4. *Arch construction commences when emergence of the saddlepoint displaces the system further from equilibrium. Insect activity is oriented by reference to a virtual singularity midway between the two pillars. (Adapted from* Information, Natural Law, and the Self-Assembly of Rhythmic Movement *(p. 80), by P.N. Kugler and M.T. Turvey, 1987, Hillsdale, NJ: Lawrence Erlbaum Associates. Reprinted by permission.)*

A saddlepoint breaks the local symmetries of neighboring deposit sites by introducing an inward bias in the direction of competing singularities. The addition of this bias adds an inward curvature to neighboring pillars that leads to the construction of an arch (Figure 14.4). Although the saddlepoint defines a local symmetry-breaking transformation in the depositing activity at the pillars, it also defines a more global symmetry-preserving transformation that relates the gradient fields of the two competing singularities. The saddlepoint defines an invariant solution that satisfies the local gradient field constraints on both pillars. The saddlepoint is a higher order attractor defining a symmetry that is invariant over the two competing gradient basins. The construction of the arch emerges out of the more global symmetry of the saddlepoint. The saddlepoint symmetry is used to coordinate the unfolding trajectories of the two local attractors defined at the tops of the pillars.

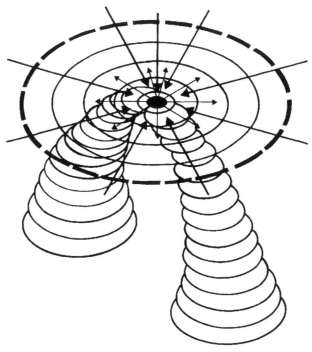

Figure 14.5. *The saddlepoint is annihilated when the arch is completed. The point of completion reverts to a standard singularity of attraction. (Adapted from* Information, Natural Law, and the Self-Assembly of Rhythmic Movement *(p. 81), by P.N. Kugler and M.T. Turvey, 1987, Hillsdale, NJ: Lawrence Erlbaum Associates. Reprinted by permission.).*

14.3.1.4   Dome construction: A return to the equilibrium mode.

The arch is completed when the two pillars meet. At this time the singularities of the two pillars merge to annihilate the saddlepoint and to create one new singularity at the top of the arch (Figure 14.5).
Gradient flows from the new singularity interact with neighboring gradient flows resulting in the emergence of an intricate pattern of new saddlepoints. These saddlepoints organize a new gradient layout that,

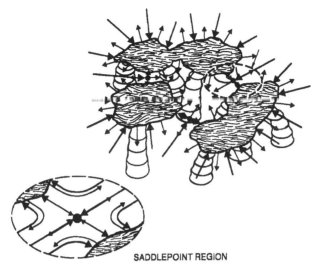

SADDLEPOINT REGION

Figure 14.6. *Dome construction results when new singularities created by completion of arches interact to generate an intricate pattern of new saddlepoints. (Adapted from <u>Information, Natural Law, and the Self-Assembly of Rhythmic Movement</u> (p. 83), by P.N. Kugler and M.T. Turvey, 1987, Hillsdale, NJ: Lawrence Erlbaum Associates. Reprinted by permission.)*

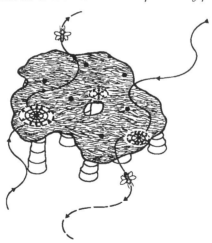

Figure 14.7. *Upon completion of the dome the building phase returns to equilibrium, beginning once again with the random flight phase. (Adapted from <u>Information, Natural Law, and the Self-Assembly of Rhythmic Movement</u> (p. 84), by P.N. Kugler and M.T. Turvey, 1987, Hillsdale, NJ: Lawrence Erlbaum Associates. Reprinted by permission.)*

in turn, provides new constraints to coordinate the construction of a "dome" (Figure 14.6).

Upon completion of the dome, the far-from-equilibrium condition is annihilated, and there is a return to the equilibrium mode. A new construction cycle begins, starting with the random deposit phase on the surface of the dome (Figure 14.7). The system begins another cycle through the sequence of construction modes (random deposit—>pillar construction—>arch construction—>dome construction—>random deposits) (Figure 14.8).

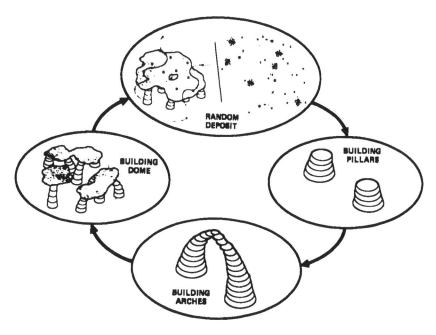

Figure 14.8. *A circular ring of building phases. Each phase is dominated by a small set of critical states that organize insect flight patterns. (Adapted from <u>Information, Natural Law, and the Self-Assembly of Rhythmic Movement</u> (p. 87), by P.N. Kugler and M.T. Turvey, 1987, Hillsdale, NJ: Lawrence Erlbaum Associates. Reprinted by permission.).*

14.3.2   Summary: An Information System

The organization of termite nest-construction activities is primarily constrained by local control of individual termites through the global influence of a pheromone field.[7] In general terms, the system self organizes via the perceptual sensitivity of individual insects to coherent, long-range information. Individual insects (and indeed, the ensemble of insects as a whole) have no explicit plan. It would be misleading to conceptualize this organized activity as being guided by any form of group intelligence (Prigogine & Stengers, 1984) or shared mental model. Nor do individual insects modify their mode of response to the pheromone field. The local character of this mode of response remains unchanged during the nest-construction cycle. Insects continue to respond in the same manner to local gradients throughout the distinctive phases of pillar construction, arch construction, dome construction, and random flight.

Particular forms of organization emerge via specific symmetry-breaking transformations that, at critical thresholds, cast group behavior into new modes. At the system level there is a perception–action cycle (Gibson, 1979/1986) that is self-writing and self-reading (Pattee, 1972). Symbolic and dynamic constraints interact to instantiate a memory of the termite activity in the form of the nest. The structure that emerges from this perception–action cycle is constrained to global forms (random deposit—> pillars—> arches—> domes—> random deposit), but within those global forms, there is room for infinite novelty. Precise long-range prediction of the temporal unfolding of sequential states or of the spatial layout of singularities is not possible.

14.4   Management and Control

A common approach to risk management is to model behavior in terms of rules. This impetus toward low-order control via a high dimensional set of informational properties is based on the expectation that a complex system can be made to work efficiently by ensuring that individuals know procedures relevant to all contingencies related to their own task. Underlying this approach is the assumption that all

---

[7]Informational constraints are defined by global flow properties of the pheromone field. These properties are organized by the number and layout of fixed points (singularities) in the pheromone field. The flow field associated with a given layout is structurally stable. Instability of the information field is associated with creation and annihilation of fixed-point properties in the pheromone field. The creation and annihilation of these fixed points is ultimately related to the amplification of a nonlinearity.

potential system states can be specified in advance. Under this assumption it should be possible to identify standard operating procedures that can guide transitions into desired states and that can prevent transitions into undesired states.

This rule-based approach, which is beguiling in its simplicity and rationality, takes various forms. Emphasis by management on specific details of operator behavior, presentation by display engineers of a mass of detailed information, and implementation by system designers of automatic safety devices are of this character. In general, these approaches can be characterized as attempts at low-order control from high levels of management, in which low-order control refers to emphasis on specific or local features of the system. These are high-dimensional strategies in the sense that, to be successful, all details of the system must be understood and manipulated by a central authority.

14.4.1 Failures of Low-Order Control

Many complex industrial systems are designed so that a mass of detailed information is available to operators in the event of a problem. A mass of information did not, however, assist operators at Three Mile Island (Perrow, 1984). If anything it distracted attention from the real problem. Furthermore, attempts at low-order control will often have unexpected consequences. The emphasis by management at Three Mile Island on the need to guard against a high-pressure incident that could rupture containment integrity provides an example. In this case, the incident that occurred resulted from loss of coolant, which demanded corrections incompatible with those required for a high-pressure event. Prior emphasis by management on a specific type of incident exacerbated the problem by encouraging an incorrect interpretation at the control interface.

Automatic safety devices, which offer another form of low-order control, often generate their own problems. They can enter into a sequence that precipitates a critical incident, as one did at the Fermi nuclear reactor in Lagoona Beach, MI, in 1966 (Perrow, 1984). In this incident, a fuel melt was caused by a metal vane that dislodged and impeded circulation of coolant. The installation of this vane had been recommended by a committee of engineers and scientists who had thought that it would enhance safety by maintaining a more even flow of coolant. Alarms and redundancy are other approaches to low-order control that can create their own problems. False signals from alarms can lead to the attitude that alarms are to be ignored. Redundant systems are often not truly redundant, as in the oxygen system of the 1970 Apollo-13 mission and, by adding to complexity, they can interfere with production schedules and can generate obscure interactions (Perrow, 1984).

14.4.2   Micromanagement and the Problems of Novelty

The apparent intractability of large-scale industrial systems has often stimulated micromanagement as one form of low-order control. The proposal of the Kemeny Commission on the nuclear accident at Three Mile Island that the safety of nuclear power operation would be enhanced by paramilitary discipline represents just such a view (Perrow, 1984). Implicit in this view is the idea that military organizations work because of tight, hierarchical control in which supervisors pay considerable attention to detail. This is a classic example of a high-dimensional, resource-intensive approach. It is a form of management that has, however, provided occasional examples of spectacular failure; the disastrous raid by aircraft from the USS Kennedy and the USS Independence against sites in Lebanon in December 1983 being just one (Wilson, 1986). Despite elaborate planning and extensive training by carrier crews to execute just such an attack, this air raid failed because higher authority specified a time of attack and targets that were not consistent with the plans that had been prepared by officers in the theater of operation nor with the exercises that had been undertaken to familiarize personnel with those plans.

A typical style of micromanagement is to formulate a set of error-based corrective procedures that rely on recognition of desired and current states. The problem facing this approach is one of novelty: Incidents in large-scale industrial systems often involve unanticipated and poorly understood interactions (Perrow, 1984). The implementation of rule-based procedures to deal with all anticipated eventualities is defeated by combinatorial complexity that can quickly lead events into novel situations for which the rule set is no longer applicable. In fact, a system can be so complex that investigating teams find it difficult to fully understand the nature of the problem even after the fact (Perrow, 1984). In practice, highly detailed sets of rules or operating procedures turn out to be brittle; they collapse precipitously where the novelty of a current state precludes its recognition.

14.4.3 The Irremedial Incompleteness of Rules

A further problem with rule-based, low-order control is that rules must be interpreted within situated action. Suchman (1987) argued that rules cannot fully specify required actions. For any specific situation, a considerable amount of implicit knowledge is required to transform a rule into action. In her terms, there is an "irremediable incompleteness" of instructions. Any actor who is following instructions is inevitably faced with the problem of ascertaining the practical significance of those instructions for situated action, and there is always room for differing interpretations. Suchman (1987) concluded that a preestablished plan as encapsulated in sets of instructions or procedures can act as a resource for action but cannot determine it.

In concert with Suchman (1987), Winograd and Flores (1986) argued that the understanding of language is grounded in an unrecognized preunderstanding. Although it is possible to extend the articulation of that preunderstanding, pursuit of that goal is aided most effectively by *breakdown*; that is, an incident in which the unexpected draws attention to an action or interaction that has previously gone unnoticed. For a large-scale industrial system with disastrous potential, breakdown is an unacceptably risky means of working through potential problems. Furthermore, Winograd and Flores (1986) argued that articulation of unrecognized preunderstanding is never ending. The attempt to articulate a preunderstanding is made in terms of language and experiences that themselves reflect unarticulated preunderstandings.

14.4.4   Summary: The Failure of Low-Order Control

The literature on critical incidents in complex industrial systems is replete with examples of failures of low-order control (Reason, 1990). Complex systems often exhibit behavior that has not been anticipated. Unpredicted events and poorly understood interactions can throw a system into a novel state for which it is difficult or even impossible to plan an appropriate set of recovery procedures (Perrow, 1984). The documented failures of large-scale industrial systems (Perrow, 1984; Reason, 1990) should stimulate careful reassessment of the rational-logical perspective that underlies this rule-based, low-order approach to risk management.

In our view, it is the essential nonlinearity of large-scale industrial systems that limits the effectiveness of the rule-based approach to system management. It is not possible, in a nonlinear system, to specify all contingencies in advance. In addition, the effects of low-order interventions cannot easily be anticipated. In particular, if the nature of the nonlinearity is not well understood, the effects of low-order control can push the system through a nonlinear region into a state that has not been anticipated. Thus, a rule-based approach to risk management may actually increase the risk of a major incident[8].

Furthermore, local human adaptation is too powerful a process to permit a rule-based system to function as planned. Individuals do not follow procedures precisely (the irremediable incompleteness problem). To the extent that individuals must interpret a rule within their own body of unarticulated knowledge (Winograd & Flores, 1986), there will inevitably be local deviations from rule-based precision. It is characteristic of nonlinear systems that even when deviations are minor (to the point of being imperceptible to the human observer), they can cascade and amplify through the system to generate major and unanticipated changes in state.

Rochlin (1989a) argued for a distinction between management and control, in which control refers to the exercise of rules developed on the basis of perfect knowledge, correct information, and a system model that encompasses all possible contingencies. Rochlin points out that attempts to establish this form of control over military operations has historically led to failure. On the other hand, a style of management that operates on more global parameters such as size of armies and standardization of tactics, and that accords local autonomy to on-site personnel, has met with considerably more success. This high-order strategy is one in which decisions about highly detailed and localized action is left to those who must implement the control actions. Rochlin's view is consistent with that of Winograd and Flores (1986) who argued that the essential concern for management is to generate contexts in which effective action can be realized.

---

[8]Recall the Fermi accident discussed earlier. Also consider the argument of Glass and Mackey (1988) regarding the reaction of chronic myelogenous leukemia to cytotoxic agents. The standard statistical distributional analysis of survival duration from time of diagnosis suggests that these cytotoxic agents have no impact on the disease. In contrast, a nonlinear analysis suggests that these agents can dramatically shorten or lengthen an individual patient's life span. In the absence of a deep appreciation of the underlying dynamics of the disease, a physician will be unable to judge whether the prescribed therapy will actually lengthen or shorten an individual patient's life.

14.5   Self-Organization in Military Systems

That large-scale nonlinear systems can be managed successfully is
evident in an analysis by Rochlin, La Porte, and Roberts (1987) of flight-
deck operations on board an aircraft carrier. A key feature in the
development of safe flight-deck procedures is an extended workup
period to prepare for full-capacity operations. After commissioning or
refit, weeks are spent in which aircraft launch and recovery rates are
gradually increased to operational levels. It seems significant that there
are no detailed operating procedures: As noted by Rochlin et al., the
only complete operating manual is the working carrier itself. From the
nonlinear perspective, this workup period is critical. It is during this
period that the microelements (individuals) self-assemble into a
functional organization.

A unique organization emerges for each carrier, but there is a
similitude or self-similar identity at the level of function that is invariant
across carriers. That is, each unique organization found in the different
U.S. Navy Carriers accomplishes the same job. The role of management
in this system is to establish global priorities and schedules, but those
who execute the multitude of various tasks to accomplish the goals set
by management are left to work out their own adaptations to local
conditions. As in the termite nest-construction system, an information
field is generated and sustained by the action of the workers reacting
individually to information in their immediate environment.

The description by Rochlin et al. (1987) of carrier operations
contrasts with the usual expectations of risk management. To the casual
observer, the procedures that emerge will appear to be ad hoc and
inefficient. Nevertheless, the system is robust to high personnel
turnover, high production demand, and the ever-present possibility of
major accidents (Rochlin, 1989a). Ironically, the system that
encompasses Navy carrier flight-deck operations, where the need for
high levels of production and the ever-present possibility of disaster go
hand in hand, is reliable and robust, specifically because of a mode of
organization that members of the Kemeny Commission might view as
distinctly nonmilitary. It is nevertheless evident that this self-organizing
mode of operation, in which autonomy for local decisions is transferred
down the hierarchy, has been a feature of many successful military
operations (Rochlin, 1989b).

14.6   Management and Control of Nonlinearity

Further insight into the relative advantages of high-order management versus low-order control can be generated by consideration of the self-organizing properties of nonlinear systems. Here understanding rests on relating descriptions at a microlevel to descriptions at a macrolevel and on elucidating the mechanisms of interaction between these two levels (Lintern & Kugler, 1991). In contrast to wholism or classical reductionism in which there is a single, privileged level of description, this path to understanding nonlinearities emphasizes the complementarity of descriptions at different levels (Pattee, 1979a, 1979b). Description at neither level is meaningful on its own.

Implicit in our discussion is the belief that the mechanisms of self-organization that drive the termite nest-construction system through a cascading series of topological forms are generic. Termite nest construction is but one example of nature's solution to the development of order within a system of loosely constrained subsystems coupled by a low-energy information field. Such systems will exhibit the dominant features of nonlinearity: large stable regions separated by small regions of instability, an absence of a shared plan at the microlevel of description, and unpredictable temporal and spatial detail nested within topological regularity. Significantly, multiple levels of description suggest opportunities for intervention at different levels; a high-order of intervention (i.e., management) applied at a macroscale or a lower order of intervention (i.e., control) applied at a microscale. At issue for large-scale industrial systems is how these contrasting strategies of management and control might be exploited to ensure safe and stable operation.

14.6.1   Order of Intervention

It would be possible, in the termite nest-construction system (or at least in a computer model of that system), to modify system response with either low- or high-order control strategies. Nest-construction activity might, for example, be disrupted by a low-order strategy in which pillars were obliterated as they neared the size that would permit arch construction. This is an error-based approach in which a pillar of a prespecified size is defined as an error. Alternatively, preferred pillar sites might be established artificially by seeding the space with appropriate deposits. Such low-order strategies are resource intensive and high dimensional in that they require constant vigilance and detailed planning. A contrasting high-order strategy for disrupting or modifying the building plan would be to destroy or bias the pheromone field. One possibility would be to generate strong air movement by mechanical means. This information-based or field-dependent strategy would require far less attention to detail. It is a low-dimensional approach that would require attention to only one global parameter.

A lesson to be drawn from the termite nest-construction systems is that information exists at two levels. Individual termites orient at the microlevel to the local gradient. For the system to work, they must be perceptually sensitive to the pheromone. In this system, that perceptual sensitivity is established presumably via adaptive evolution, but in human-machine systems, sensitivity of operators to critical information will often be developed via perceptual learning. Information also exists at a macrolevel in the form of a gradient field organized by the deposit sites as attractors and by interactions between deposit sites as virtual attractors. Constraints on the termite system (e.g., number of participating insects) occur naturally, but in a human-machine system, natural constraints are supplemented by management policies. An implication of our analyses, consistent with Rochlin's distinction between management and control, is that managers and individual workers must attempt to influence the system at different levels.

14.6.2 Summary: Open, Nonlinear Systems

In any system, the micro- and macrolevels discussed in reference to open systems will be found in the process to be managed, and in the control, prediction, or management that must be applied to that process. One implication from our analysis of open, nonlinear systems is that the divergence of the system from the current global state of stability will be constrained primarily by higher order processes such as fields and gradients. In general, system stability or orderly transition to new states requires that constraints on the dynamic process be implemented at a higher order than the order of the process to be controlled. As is consistent with Rochlin's (1990) view, responsibility for system stability lies with management at the macro level while individual workers remain responsible for detailed implementation via control at the micro level. Given appropriate management of the global level descriptors, it should be possible to maintain robust and stable system performance via the activities of individuals workers who are adequately sensitive to their local conditions.

14.7   Information for Management and Control

The nonlinear perspective is consistent with the view that effective management will seek to transfer decision making to the lower levels of the organizational hierarchy where information is likely to be most detailed and most accurate (Roberts, 1989). An effective approach to risk management will encapsulate a high order of control in which management is tuned to information about the global state of the system, but will rely on the sensitivities and experiential knowledge of operators at the interface to detect and to adapt to local fluctuations. The term *information* is used here in Gibson's (1979/1986) specificational sense: An invariant and unique perceptible property (or an invariant combination of perceptible properties) specifies the character of an object or event.

Managers and controllers must be sensitive to critical information at their own level of influence; information will be of a different order for managers versus controllers. The primary role of management in this scheme is to establish global constraints such as emphases on safety, timeliness, and productivity. To be consistent with Gibson's (1979/1986) sense of information, the invariant properties (or invariant combinations of properties) that specify such constraints must be potentially perceptible by the appropriately attuned and attending observer (or they must be connected lawfully to perceptible properties). In addition, those properties must be substantial, not merely conceptual (Turvey, 1992).

This view may be compared to one that would have organized activity for multiple participants emerging from shared plans or shared models (Fischhoff & Johnson, 1990). In contrast to Fischhoff and Johnson (1990), we believe that self-organization is more useful than shared models as a unifying construct. Although human action is probably influenced to some degree by common knowledge of overall goals and plans, we believe that shared goals and plans have a much less potent role to play than the mechanisms of self-organization. As in the case of termite nest construction, the notion of a shared plan or model as the principal constraint on organization within complex industrial systems is likely to be misleading (see also Suchman, 1987).

A relevant example consistent with the nonlinear perspective is found in Klein's (1989) analysis of fireground commanders, whose task it was to combat forest fires. Klein argued that situational recognition is the primary decision process used by fireground commanders to guide them toward appropriate strategies for unique problems. Here it is apparent that a low-order strategy of directing resources to specific locations as they start to burn will generally not be optimum. The fire itself is controlled by higher order processes such as wind and extent and combustibility of material. A fireground commander's task is to establish control at a higher order: that is, to develop a strategy that takes account of the global processes driving the fire. Strategic construction of firebreaks would constitute one such higher order strategy. If situational recognition is the key process in decisions of this type, sensitivity to higher order information would seem to be crucial. In contrast, firefighters at the location of the fire would need to be sensitive to both the intent of the instructions they receive from their commander and to the changing conditions in their locality so that they could implement their commander's strategy within their own situated context.

There is some evidence that individual workers can become sensitive to critical information that signals an imminent breakdown. Reason (1991), in a discussion of work by La Porte and Consolini (1989), described how air traffic controllers would assemble around a fellow worker who had an unusually high number of aircraft to control. Information that the loaded individual would normally acquire him- or herself is offered at critical moments without request. When the load eases, this impromptu support group disassembles without explicit recognition that the impending crisis is over or even that it existed.

For the present discussion, it is useful to view a complex industrial system as one in which the microelements (individual workers) interact with the local conditions of a global information field. One implication of this view is that operators at a control interface must be given access to significant low order (local) information about the state and activity of the system, and they must be given opportunities to fully assimilate the meaning and importance of that information (Roberts, 1989). Another implication of this view is that managers must also be given access to high-order (global) information, and must have the opportunity to assimilate its meaning and importance.

The development of sensitivity to information (of either order) is a matter of training, although it is training of a special sort. The workup period on a carrier flight deck can be viewed as an opportunity to build and to explore the (information) environment of the workspace. It develops a form of experiential knowledge (Rochlin, 1989b) via a process identified by Lave and Wenger (1991) as legitimate peripheral participation. In essence, training must provide opportunities for new participants to increase their involvement in a full range of tasks as they move gradually toward full participation in the activity. During this process, a new participant must become sensitive to information that specifies system state and critical system behavior in either localized or global terms as is appropriate for the level of intervention that can be implemented by that participant. This may be accomplished by on-the-job familiarization of the type described by Rochlin et al. (1987), by legitimate peripheral participation in communities of practice as described by Lave and Wenger (1991), or by special training scenarios that include simulation of relevant information flows (Lintern, 1991).

14.8 Regions of Instability

There is an after-the-fact impression that events leading to any particular incident (e.g., the destruction of Iran Air Flight 655 by the USS Vincennes or the loss of the Challenger) could have occurred anywhere and at anytime. It was just sheer bad luck that they did not happen in more benign circumstances. Some might argue that such incidents are precipitated by quirks of fate and are essentially beyond control. An alternate view is that certain designs and certain attitudes generate potentially dangerous potentialities (Rochlin, 1989b, 1990). There are problems waiting to surface given the precipitating circumstances; in Reason's (1990) terms, latent errors. The precipitating circumstances may be quite ordinary. A maintenance worker might drop a light bulb, or a mechanical valve may leak (Perrow, 1984). Events precipitated by such ordinary events may be characterized as normal accidents.

The question arises as to whether it is possible to identify systems that contain latent errors or are prone to normal accidents. Perrow's (1984) structural analysis identified complexity and coupling as critical dimensions. Rochlin (1989b) argued that a lack of opportunities for controllers or managers to develop experiential knowledge through exploration and assessment of potential scenarios can cause significant problems. These are useful analyses that identify characteristics of potentially dangerous systems, but they downplay the dynamic nature of the system that may leave it more prone to failure at some times versus others.

Specifically, open systems are nonstationary; that is, they evolve and change over time. Prigogine's theory of dissipative structures highlights the fact that evolving fields (information at the macrolevel) modify the potentialities of open systems to change states as a result of minor (or normal) fluctuations. In what follows, we supplement the analyses of

Perrow (1984) and Rochlin (1989a, 1990) by pursuing the observation that nonlinear systems offer characteristic signatures as they approach an instability (Schmidt, Carello, & Turvey, 1990). From an identification of these signatures it may be possible to anticipate critical regions in the evolving conditions surrounding large-scale industrial systems that offer increased likelihood of disastrous state transitions. From there it may be possible to identify effective management interventions that will transition a system into or maintain it in the desired state.

14.8.1 Information for Control of Action

Following the lead of Gibson (1979/1986), identification and characterization of information for perception has become a significant issue in the study of human activity (Cutting, 1986; Lintern, 1991; Mark, 1987; Warren & Whang, 1987). In Gibson's (1979/1986) analysis, perceivers are sensitive to information that specifies the state of the environment and their interaction with it. The particular properties to which an observer is sensitive are at an actor-relevant scale. As we argued earlier, observers may become attuned to these properties through training or experience.

By Gibson's thesis, there is a reciprocity between perception and action; information that specifies an environmental state can be used for control. Within the context of an individual perceiver-actor, time-to-contact information (Lee, 1976; Lee & Reddish, 1981) specifies imminence of a collision, but that same time-to-contact information might be used as the basis for control of action that will avoid collision. Included in discussions of time-to-contact, is the notion of criticality. Perceiver-actors become exquisitely attuned to the critical boundary values of time-to-contact that separate collision from noncollision. The appropriate adaptive strategy is to maintain this information at values on the noncollision side of the boundary.

14.8.2 Signatures of Impending Instability

The behavior of open, nonlinear systems has been notoriously difficult to predict. In terms of local behavior, there is no change. Global state transitions occur suddenly and seemingly unannounced. However, there are subtle signals in the relaxational dynamics of nonlinear systems that cast an anticipatory shadow to announce the impending instability. It is this type of global information that will inform a rider in advance of whether a horse is about to break into a canter or a gallop.

There may be a relatively rapid increase in the amplitude of fluctuations, or there may be a critical slowing down in response of a system as it returns to equilibrium following a perturbation. These are changes characteristic of an instability that is about to amplify and to launch the system into a new state.

14.8.3 Information for Control of Instability

One difficulty for risk management is the discrimination of a stable regime from one that is approaching instability. In large-scale industrial systems there is generally no prior record of the conditions that will cast the system into an undesired state. Thus, system managers need to exploit current information in preference to prior records or memories of prior experiences. From the open systems perspective, effective management might be established through interpretation of the information in the transient dynamics leading up to a change of state. Specifically, information about the imminence of a change of state and the nature of the change will be contained in the small temporal and spatial regions surrounding the impending instability.

Managers of large-scale industrial systems might become attuned to critical values of the information in relaxational dynamics that specify impending instabilities, for example, critical changes in the period of the transient following a perturbation or critical increases in amplitude of fluctuations in system response. We suspect that managers may develop sensitivity to such criticalities via the same learning process that perceiver-actors become sensitive to critical values of time-to-contact information. It is possible that competent decision makers such as fireground commanders attend to and control these types of criticalities when they monitor the progress of a fire (Klein, 1989). The management problem in this case is to maintain information on the nontransitional side of the critical boundary.

We assume that critical information boundaries are present in all large-scale systems. For example, fluctuations in stock market indexes may carry information about impending instabilities. It should, therefore, be possible in principle to predict when a perturbation such as a shift in interest rates will precipitate a major adjustment in stock market values. This does, however, point to one of the difficulties of using relaxational dynamics in transients to predict trends. It is often difficult to distinguish critical from noncritical changes. Although it is known that perceivers are exquisitely sensitive to time-to-contact information (Lee, Young, Reddish, Lough, & Clayton, 1983), it is not clear that managers can develop appropriate sensitivity to the global criticalities that specify the onset of instability in large-scale systems. However, we speculate that development of management skill with large-scale industrial systems is, in some part, derived from sensitivity to and calibration of information in transient dynamics. For example, expert fireground commanders may have learned to calibrate the information contained in the speed with which their strategies have some impact on a fire.

There are, nevertheless, many large-scale systems for which there seem to be no experts who are adequately sensitive to the criticalities in the transient dynamics: a nuclear process, a country on the verge of revolution, or an economy on the verge of hyperinflation. In addition, in some large-scale systems, managers appear insensitive to the fact that specific control techniques are not having the desired effects. This may be due to the limited opportunities for managers to experience scenarios that generate information at this scale. In contrast, experience with time-to-contact information is acquired in a wide range of normal human activities. Opportunities for enhancing sensitivity and for recalibration arise as we walk, run, drive, play a racquet sport, or catch a ball (Lee & Young, 1985), and further opportunities can be provided via special training scenarios (Young & Lee, 1987). Thus, the open systems perspective presents a major challenge and a significant opportunity. We might advance our ability to manage risks in large-scale industrial systems by first identifying the dimensions of information that can signal impending instability, and then by developing training scenarios that could sensitize managers to those dimensions of information and to the critical values in the transient dynamics.

The proposal we present here should not be taken as an "analogical" extension of principles underlying organization in open chemical and biological systems. To be consistent with the materialist ontology of the ecological approach to perceiving and acting (Turvey, 1992) it is essential to treat the nonlinearities of complex industrial systems as substantial (versus conceptual). Thus, control parameters and order parameters must be lawfully connected to potentially perceptible properties. These control and order parameters for complex industrial systems has not yet been identified, but we speculate that control parameters may be found in management emphases on safety and productivity, whereas order parameters will take the form of safety and productivity outcomes. It remains a challenge to proceed beyond speculation in identifying control and order parameters. As a first step, an event audit within the workplace might suggest the nature of important control parameters. From there it would be necessary to establish the effects of variation in the control parameters and also to identify the invariant perceptual properties that are lawfully related to changes in those properties and that can be used to identify proximity to critical transitions.

14.9   Conclusion

Large-scale industrial systems have the capacity to generate subtle and potentially disastrous interactions. The standard engineering solution to this problem is to create a closed (completely defined) rule-governed system. In practice, this approach will fail because faults are often unique (Rasmussen & Vicente, 1989). The combinatorial complexity inherent in large-scale systems can quickly depart from any anticipated situation, and operators are left with little useful information for effective control. Automatic safety devices can be fitted, and standard operating procedures specified, but these are useful only for those precisely constrained error states that have been anticipated. Ad hoc backfitting of safety devices and development of new procedures to cover the latest accident (Perrow, 1984; Rasmussen, 1988) are inadequate. Clearly, a more robust approach is needed. In our view, attempts to create closed, rule-governed systems are doomed to failure. Biological systems in general, and human organizations in particular, remain open to new forms of order. A realistic approach to risk management must take account of the fact that potentialities are unbounded.

Failures in management and control of large-scale industrial systems stem partly from the fact that their open thermodynamic nature is not fully appreciated. Open thermodynamic systems are capable of never-ending novelty. Their nonlinear nature makes them prone to unanticipated state transitions. Of concern is how such systems transition between states and how impending state transitions can be managed. Because of their nonlinear nature, attempts to generate new states in these systems or to maintain current states carry considerable dangers. Low-order control by managers who do not directly experience local conditions at the human-machine interface can have entirely unexpected consequences, and the transitions to new states can be sudden. In particular, a system that is near a nonlinear transition will respond quickly and dramatically to control inputs that will, at other times, generate a barely noticeable ripple in system behavior.

The information that would reveal an impending instability may not be available to the operator; that is, the processes driving errors or faults may not reveal themselves at the operator-system interface (Rasmussen, 1988). In open systems that have the possibility for self-organization, global information carried in the relaxational dynamics of transients can signal an impending instability. This high-order information might be detected at the management level. It could provide an essential, and sometimes only, source of information for maintaining stability of a large-scale industrial system. However, we do not yet fully understand the nature of the information that should be monitored. A challenge to the approach we have outlined here is to transform intuitions about the nature of the high-order information available to management into objective specification, and to show how managers might be oriented to the criticalities in the information that signals an impending nonlinear transition.

Acknowledgment

Preparation of this chapter was supported in part by a grant from the Veteran's Administration #1-5-26689, Vet Admin V578P-3558. Kim Vicente, Gary Riccio, and Thomas Stoffregen reviewed earlier drafts of this chapter. Aaron Contorer assisted our understanding of self-organization by creating a simulation of the nest-building system that reveals graphically the evolutionary processes discussed in this chapter.

14.10 References

Atkins, P. W. (1984). *The second law.* New York: W.H. Freeman.
Babloyantz, A. (1986). *Molecules, dynamics and life: An introduction to self-organization of matter.* New York: Wiley-Interscience.
Bruinsma, O. II., & Leuthold, R.H. (1977). An analysis of building behavior of the termite macrotermes subhyalinus. *Proceedings of the VIII Congress IUSSI* (pp. 257–258) Wargeningen.
Chance, B., Pye, E. K., Ghosh, A. M., & Hess, B. (Eds.). (1973). *Biological and biochemical oscillators.* New York: Academic Press.
Cutting, J. E. (1986). *Perception with an eye for motion.* Cambridge, MA: MIT Press.
Davies, P. (1988). *The cosmic blueprint: New discoveries in nature's creative ability to order the universe.* New York: Simon and Schuster.
Davies, P. (1989). *The new physics.* Cambridge, MA: Cambridge University Press.

Deneubourge, J.L. (1977). Application de l'ordre par fluctuation a la description de certaines etapes de la construction du nid chez les termites. [Application to the order by fluctuations to the description of some stages in the building of the termintes' nest]. *Insectes Sociaux, Journal International pour l'Etude des Arthropodes Sociaux, 24*, 117–130.

Fischhoff, D., & Johnson, S. (1990). The possibility of distributed decision making. In Committee on Human Factors (Eds.), *Distributed decision making: Report of a workshop* (pp. 25–58). Washington, DC: National Academy Press.

Gibson, J.J. (1986). *The ecological approach to visual perception.* Boston: Houghton Mifflin. (Original work published 1979)

Glansdorff, P., & Prigogine, I. (1971). *Thermodynamic theory of structure, stability, and fluctuation.* New York: Wiley-Interscience.

Glass, L., & Mackey M. C. (1988). *From clocks to chaos: The rhythms of life.* Princeton, NJ: Princeton University Press.

Glieck, J. (1987). *Chaos: Making a new science.* New York: Viking.

Grasse, P. P. (1959). La reconstruction du nid et les coordination interindividuelles chez Bellicositermes natalensis et cubitermes sp. La theorie de la stigmergie: essai d'interpretation des termites constructeurs. [The rebuilding of the nest the interpersonal coordination The theory of the stigmergie: essay on the interpretation of the behavior of the termite builders]. *Insectes Sociaux, Journal International pour l'etude des Arthropodes Sociaux, 6*, 41–83.

Klein, G. A. (1989). Recognition-primed decisions. In W. R. Rouse (Ed.), *Advances in man-machine system research* (Vol. 5, pp. 47–92). Greenwich, CT: JAI Press.

Kugler, P. N., & Turvey, M. T. (1987). *Information, natural law, and the selfassembly of rhythmic movements.* Hillsdale, NJ: Lawrence Erlbaum Associates.

La Porte, T. R., & Consolini, P. M. (1989, August). *Working in practice but not in theory: Theoretical challenges of high reliability organizations.* Paper presented at the Annual Meeting of the American Political Science Association, Washington, DC.

Lave, J., & Wenger, E. (1991). *Situated learning: Legitimate peripheral participation.* Cambridge, MA: Cambridge University Press.

Lee, D. N. (1976). A theory of visual control of braking based on information about time-to-collision. *Perception, 5*, 437–459.

Lee, D. N., & Reddish, P. E. (1981). Plummeting gannets: A paradigm of ecological optics. *Nature, 293* (5830), 293–294.

Lee, D. N., & Young, D. S. (1985). Visual timing of interceptive action. In D. J. Ingle, M. Jeannerod, & D. N. Lee (Eds.), *Brain mechanisms and spatial vision* (pp. 1–30). Boston: Martinus Nijhoff Publishers.

Lee, D. N., Young, D. S., Reddish, P. E., Lough, S., & Clayton, T. M. H. (1983). Visual timing in hitting an accelerating ball. *Quarterly Journal of Experimental Psychology, 35A,* 333–346.

Lintern, G. (1991). An informational perspective on skill transfer in human-machine systems. *Human Factors, 33,* 251–266.

Lintern, G., & Kugler, P. N. (1991). Self organization in connectionist models: Associative memory, dissipative structures, and Thermodynamic Law. *Human Movement Science, 10,* 447–483.

Mark, L. S. (1987). Eyeheight-scaled information about affordances: A study of sitting and stair climbing. *Journal of Experimental Psychology: Human Perception and Performance, 13,* 361–370.

Nicolis, G., & Prigogine, I. (1977). *Self-organization in non-equilibrium systems.* New York: Wiley.

Nicolis, G., & Prigogine, I. (1985). *Exploring complexity.* Munchen: Piper.

Pattee, H. H. (1972). Physical problems of decision-making constraints. *International Journal of Neuroscience, 3,* 99–106.

Pattee, H. H. (1979a). The complementarity principle and the origin of macromolecular information. *BioSystems, 11,* 217–226.

Pattee, H. H. (1979b). Complementarity vs. reduction as explanation of biological complexity. *American Journal of Physiology, 236(5),* 241–246.

Perrow, C. (1984). *Normal accidents.* New York: Basic Books.

Prigogine, I. (1947). *Etude thermodynamics des processus irreversibles.* Liege: Desoer.

Prigogine, I. (1962). *Introduction to thermodynamics of irreversible processes.* New York: Wiley-Interscience.

Prigogine, I. (1967). Structure, dissipation and life. In M. Marios (Ed.), *Theoretical physics and biology* (pp. 23–52). Amsterdam: North-Holland.

Prigogine, I. (1980). *From being to becoming: Time and complexity in the physical sciences.* San Francisco: W. H. Freedman.

Prigogine, I., & Stengers, I. (1984). *Order out of chaos.* New York: Bantam Books.

Rasmussen, J. (1988, June). *Coping safely with complex systems.* Paper presented at the American Association for the Advancement of Science, AAAS Annual Meeting, Boston, MA.

Rasmussen, J., & Vicente, K. J. (1989, July). *Cognitive engineering: An ecological frontier.* Paper presented at the 3rd International Conference on Perception and Action, Miami University, Oxford, OH.

Reason, J. (1990). *Human error.* Cambridge, MA: Cambridge University Press.

Reason, J. (1991). The contribution of latent human failures to the breakdown of complex systems. *BASI Journal, 9, 3–17.*

Roberts, K. H. (1989). New challenges in organizational research: High reliability organizations. *Industrial Crisis Quarterly, 3,* 111–125.

Rochlin, G. I. (1989a). Informal organizational networking as a crisis-avoidance strategy: US naval flight operations as a case study. *Industrial Crisis Quarterly, 3,* 159–176.

Rochlin, G. I. (1989b). *The case for experiential knowledge.* Unpublished paper presented at the Second International Workshop on Safety Control and Risk Management, Karlsbad, Sweden.

Rochlin, G. I. (1990, October). *Iran air flight 655 and the USS Vincennes: Complex, large-scale military systems and the failure of control.* Paper presented at the Conference on Large-Scale Technological Systems, Berkeley, CA.

Rochlin, G. I., La Porte, T. R., & Roberts, K. H. (1987, Autumn). The self-designing high-reliability organization: Aircraft carrier flight operations at sea. *Naval War College Review,* pp. 76–90.

Schmidt, R. C., Carello, C., & Turvey, M. T. (1990). Phase transitions and critical fluctuations in the visual coordination of rhythmic movements between people. *Journal of Experimental Psychology: Human Perception and Performance, 16,* 227–247.

Suchman, L. A. (1987). *Plans and situated actions: The problem of human machine communication.* Cambridge, MA: Cambridge University Press.

Turvey, M. T. (1992). Affordances and prospective control: An outline of the ontology. *Ecological Psychology, 4,* 173–187.

Tyson, J. J. (1976). *The Belousov–Zhabotinski reaction.* Berlin: Springer-Verlag.

Warren, W. H., & Whang, S. (1987). Visual guidance of walking through apertures: Body-scaled information for affordances. *Journal of Experimental Psychology: Human Perception and Performance, 13,* 371–383.

Wilson, G.C. (1986). *Super carrier.* New York: MacMillan.

Winfree, A. T. (1974). Rotating chemical reactions. *Scientific American, 23,* 82–95.

Winograd, T., & Flores, F. (1986). *Understanding computers and cognition:*

A new foundation for design (paperback ed.). Reading, MA: Addison-Wesley.

Yates, F. E. (Ed.). (1987). *Self-organization: The emergence of order.* New York: Plenum Press.

Young, D. S., & Lee, D. N. (1987). Training children in road crossing skills using a roadside simulation. *Accident Analysis and Prevention, 19,* 327–341.

Zhabotinski, A. M. (1974). *Self-oscillating concentrations.* Moscow: Nauka.

Author Index

Lakoff, G., 214, 215, 216, *231*
Lam, N., 305, 306, *319*
Lambiotte, J. G., 377, *389*
Landwehr, K., 26, *32*
Langeweische, W., 72, 73, 74, 91, *100*
LaPorte, T. R., 398, 399, *414,* 436, 440, 441, *447, 449*
Larish, J. F., 7, *30,* 74, 78, 79, 87, 89, 96, *98, 100*
Larson, C. E., 396, *414*
Larson, C. O., 377, *389*
Latane, B., 376, 379, *389*
Laughlin, P. R., 374, 377, *389*
Laurent, M., 25, 26, *30,* 112, *120*
Lave, J., 366, 368, *389, 390,* 398, *414,* 440, 441, *447*
Lawrence, L., 227, *230*
Lawson, D., 396, *413*
Lawton, D. T., 281, *283*
Leberl, F. W., 305, *322*
Lee, D. N., 16, 18, 21, 23, 25, *32, 35,* 40, 48, 49, *62, 63,* 70, 73, 90, 92, *100,* 111, 112, 113, *120, 121,* 126, 161, *178, 179,* 189, 190, 198, 199, *205,* 211, 228, *231,* 442, 443, 444, *447, 448, 450*
Leibowitz, H., 5, 6, 12, 27, *32, 34*
Leifer, L., 202, *205*
Leiser, D., 236, *252*
Lenzen, T., 258, *283*
Leplat, J., 224, *231*
Lesh, R., 373, *389*
Lestienne, F. G., 161, 162, *178, 179*
Leuthold, R. H., 423, *446*
Levitt, T, 281, *283*
Lewin, K., 10, *32*
Lewis, C., 216, *230*
Lewis, P., 305, *322*
Lilienthal, 163, 165, 166, *178*
Lillesand, T. M., 290, *321*
Lincoln, J. E., 129, *177*
Lindsay, R. W., 218, 225, 226, *231*
Lintern, G., 7, *30,* 74, 91, 94, 96, 97, *98, 100,* 135, *179,* 437, 441, 442, *448*
Lishman, J. R., 18, 25, *32,* 48, *62,* 112, *120,* 161, *178, 179*
Liu, Y.-T., 91, 94, *100*
Lobeck, A. K., 292, 294, *321*
Lock, B. F., 258, *283*
Lombardo, T. J., 14, *32*
Longuet-Higgins, H. C., 197, *205*
Lough, S., 25, *32,* 49, *62,* 112, *120,* 443, *448*
Lusk, S., 147, *179*
Lyman, B., 27, *34*
Lynn, S. J., 336, *357*

M

Mace, W. M., 190, *205,* 370, 379, *390*
MacGregor, D., 325, *357*
Mackey, M. C., 435, *447*
Magee, L. E., 166, 167, *178*
Malinowski, B., 399, *414*
Mangold, S. J., 78, *100*
Manos, D., 396, *413*
Mark, D. M., 307, 310, 314, 318, *321*
Mark, L. S., 442, *448*
Mark, R. K., 313, 318, *319*
Marmo, G., 45, *62*
Marr, D., 198, *205*
Martin, D. J., 156, 172, *179*
Martin, E. A., 136, 147, 150, 151, 176, *179*
Martin, E. J., 39, 40, 48, 49, 56, 59, 60, 61, *62, 63,* 154, 172, 173, *180*
Matlin, M., 3, *32*
Mayer, R. E., 381, *389*
McCloskey, M., 216, *231*
McCormick, E., 4, *32, 34*
McFarren, M. R., 363, 364, *389*
McGinnis, P. M., 45, *62, 63*
McGrath, J. E., 375, *389*
McLeod, R., 13, 25, 26, *32*
McMillan, G. R., 129, 131, 136, 140, 147, 149, 150, 151, 161, 172, 176, *177, 179,* 179
McNamara, T. P., 236, *252*
McNaughton, G. B., 65, *100*
McNeese, M. D., 363, 364, 373, 374, 377, 378, *389, 391*
McRuer, D. T., 146, *179*
Meacham, J. A., 359, 367, *390*
Messmore, J. A., 294, *321*
Mestre, D. R., 95, *102,* 106, *121,* 139, *181*
Metzler, J., 333, *358*
Michaels, C. F., 40, *63,* 399, *414*
Michaelsen, J. C., 305, 306, *319*
Middendorf, M. S., 140, 161, *179*
Middleton, D., 367, 377, *387, 388*
Miller, M., 65, *100,* 212, *230*
Mintzer, O., 294, *321*
Mitchell, T. R., 328, 332, *356*
Mitroff, I., 270, *283*
Miura, T., 21, *32, 33*
Moen, J. B., 267, *283,* 368, *388, 389*
Monmonier, M., 237, *252*
Moore, J. L., 366, 368, 369, 370, 375, 383, 385, *388*
Moorehead, I., 28, *31*
Morisawa, M., 307, *321*
Morita, T., 21, *32, 33*
Morris, M. W., 95, *102,* 106, *121,* 139, *181*
Morris, N. M., 333, *358*
Mulder, J. A., *99*
Mullen, B., 336, *356*

Subject Index